T0213962

Undergraduate Texts in Mathematics

Undergraduate Texts in Mathematics

Undergraduate Texts in Mathematics are generally aimed at third- and fourth-year undergraduate mathematics students at North American universities. These texts strive to provide students and teachers with new perspectives and novel approaches. The books include motivation that guides the reader to an appreciation of interrelations among different aspects of the subject. They feature examples that illustrate key concepts as well as exercises that strengthen understanding.

More information about this series at http://www.springer.com/series/666

Béla Bajnok

An Invitation to Abstract Mathematics

Second Edition

 Springer

Béla Bajnok
Department of Mathematics
Gettysburg College
Gettysburg, PA, USA

ISSN 0172-6056 ISSN 2197-5604 (electronic)
Undergraduate Texts in Mathematics
ISBN 978-3-030-56176-5 ISBN 978-3-030-56174-1 (eBook)
https://doi.org/10.1007/978-3-030-56174-1

Mathematics Subject Classification: 00, 01, 03

This Springer imprint is published by the registered company Springer Nature Switzerland AG
The registered company address is: Gewerbestrasse 11, 6330 Cham, Switzerland

Preface to Instructors

What Kind of Book Is This?

It has been more than three decades since the first so-called transition book appeared on the mathematics shelves of college bookstores, and there are currently several dozen such books available. The aim of these books is to bridge the gap between the traditional lower-level courses, primarily calculus, and the upper-level courses that require deeper understanding and maturity, such as modern algebra and real analysis. Thus, the main focus of transition books is on the foundations of abstract mathematics, giving a thorough treatment of elementary logic and set theory and introducing students to the art and craft of proof writing.

While this book certainly hopes to provide students with a firm foundation for the upper-level courses of an undergraduate mathematics program, it is not geared solely toward students who intend to major in mathematics. It is the disappointing reality at many institutions that some of the most able students are not considering mathematics as a possible major; in fact, coming out of a standard calculus sequence, most students are not familiar with the true nature of this beautiful subject. Therefore, an important mission of the book is to provide students with an understanding and an appreciation of (abstract) mathematics, with the hope that they choose to study these topics further.

Recognizing that not all our students will have the opportunity to take additional courses in mathematics, this textbook attempts to give a broad view of the field. Even students majoring in mathematics used to complain that they were not given an opportunity to take a course on "mathematics" without an artificial division of subjects. In this textbook, we make an attempt to remedy these concerns by providing a unified approach to a diverse collection of topics, by revisiting concepts and questions repeatedly from differing viewpoints, and by pointing out connections, similarities, and differences among subjects whenever possible. If, during or after reading this book, students choose to take further courses in mathematics, then we have achieved our most important goal.

In order to provide students with a broad exposure to mathematics, we have included an unusually diverse array of topics. Beyond a thorough study of concepts that are expected to be found in similar books, we briefly discuss important milestones in the history of mathematics and feature some of the most interesting recent accomplishments in the field. This book aims to show students that mathematics is a vibrant and dynamic human enterprise by including historical perspectives and notes on the giants of mathematics and their achievements; by mentioning more recent results and updates on a variety of questions of current activity in the mathematical community; and by discussing many famous and less well-known questions that have not yet been resolved and that remain open for the mathematicians of the future.

We also intended to go beyond the typical elementary text by providing a more thorough and deeper treatment whenever feasible. While we find it important not to assume any prerequisites for the book, we attempt to travel further on some of the most enchanting paths than is customary at the beginning level. Although we realize that perhaps not all students are willing to join us on these excursions, we believe that there are a great many students for whom the rewards are worth the effort.

Another important objective—and here is where the author's Hungarian roots get truly revealed—is to center much of the learning on problem-solving. George Pólya's famous book *How to Solve It*[1] introduced students around the world to mathematical problem solving, and, as its editorial review says, "show[ed] anyone in any field how to think straight." Paul Halmos—another mathematician of Hungarian origin[2]—is often quoted[3] about the importance of problems:

> The major part of every meaningful life is the solution of problems; a considerable part of the professional life of technicians, engineers, scientists, etc., is the solution of mathematical problems. It is the duty of all teachers, and of teachers of mathematics in particular, to expose their students to problems much more than to facts.[4]

Our text takes these recommendations to heart by offering a set of carefully chosen, instructive, and challenging problems in each chapter.

It is the author's hope that this book will convince students that mathematics is a wonderful and important achievement of humankind and will generate enough enthusiasm to convince them to take more courses in mathematics. In the process, students should learn how to think, write, and talk abstractly and precisely—skills that will prove immeasurably useful in their future.

---

[1] Originally published in hardcover by Princeton University Press in 1945. Available in paperback from Princeton University Press (2004).

[2] Halmos, author of numerous prize-winning books and articles on mathematics and its teaching, is also known as the inventor of the $\square$ symbol, used to mark the end of proofs, and the word "iff," a now-standard abbreviation for the phrase "if, and only if."

[3] For example, by a report of the Mathematical Association of America Committee on the Teaching of Undergraduate Mathematics (Washington, D.C., 1983) and by the *Notices* of the American Mathematical Society (October 2007, page 1141).

[4] "The Heart of Mathematics," *American Mathematical Monthly* 87 (1980), 519–524.

How Can One Teach from This Book?

Can abstract mathematical reasoning be taught? In my view, it certainly can be. However, an honest answer would probably qualify this by saying that not all students will be able (or willing) to acquire this skill to the maximal degree. I often tell people that, when teaching this course, I feel like a ski instructor: I can show them how the pros do it and be there for them when they need my advice, praise, or criticism, but how well they will learn it ultimately depends on their abilities, dedication, and enthusiasm. Some students will be able to handle the steepest slopes and the most dangerous curves, while others will mostly remain on friendlier hills. A few might become Olympic champions, but most will not; however, everyone who gives it an honest effort will at least learn how to move forward without falling. And, perhaps most importantly, I hope that, even though some occasionally find the training frightening and difficult, they will all enjoy the process.

This second edition of the book contains twenty-two chapters and seven appendices arranged in four parts. Each chapter consists of a lecture followed by about a dozen problems. I am a strong believer in the "spiral" method: topics are often discussed repeatedly throughout the book, each time with more depth, additional insights, or different viewpoints. The chapters are written in an increasingly advanced fashion; the last ten chapters (and especially the last three or four) are particularly challenging in both content and language. The lectures and the problems build on one another; the concepts of the lectures are often introduced by problems in previous chapters or are extended and discussed again in problems in subsequent chapters. (The LaTeX command "ref" appears more than one thousand times in the source file.) Therefore, if any part of a lecture or any problem is skipped, this should be done with caution.

The book is designed both for a one-semester course and a two-semester sequence; the latter choice will obviously allow for a more leisurely pace with opportunities for deeper discussions and additional student interactions. The first three parts—What's Mathematics?; The Foundations of Mathematics; and How to Prove It—should probably be covered in any transition course. The last part—Advanced Math for Beginners—aims to give students a fuller view of mathematics. The ten chapters in this part have an approximate dependency chart as follows:

$$13 \rightarrow 14 \rightarrow 15$$
$$16 \rightarrow 17 \rightarrow 18$$
$$19 \rightarrow 20$$
$$21 \rightarrow 22$$

These four branches are largely independent of one another and can be covered at will.

The heart and soul of this book is in its 329 problems (most with multiple parts); the lectures are intended to be as brief as possible and yet provide enough information for students to attack the problems. I put considerable effort into keeping the number of problems relatively small. Each problem was carefully chosen to clarify a concept, to demonstrate a technique, or to enthuse. There are very few routine problems; most problems will require relatively extensive arguments, creative approaches, or both. Particularly in later chapters, the problems aim for students to develop substantial insight. To make even the most challenging problems accessible to all students, hints are provided liberally.

Many of the problems are followed by remarks aimed at connecting the problems to areas of current research with the hope that some of these notes will invite students to carry out further investigations. The book also contains several appendices with additional material and questions for possible further research. Some of these questions are not difficult, but others require a substantial amount of ingenuity— there are even known open conjectures among them; any progress on these questions would indeed be considered significant and certainly publishable. I feel strongly that every undergraduate student should engage in a research experience. Whether they will go on to graduate school, enroll in professional studies, or take jobs in education, government, or business, students will benefit from the opportunities for perfecting a variety of skills that a research experience provides.

In order to reach a larger audience with this second edition, I made several changes to the first edition. Much of the material was reorganized and streamlined to better match the outline of typical transition courses. In particular, the chapter on functions was moved earlier (and thus received a less abstract treatment); the chapters on mathematical structures moved further back and were augmented by a new separate chapter on groups; and some of the material on the culture, history, and current directions of mathematics were placed into (much enhanced) appendices. This reorganization of course meant that some of the material in other parts of the book had to be rewritten. The second edition includes some of the new developments of the past few years, such as the Ternary Goldbach Conjecture (now theorem) and the improved lower bound on the chromatic number of the plane. Furthermore, to make this admittedly challenging book more approachable, new problems were included and additional guidance (in the form of hints for ways to get started or warnings of potential dead-ends) was added.

Allow me to add a few notes on my personal experiences with this book. I taught courses using this text dozens of times, but find it challenging each time. The approach I find best suited for this course is one that maximizes active learning and class interaction (among students and between students and myself). Students are asked to carefully read the lecture before class and to generate solutions to the assigned problems. Our organized and regular out-of-class "Exploratorium" sessions—where students work alone or with other students in the class under the supervision of teaching associates—seem particularly beneficial in helping students prepare for class. I spend nearly every class by asking students to present the results of their work to the class. As I tell them, it is not necessary that they have completely correct solutions, but I expect them to have worked on all of the problems before class to the best of their ability. I try to be generous with encouragement, praise, and constructive criticism, but I am not satisfied until a thorough and complete solution is presented for each problem. It is not unusual for a problem to be discussed several times before it gets my final PFB ("Perfect for Béla") approval.

Without a doubt, teaching this course has been one of the most satisfying experiences I have had in this profession. Watching my students develop and succeed, perhaps more so than in any other course, is always a superbly rewarding adventure.

Preface to Students

Who Is This Book Written For?

This book is intended for a broad audience. Any student who wishes to learn and perfect his or her ability to think and reason at an advanced level will benefit from taking a course based on this textbook. The skills of understanding and communicating abstract ideas will prove useful in every professional career: law, medicine, engineering, business, education, politics, science, economics, and others. The ability to express oneself and to argue clearly, precisely, and convincingly helps in everyday interactions as well. Just as others can see if we look healthy physically, they can also assess our intellectual fitness when they listen to our explanations or read our writings. Abstract mathematics, perhaps more than any other field, facilitates the learning of these essential skills.

An important goal of this book, therefore, is to help students become more comfortable with abstraction. Paradoxically, the more one understands an abstract topic or idea, the less abstract it will seem! Thus, the author's hope is that his *Invitation to Abstract Mathematics* is accepted, but that, by the time students finish the book, they agree that there is no need for the word "abstract" in the title—indeed, this book is (just) about mathematics.

The prerequisites to the text are minimal; in particular, no specific knowledge beyond high school mathematics is assumed. Instead, students taking this course should be willing to explore unusual and often difficult topics and be ready to face challenges. Facing and overcoming these challenges will be students' ultimate reward at the end.

How Can One Learn from This Book?

Welcome to abstract mathematics! If you are like 99 percent of the students who have taken a course based on this book, you will find that the course is challenging you in ways that you have not been challenged before. Unlike those in your previous mathematics courses, the problems in this book will not ask you to find answers using well-described methods. Instead, you will face problems that you have not seen before, and you will often find yourself puzzled by them for hours, sometimes days. In fact, if you do not need to struggle with the concepts and problems in this book, then this is the wrong course for you since you are not being challenged enough to sharpen your mind. (You should not worry too much about not being challenged though!)

My recommendations to you are as follows. If your instructor gives a lecture, make sure you understand what is being said by asking questions—your classmates will be grateful too. At home, read the relevant material *slowly*. Again, if anything is not completely clear, ask. Next, attack the assigned problems. Do not say that I did not warn you: these problems are hard! In almost all cases, you will need several attempts before you find a solution. It can happen that you spend days without any progress on a given problem, or that the solution you discover at midnight will prove wrong when you try to write it up the next morning. If this happens, *don't panic!* And, *don't give up!*

If you feel you do not know where to start, make sure that you understand what the problem is asking. Look at special cases. Draw illustrative diagrams. Try to turn the question into a simpler question and solve that first. If all these fail, give yourself a break (by moving to another problem) and come back to the beast at a later time.

If you are not sure whether your solution is valid, explain it to others. If they are not convinced or cannot follow your argument, it often means that you have a gap in your proof. Making a jump from one statement to the next without being able to furnish the details means that your work is incomplete.

Even if you succeed in solving a problem, I recommend that you consult with other students in your class. It is always beneficial to discuss your work with others; you might find it interesting to listen to the thoughts of a variety of people and to compare different approaches to the same questions. You might learn more from these brainstorming sessions than by working alone—and, for all but the most antisocial people, it is a lot more fun too!

I am absolutely convinced that your work will pay off, and I hope that you will find it enjoyable as well!

Acknowledgments

I am fortunate to have grown up in a country with a long tradition of superb mathematics education. The legendary culture of mathematics in Hungary[1] was influenced by the many giants who took an active role in education at all levels; for example, Paul Erdős, the most prolific mathematician in history, was a frequent visitor to elementary schools to talk about "adult" mathematics. I remain greatly influenced by the excellent education I received in Hungary, and I am especially grateful to Professors Róbert Freud, Edit Gyarmati, Miklós Laczkovich, and Lajos Pósa who are most responsible for my love of mathematics.

This book is a result of many years of teaching a transition course at Gettysburg College. I would like to thank all the students who used earlier versions of the book during the past—their feedback was invaluable. I am particularly thankful to my colleagues Darren Glass, Benjamin Kennedy, and Keir Lockridge for having used the first edition of this book; the second edition incorporates their many insightful observations and recommendations. I also greatly benefited from the advice on both content and style provided by other friends and colleagues, including Matthias Beck, Steve Berg, Joseph Bonin, Jill Dietz, Róbert Freud, Klaus Peters, Máté Wierdl, and Paul A Zeitz. My students Peter Francis, Emma Gruner, and Hoang Anh Just and freelance editor Martha Masterson Francis offered to carefully read the draft of the second edition—I am very thankful to them for the astute comments and suggestions. I also wish to express my gratitude to my editor Loretta Bartolini for her attentive handling of the manuscript, to the UTM series editors for helpful suggestions, and to everyone else at Springer who assisted with the production of this book.

I would be interested in hearing your opinions, whether you are an instructor, a student, or a casual reader of this book. Also, please let me know of any mistakes or typos. I can be contacted at bbajnok@gettysburg.edu. Thanks!

---

[1] See, for example, "A Visit to Hungarian Mathematics," by Reuben Hersch and Vera John-Steiner, in *The Mathematical Intelligencer*, Volume 15, (2), (1993) 13–26.

Contents

Part IV Advanced Math for Beginners

Part I
What's Mathematics?

Chapter 1
Let's Play a Game!

We start our friendship with abstract mathematics with an example that illuminates how concrete quantitative problems may turn into abstract mathematical situations and how mathematicians might be led from specific questions to the development of highly abstract concepts and discoveries.

We should admit right away that not all abstractions have an immediate application; indeed, many branches of mathematics were born and developed independently of the "real world." Some of these areas later found critical uses; for example, number theory, dubbed by Carl Friedrich Gauss (1777–1855) as the "Queen of Mathematics" but long considered arcane, is now playing major roles in computer technology. Other areas, particularly those recently developed, are still waiting for applications.

Our example comes from the interesting, newly-developing field of combinatorial game theory. We have simplified the problem as much as possible to reduce technicalities and make our computations simpler. Yet this particular game—and the more challenging (yet, more interesting) games introduced in the problems at the end of this chapter—will enable us later on to discuss some of the most fundamental elements of abstract mathematics, both its objects of study (for example, the mathematical structures of an Abelian group and of a field) and its methods (logic, quantifiers, proof techniques, and more). In fact, at the end of our journey, once we develop the necessary tools, we will return to these games to provide a more thorough and far-reaching analysis to see how we can compare games to one another and decide, for example, when one game is more "advantageous" for a particular player than another game! This will lead us to a deeper understanding of numbers (zero, positive and negative integers, rational numbers, real numbers, and surreal numbers). As the title of Chapter 22 declares, games are "value"-able!

To begin, let us consider the following situation. Suppose that there is a group of competing companies, all trying to gain control of a certain segment of the market. Each company has a number of possible courses of action that it can carry out (such as introducing or terminating a product, changing prices, etc.). These options

© Springer Nature Switzerland AG 2020

B. Bajnok, *An Invitation to Abstract Mathematics*, Undergraduate Texts in Mathematics, https://doi.org/10.1007/978-3-030-56174-1_1

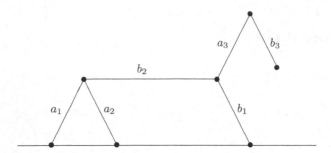

Fig. 1.1 Diagram of the six options for companies A and B arranged to form a "horse"

typically have a limiting effect on one another; in other words, once an option (or a set of options) has been carried out by a certain company (or companies), it prevents another company (or perhaps the same company) from exercising some of its options later.

In our specific example we have two companies, A (Apple Core Corp.) and B (Blue Ink, Inc.), with company A having three available options, which we denote by a_1, a_2, and a_3, and company B having another three available options, denoted by b_1, b_2, and b_3. Let us assume that the restrictions are as follows:

- b_3 is available only if a_3 is still available;
- a_3 is available only if at least one of b_1 or b_2 is still available; and
- b_2 is available only if at least one of a_1, a_2, or b_1 is still available.

We assume that the two companies take turns carrying out their options, with each company having to exercise one of its options when it is its turn (each option can only be performed once). The first company unable to exercise any of its options loses. Our question: Which of the two companies is in a better position when the game starts?

It is helpful to model our game with a figure. In Figure 1.1 the options available to for companies A and B are illustrated by line segments. A move will be reflected by removing the corresponding segment from the figure, and the restrictions listed above will be seen in the fact that, when certain segments are removed, then some other segments get disconnected from the base. For example, the first restriction means that if we remove the "neck" of the horse, then its "head" gets cut off. It is easy to check that our figure reflects all three of our restrictions and no others. As we play our game, we will regard segments disconnected from the ground as unavailable options.

Let us see now how we can analyze this game that we call *Arion* (after the horse with the same name in Greek mythology). Note that we did not specify above which company, or "player" from now on, starts the game. Is the winner going to depend on that? In many games it certainly does.

We will see, however, that no matter who starts *Arion*, A has a way to win against B no matter how B plays. To see this, we check all possibilities as follows.

Fig. 1.2 The result after A removes option a_3

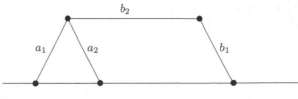

Fig. 1.3 The decision tree when A starts with option a_3

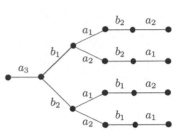

Assume first that A starts. One (and, as we will see shortly, the best) choice for A is to remove a_3 first. Then b_3 becomes unavailable for B, and the game reduces to that of Figure 1.2.

Now both A and B have two available options, and it is B's turn. No matter how B moves, the game will terminate in two more rounds, and it will be B who will first run out of options. Thus we see that, if A starts, then for *any* sequence of moves by B, A has *some*—maybe not always the same—way of responding that guarantees a win for A. We describe this situation by saying that A has a *winning strategy*.

What if B has the chance to move first? If B starts by removing b_3, then A can respond by removing a_3, and the game reduces to the one that we just analyzed, and B loses. The other two initial options are even worse for B, since A will again respond by removing a_3, and the game will terminate in just one more round by B running out of options. Again, A has *some* winning sequence of moves for *every* sequence of moves by B. We see, therefore, that no matter who starts our game, A will be able to win no matter how B plays. We can say that this game is a *win for A*.

We can similarly analyze larger games, but it is not hard to imagine that our arguments become complicated as the number of options and restrictions increases. There is a more systematic approach that, though only after some rather tedious work, makes the conclusion clear. This approach uses so-called *decision trees*.

Part of the decision tree of our game, showing all possible plays that start with A starting the game by removing a_3, is shown in Figure 1.3. The figure is read from left to right and shows all possible moves for each round. We see that A can win the game in all cases.

In comparison, consider Figure 1.4, which shows all possible plays if A starts by removing a_1.

We see from the tree that in this case B has a winning strategy. Namely, if B removes b_3, then A loses whether it plays a_2 or a_3 (in the latter case B should respond by removing b_2). The situation is similar if A's initial move is a_2; B again will win if it responds by b_3. Thus, B has *some* winning strategy against *some* initial

Fig. 1.4 The decision tree
when A starts with option a_1

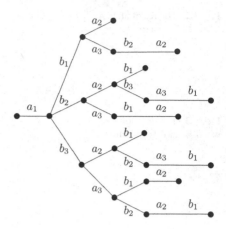

moves on A's part. However, we saw that the game is a win for A; that is, if A plays
optimally, the game is won by A regardless of who starts the game or how B plays.

The game that we just analyzed is an instance of the so-called *Hackenbush
games*. Naturally, not all competitive situations can be modeled by *Hackenbush*
games. The field of *combinatorial game theory* attempts to analyze and evaluate
various kinds of games as well as to develop a theory that applies to all games.

We will often employ games to introduce abstract topics and methods in this
book. We introduce some of these games in the problem set below.

Problems

1. Draw diagrams that represent the following *Hackenbush* games. Assume that
   the available options for companies A and B are a_1, a_2, a_3 and b_1, b_2, b_3,
   respectively, subject to the restrictions listed below. Use decision trees or other
   arguments to decide which player has a winning strategy.

   (a) • b_3 is available only if a_3 is still available;
       • a_3 is available only if at least one of a_1 or a_2 is still available;
       • b_2 is available only if at least one of a_2 or b_1 is still available; and
       • a_2 is available only if at least one of a_1 or b_1 is still available.
       (Hints: Prove that the game is a win for A. Note that in order to prove this
       claim, it suffices to verify that A has *some* sequence of winning moves no
       matter what B's moves are. In particular, when A starts, it suffices to exhibit
       one initial move that leads to a win regardless of how B responds, but when
       B starts, we need to consider what A's response is to *each* of B's initial
       moves.)

(b) • b_3 is available only if b_2 is still available;
  • b_2 is available only if at least one of a_1 or a_2 is still available;
  • a_3 is available only if at least one of a_2 or b_1 is still available; and
  • a_2 is available only if at least one of a_1 or b_1 is still available.

2. The *Divisor* game is played by two players who choose a positive integer n and
   then take turns naming a positive divisor of n with the condition that they cannot
   name a multiple of a number named earlier (including the number itself). The
   player who is forced to name the number 1 loses (and the other player wins). For
   example, if the originally chosen number is $n = 9$, then the first player can name
   3 or 9 (the third choice, being 1, would result in an immediate loss). If the first
   player chooses 3, then the second player loses immediately as the only number
   left to be named is 1; therefore, the first player has a winning strategy. (If the
   first player chooses 9, then the second player will choose 3, after which the first
   player is forced to name 1 and will lose the game. Thus, the first player must start
   the game by naming 3 in order to win.)
   Use decision trees or other arguments to decide which player has a winning
   strategy if the initial number is

   (a) $n = 8$;
   (b) $n = 10$;
   (c) $n = 12$.

   Remark Later in the book we return to the *Divisor* game for a more general
   analysis.

3. Two players, High and Low, play the following game on the "board" shown in
   Figure 1.5. Given a coin on the top node, they take turns to move the coin one

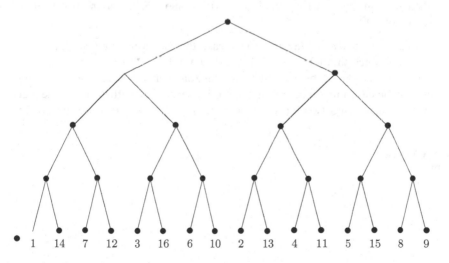

Fig. 1.5 The board for the game of High and Low

node downward until the coin arrives at one of the sixteen terminal nodes. High's goal is that the game ends with the coin at a node with as high a value as possible, while Low's aim is that the target node has as low a value as possible. Assuming that both of them play optimally, what is the outcome (the value of the last node) of the game if

(a) High starts?
(b) Low starts?

(Hints: It may be helpful to think "backwards": For example, supposing first that High starts, how will the game end if after three moves the coin arrives at the node above nodes 1 and 14? How will it end if the coin is at the node above nodes 7 and 12? Given your answers to these two questions, what should High's choice be if the coin is at the left-most node of the middle row?)

4. The game *Capture* is played by two players, First and Second, who take turns to move toward one another on a narrow bridge. Initially, the two players are at opposite ends of the bridge, at a distance of n feet from each other. (Here n is an arbitrary positive integer.) When it is their turn, they are allowed to jump 1, 2, or 3 feet toward each other. The game starts by First making a move first; the game is over when one player is able to capture (jump on top of) the other player.

(a) Which player has a strategy to capture the other player if $n = 8$?
(b) What if the initial distance is $n = 21$ feet?
(c) Generalize the problem for all positive integers n.
(d) Generalize the problem further to the case in which the initial distance is n feet and the players are allowed to jump any (positive) integer number of feet up to k feet. (Here n and k are arbitrary positive integers.)

Remark This game—with $n = 21$ and $k = 3$, as in part (b)—was a featured challenge on the American reality television show *Survivor* in the "Thailand" season of 2002.

5. The game *Cutcake* is played using a cake that has a rectangular shape and is scored by lines parallel to its edges, as shown in Figure 1.6.

The game is played by two players, Horizontal and Vertical, who take turns cutting the cake into pieces along the scoring, each time cutting one of the pieces created earlier into two pieces. Horizontal is only allowed to make horizontal

Fig. 1.6 A cake of size
3-by-5

cuts, and Vertical is to make vertical cuts. The game ends when one player is unable to move; this player is then the loser, and the other player is the winner.

(a) Which player has a winning strategy on the cake illustrated above if Horizontal moves first?

(b) Which player has a winning strategy on the same cake if Vertical moves first?

(c) Repeat parts (a) and (b) for a cake of size 4-by-7 (i.e., the cake is scored by three horizontal lines and six vertical lines).

Remark We will periodically return to *Cutcake* to provide further analysis.

6. The classical game *Nim* is played with a finite collection of heaps where each heap consists of some chips (or coins, matches, etc.). The notation $N(n_1, n_2, \ldots, n_m)$ stands for the game played with m heaps, containing $n_1, n_2, \ldots, n_m$ chips, respectively; here m is a positive integer and $n_1, n_2, \ldots, n_m$ are (not necessarily distinct) nonnegative integers. Two players, named First (making the first move) and Second, take turns to select a single heap and to remove any number of chips from it. They are allowed to remove every chip from their chosen heap if they so choose but must remove at least one chip. The player who is unable to move loses (and the other player wins).

The *Nim* game $N(n_1)$, consisting of a single heap, is quite obvious to analyze: If $n_1 = 0$, then First loses immediately (and thus Second wins); if $n_1 > 0$, then First wins by removing all chips (and thus Second loses).

(a) Which player has a winning strategy for the two-heap game $N(5, 8)$?

(b) Generalize part (a) for any two-heap game $N(n_1, n_2)$.

(c) Which player has a winning strategy for the three-heap game $N(2, 3, 5)$?

(d) Which player has a winning strategy for the three-heap game $N(5, 7, 9)$?

7. (a) The game *Quattro* is played by two players, First and Second (with First making a move first) on a game board that has four points marked; no three of the points are collinear (are on the same straight line). The players take turns, each time connecting two of the four points by a straight line segment. They are allowed to connect any two of the four points that have not been connected yet, except that they are not allowed to form a triangle (a triangle is formed when each of three original points is connected with the other two; triangles whose vertices are not all among the four given points are okay). The first player unable to move loses. Draw the decision tree for this game and determine which player has a winning strategy.

(b) A modified version of *Quattro* is played on the same board, but this time the players' goal is to create a triangle; the player first able to do so wins the game. Determine which player has a winning strategy.

(Hints: The key to being able to provide a complete analysis to these games, without having to discuss a huge number of possible move sequences, is in understanding how certain positions are essentially the same. (The formal word for this equivalence is that the positions are *isomorphic*.) Clearly, all first moves by First are essentially the same, so it is sufficient to assume that the move connects points 1 and 2, for example. Second can then have two choices: either draw in a segment that is adjacent to the segment drawn by the first player or draw one that is not. Therefore, we only have to distinguish between two cases: the segment can connect points 2 and 3, or 3 and 4. Continuing the analysis this way greatly reduces the number of cases to be considered.)

8. (a) The game *Two-Color Quattro* is similar to *Quattro* (cf. the previous problem) except that the two players use different colors to draw their line segments: player Red draws his connections with red, and player Green draws hers with green. They are allowed to connect any two of the four points that have not been connected yet, except that they are not allowed to form a mono-chromatic triangle (a triangle whose three sides are drawn with the same color). The first player who is forced to create a monochromatic triangle loses; if the game ends without either player creating a monochromatic triangle, then we say that the game ends in a draw. Without loss of generality, assume that Red makes the first move. Does either player have a winning strategy or does the game end in a tie?

   (b) The game *Two-Color Penta* is just like *Two-Color Quattro*, but it is played on a board with five points. Is it possible that, perhaps not playing optimally, the two players end the game in a tie?
   (Hints: Do not attempt to analyze who has the winning strategy in *Two-Color Penta* as it is quite complicated! Note that the question is simply asking whether it is possible to make all connections among the five points with five red segments and five green segments without a mono-chromatic triangle.)

   (c) Is a tie possible in *Two-Color Hexi* (the similar game on six points)?
   (Hints: Prove that a tie is not possible. Labeling the six points A, B, C, D, E, and F, start by considering the five edges AB, AC, AD, AE, and AF. What are the possibilities for their colors? Explain why each possibility leads to a monochromatic triangle once all segments are drawn.)

   Remark A group of researchers at McMaster University in Canada showed in the 1970s that this game, sometimes referred to as *SIM*, is a win for the second player, but the winning strategy is very complicated.

9. (a) The game *Acrostic Twins* is played on a rectangular grid where each square contains a coin; in the initial position, each coin is showing "heads." Two players, First and Second (as usual, First is to move first) take turns; each move entails (i) selecting a coin that shows heads, (ii) selecting another coin to the left of it in the same row or above it in the same column (this second coin may show either heads or tails), and (iii) turning over both coins. The first player unable to move loses (and the other player wins).
   For each positive integer m and n with $m \leq 5$ and $n \leq 5$, decide which player has a winning strategy on an m-by-n board.

   (b) The game *Turning Corners* is similar to *Acrostic Twins*, but here a move consists of turning over the four corners of a rectangle (whose edges are parallel to the sides of the board) with the condition that the coin at the lower right must show heads. Again, the initial configuration shows all heads, and the game ends when a player is unable to move (this player then loses and the other wins).

For each positive integer m and n with $m \leq 5$ and $n \leq 5$, decide which player has a winning strategy on an m-by-n board.

Chapter 2
What's the Name of the Game?

Abstract mathematics deals with the analysis of mathematical concepts and state-ments. The concepts of mathematics have a precise and consistent meaning, and its results, once established, are not subject to opinions or experimental verification and remain valid independently of time, place, and culture—although their perceived importance might vary. In this chapter we discuss mathematical concepts; in Chapter 3 we study mathematical statements.

Mathematical concepts are usually introduced by *definitions*. A definition needs to say unambiguously what the meaning of the newly introduced concept is and has to be expressed in terms of previously introduced concepts only.

As an example, consider the definition of primes, which can be stated as follows:

Definition 2.1 *An integer is a* prime *if it has exactly two positive divisors.*

For example, 2, 3, 5, and 7 are prime numbers, since their two positive divisors are 1 and themselves. Note that our definition allows for the possibility of negative primes as well; indeed, -2, for example, is a prime as it has two positive divisors (1 and 2). But 4, 6, 8, and 9 (and their negatives) are not primes, as they have more than two positive divisors; such integers are called *composites*. Note that 1 and -1 are not primes either, since they have only one positive divisor, and neither is 0 a prime, as it has infinitely many positive divisors. We can categorize integers according to how many positive divisors they have. Letting $d(k)$ denote the number of positive divisors of the integer k, we can say that

- if k is 1 or -1, then $d(k) = 1$;
- if k is a prime number, then $d(k) = 2$;
- if k is a composite number, then $d(k)$ is greater than 2 but finite; and
- if $k = 0$, then $d(k)$ is infinite.

Primes are the basic building blocks of integers. According to the *Fundamental Theorem of Arithmetic* (which we will prove in Chapter 11), every integer n with $n \geq 2$ is either a prime or can be expressed as a product of primes; furthermore, this

© Springer Nature Switzerland AG 2020
B. Bajnok, *An Invitation to Abstract Mathematics*, Undergraduate Texts in
Mathematics, https://doi.org/10.1007/978-3-030-56174-1_2

factorization into primes is essentially unique (that is, there is only one factorization if we ignore the order of the prime factors or the possibility of using their negatives). Note that this uniqueness is one reason why we do not consider 1 to be a prime; if it were, then even the number of prime factors would not be unique (e.g., we could factor 6 as $6 = 2 \cdot 3$ or $6 = 1 \cdot 2 \cdot 3$ or $6 = 1 \cdot 1 \cdot 2 \cdot 3$, etc.). We will soon see other important properties of primes. An excellent source of information for primes and their properties (including current prime records) can be found at www.primes.utm.edu.

The wording of a definition needs to be chosen carefully. For example, none of the descriptions

- integers that are divisible by 1 and themselves,
- integers that have no divisors between 1 and themselves, or
- integers that do not factor into a product of other integers

provides a correct definition for primes. The first description applies to every integer (e.g., 6 is certainly divisible by both 1 and 6); the second includes 1 (it has no divisors between 1 and itself); and the last condition is not satisfied by any number, since every integer factors (e.g., $7 = 1 \cdot 7$ or $7 = (-1) \cdot (-7)$).

However, there might be several alternative ways to define primes correctly. For example, it is easy to see that the following would work.

Definition 2.1a An integer is a *prime* if it is different from 1 and -1 and it cannot be factored into a product of two integers without one of them being 1 or -1.

To be able to claim that Definition 2.1a is *equivalent* to Definition 2.1 requires us to verify that Definition 2.1a is satisfied by primes but not by integers that are not primes. This can be easily accomplished, as follows. Definition 2.1a explicitly excludes 1 and -1, and it also excludes 0 since we can factor 0 as, say, $0 = 0 \cdot 2$. Furthermore, it also excludes composite numbers, since for a composite number k to have more than two positive divisors means that k has factorizations different from $k = 1 \cdot k$ or $k = (-1) \cdot (-k)$ (only one of k or $-k$ is positive). Conversely, we see that for a prime number p to have exactly two positive divisors means that p cannot be factored in any way other than $p = 1 \cdot p$ or $p = (-1) \cdot (-p)$. Therefore, Definition 2.1a is satisfied by primes and nothing but the primes, so it is equivalent to Definition 2.1.

It is considerably more difficult to see that the following is also a valid definition for primes.

Definition 2.1b (Euclid's Principle) *An integer is a* prime *if it is different from 0, 1, and -1, and it cannot divide a product of two integers without dividing at least one of them.*

One can prove that Definition 2.1b is equivalent to Definition 2.1, that is, that the two definitions are satisfied by the same set of integers. Since Definition 2.1b explicitly excludes 0, 1, and -1, one only needs to prove that it also excludes composite numbers but includes all primes. We can illuminate why composite numbers don't satisfy Definition 2.1b by considering, for example, $p = 20$: it does not divide 4 or 5, but it divides the product of 4 and 5. We can easily adapt this argument for an *arbitrary* composite number—see Problem 3. The claim that a prime number cannot divide a product of two integers without dividing at least one of them was stated first by the Greek mathematician Euclid of Alexandria about twenty-three hundred years ago. For example, for $p = 2$, this claim states that the product of two integers can only be even if at least one of the factors is even. While this may be quite clear, the claim for an arbitrary prime is considerably more difficult to verify—we will provide a proof in Chapter 10.

A definition is given in terms of other concepts, and we assume that, by the time we state our definition, we are familiar with their meanings. This can happen in two ways: either these concepts have been previously defined, or we agree to understand their meaning without definitions. Concepts that are assumed to be understood without definitions are called *fundamental concepts* or *primitives*. Every branch of mathematics will have to have some fundamental concepts, as we cannot reduce our concepts to previous terms endlessly.

At this point, it might be desirable for us to list all fundamental concepts so that we can rely on them as we proceed. We will not be following such a "fundamentalist approach," however. Although it may sound tempting to build our theory on such a solid foundation, it turns out that this approach is not feasible; it would be too cumbersome, too lengthy, and too limiting (not to mention quite boring for most people). For now, let us assume that the following concepts are primitives:

- *natural number, integer, rational number, real number*, and *complex number*;
- *equality, addition*, and *multiplication of two real (rational, etc.) numbers*;
- *positive real number*;
- *set* and *element of a set*;
- *point, line, plane*, and *intersection of lines and planes*;
- *segment, angle*, and *congruent segments and angles*;
- *distance between two points*.

Note that the choice of primitives is unavoidably arbitrary. For example, we listed the term "positive real number" as a primitive (and then can use it to define the term "negative real number"; see Problem 1). We could have, instead, listed "negative real number" as a primitive (and then could use that to define "positive real number").

Another reason for our fluidness with the list of primitives is that, depending on our desire to avoid shortcuts, we could reduce the number of primitives on our list. For example, instead of assuming that the natural numbers, the integers, the rationals, the reals, and the complex numbers are all primitives, it would suffice to just list two numbers: 0 and 1. We could then provide formal definitions for all other numbers. We may use the concepts of 1 and addition to define all positive integers: 2 can be defined as $1 + 1$, after which 3 can be defined as $2 + 1$, and so on. Then we could use the positive integers and zero to define the negative integers; from the set of all integers we might define the set of rational numbers, from there the real numbers, and finally the complex numbers. We will not provide this development here; however, we return to a more formal development of the number sets in Chapter 21 where even 0 and 1 will be defined!

Our definition of primes in Definition 2.1 uses only one other mathematical concept that needs definition, the concept of divisors. This can be done as follows:

Definition 2.2 *Given two integers a and b, we say that a is a* divisor of *(or divides) b (or b is* divisible *by a) whenever there is an integer c for which a · c = b. If a is a divisor of b, we write a|b.*

For example, $3|6$ and $6|6$, but $6 \nmid 3$. Also, $5|0$, since $5 \cdot 0 = 0$; in fact, $0|0$, since (for example) $0 \cdot 7 = 0$. But $0 \nmid 5$, since (as we will prove in Chapter 14) there is no integer c for which $0 \cdot c = 5$ because for every real number c, we have $0 \cdot c = 0$. Again, we need to be very careful with the precise meaning of our definition. According to Definition 2.2, saying that a is a divisor of b is not quite equivalent to saying that the fraction b/a is an integer: 0 is a divisor of 0, but $0/0$ is not an integer!

We should point out that we listed only the addition and multiplication of *two* numbers. Can we give definitions for operations of three or more terms/factors? How can we define, for example, $2 + 3 + 5$? Or how about 2^7? The answer is provided by *recursive definitions*. For example, we define $2 + 3 + 5$ as the sum of $2 + 3$ (which is understood as a primitive) and 5. The power 2^7 can be defined as the product of 2^6 and 2, where 2^6 is defined as the product of 2^5 and 2, and so on; we can trace this back to 2^2 that can be defined as the primitive concept $2 \cdot 2$. We can formalize this, as follows:

Definition 2.3 *Let $a_1, a_2, a_3, \ldots, a_n$ be a list of real numbers. The sum*

$$a_1 + a_2 + \cdots + a_n,$$

denoted by $\Sigma_{i=1}^{n} a_i$, is defined as follows:

1. *$\Sigma_{i=1}^{1} a_i = a_1$; and*
2. *$\Sigma_{i=1}^{n} a_i = (\Sigma_{i=1}^{n-1} a_i) + a_n$ for any integer $n \geq 2$.*

Note that in our definition of the sum of an arbitrary number of terms, we used only the primitive concept of the sum of two terms (and the method of recursion). Our definition is complete: it defines the sum of an arbitrary number of terms (the "sum" of a single term had to be defined separately). The product $\prod_{i=1}^{n} a_i$ of an arbitrary positive integer number of factors and the positive integer power a^n of a number a can be defined similarly (cf. Problem 4). The issue of having to provide definitions using only previously understood concepts is particularly relevant to such recursive definitions.

The idea of recursive definitions is to use the previous term to define the current term. We often use this method to describe sequences. For example, the power sequence 1, 2, 4, 8, 16, ... can be described (or defined) recursively by $a_1 = 1$; and for $n \geq 2$, $a_n = 2a_{n-1}$.

Recursions can even use more than just the previous term; for example, if the terms of the sequence are defined using the previous two terms, the recursion is said to have *order* 2. Consider, for example, the sequence defined as follows: $F_1 = 1$,

$F_2 = 2$, and, for $n \geq 3$, $F_n = F_{n-1} + F_{n-2}$. This sequence is called the *Fibonacci sequence* after the Italian merchant and mathematician Leonardo of Pisa (c. 1170–1250) who was known as Fibonacci. Is this second-order recursion for this sequence of numbers well defined? Using the recursion, we find that $F_3 = F_2 + F_1 = 2 + 1 = 3$, $F_4 = F_3 + F_2 = 3 + 2 = 5$, and $F_5 = F_4 + F_3 = 5 + 3 = 8$. Continuing this way, we arrive at the sequence

$$1, 2, 3, 5, 8, 13, 21, 34, 55, 89, 144, 233, \ldots$$

We see that each term in the sequence is computed easily from the previous terms; however, this method is not efficient if one only needs to find a particular term in the sequence as we need to compute all previous terms. An explicit formula exists too: in Chapter 11 we will prove that

$$F_n = \frac{1}{\sqrt{5}} \left[\left(\frac{1 + \sqrt{5}}{2} \right)^{n+1} - \left(\frac{1 - \sqrt{5}}{2} \right)^{n+1} \right]$$

for every natural number n. It is surprising that, while the formula involves irrational numbers, it claims to give an integer value (namely, F_n) for every n. The number

$$\frac{1 + \sqrt{5}}{2} = 1.61803398875\ldots$$

is a famous number, sometimes referred to as the *golden ratio*; the other irrational base involved in the above expression, $\frac{1 - \sqrt{5}}{2}$, is the negative of its reciprocal. Since $\frac{1 - \sqrt{5}}{2}$ has absolute value less than 1, its powers become very small for large values of its exponent. This fact enables us to compute F_n even faster: namely, F_n is simply the closest integer to

$$\frac{1}{\sqrt{5}} \left(\frac{1 + \sqrt{5}}{2} \right)^{n+1} .$$

This explains why the ratio of two consecutive Fibonacci numbers is close to the golden ratio; indeed, it can be verified that the sequence

$$\frac{2}{1}, \frac{3}{2}, \frac{5}{3}, \frac{8}{5}, \frac{13}{8}, \frac{21}{13}, \frac{34}{21}, \frac{55}{34}, \ldots$$

approaches exactly $\frac{1 + \sqrt{5}}{2}$. Fibonacci numbers appear in surprising places in mathematics, nature, music, art, architecture, and elsewhere.

Problems

1. Write a precise definition for the following concepts. Use only the primitives listed and the concepts defined in this chapter. (Once you define a concept, you may choose to use it in the definition of a subsequent concept in the list below.)

   (a) An *even* integer;
   (b) an *odd* integer;
   (c) a *negative* real number;
   (d) the *negative of a real number*;
       (Hint: The phrases "negative real number" and "negative of a real number" refer to two separate concepts; for example, the negative of the real number -6 (namely, 6) is not a negative real number!)
   (e) the *reciprocal of a nonzero real number*;
   (f) a real number being *greater than* another;
   (g) a real number being *less than* another;
   (h) two positive integers being *relatively prime*;
   (i) three positive integers being *relatively prime*;
   (j) three positive integers being *pairwise relatively prime*;
   (k) a *perfect square* number;
   (l) the *square root(s)* of a nonnegative number;
   (m) the *absolute value* $|x|$ of a real number x;
       (Hint: It is best to separate cases when $x \geq 0$ and when $x < 0$.)
   (n) the *floor* $\lfloor x \rfloor$ of a real number x;
       (Hint: Avoid yet-to-be-defined decimal representations as they may be misleading; for example, $\lfloor 3.9999 \ldots \rfloor = 4$, not 3!)
   (o) the *ceiling* $\lceil x \rceil$ of a real number x;
   (p) a *circle* of a given radius and center;
   (q) two lines being *parallel*.

2. Definition-like sentences are often used to define notations that appear in mathematics books and other works. For example, as we have already mentioned, for a given integer k, the quantity $d(k)$ denotes the number of positive divisors of k.

   The notations $N(a)$, $K(a, b)$, $M(a, c)$, $P(b, c)$, and $R(a, b, c)$ are taken from an imaginary mathematics paper. Their definitions, in a random order, are stated below; there are also three additional definitions without corresponding notations. Match the notations with their corresponding definitions. Assume that all variables denote integers. Also compute the values of $N(10)$, $K(2, 3)$, $M(2, 3)$, $P(2, 3)$, and $R(1, 2, 3)$.

   (a) $a^3 + b^3$;
   (b) the least integer b for which $b^3 > a$;
   (c) the largest integer b for which $b^3 > a$;
   (d) $a^3 + b^3 + c^3$;

(e) the least integer a for which $b^3 > a^3 + c^3$;
(f) the largest integer a for which $b^3 > a^3 + c^3$;
(g) the least integer b for which $b^3 > a^3 + c^3$;
(h) the largest integer b for which $b^3 > a^3 + c^3$.

3. (a) Decide which of the following descriptions provide a correct definition for primes. Justify your answer.

    i. An integer is a prime if it has exactly four divisors.

    ii. An integer is a prime if it is different from 1 and -1 and it has at most four divisors.

    iii. An integer p is a prime if it is different from 1 and -1 and it cannot be factored into a product of two integers without one of them being $|p|$.

    iv. An integer p is a prime if it is different from 1 and -1 and it cannot be factored into a product of two integers with both factors having absolute value less than $|p|$.

    v. An integer p is a prime if it is different from 0, 1, and -1 and it cannot be factored into a product of two integers with both factors having absolute value less than $|p|$.

    vi. An integer is a prime if it is different from 1 and -1 and it cannot divide a product of two integers without dividing at least one of them.

(b) Explain why any integer satisfying Definition 2.1b also satisfies Definition 2.1.

(Hints: We only need to verify that an integer p satisfying Definition 2.1b cannot be composite. Note that if p has a positive divisor a, then p divides the product of the integers a and p/a. Explain why p dividing a can only happen if $a = |p|$ and why p dividing p/a can only happen if $a = 1$. This then means that an integer p satisfying Definition 2.1b cannot have more than two positive divisors and thus cannot be composite.)

4. Recall that we consider multiplication of *two* real numbers as a primitive.

(a) Define the product of three real numbers.
(b) Define the product of five real numbers.
(c) Give a recursive definition for the product of an arbitrary (positive integer) number of real numbers.
(Hint: Try adapting Definition 2.3.)
(d) Give a recursive definition for an arbitrary positive integer exponent of a real number.

Remark Utilizing the definition of products from part (c), it is entirely possible to define a^n explicitly as $\prod_{i=1}^{n} a$ for any real number a and positive integer n. However, this problem is asking you to provide a recursive definition.

5. (a) Verify that the number 8, 191 is prime.

   (Hints: Explain how this can be done by performing 8, 190 divisions by making a general statement of the form "If a number n is not divisible by $\ldots$, then it is prime." Much better: Explain why it suffices to use 89 divisions (again make a general statement). Can the primality of 8, 191 be established with even fewer divisions?)

   (b) Describe an efficient method that finds all positive prime numbers up to 100. How about up to 1, 000?

6. In this problem we study two of the most famous number sequences: Mersenne numbers and Fermat numbers. (Our approach here will be experimental; a more precise treatment—with proofs—will be provided in Chapter 11.)

   (a) The n-th *Mersenne number* can be defined recursively by $M_0 = 0$ and

   $$M_n = n + \sum_{i=0}^{n-1} M_i$$

   for $n \geq 1$.

      i. Compute M_n for $0 \leq n \leq 5$.

      ii. Find a recursive definition for M_n that has order 1, that is, the recursive formula for M_n only involves M_{n-1} (and possibly some constants).

      iii. Find an explicit formula for M_n, that is, a formula that involves n only (and possibly some constants).

   Remarks The number M_n is called the n-th *Mersenne number*, named after the French monk Marin Mersenne (1588–1648) who first studied them. It is a very intriguing problem to find those Mersenne numbers that are prime. At the time of writing, there are only fifty-one Mersenne primes known. The one most recently found is the largest of them, having close to twenty-five million decimal digits. For more information on the *Great Internet Mersenne Prime Search (GIMPS)*, see www.mersenne.org. We will study Mersenne numbers in Chapters 3 and 4 in more detail.

   (b) The n-th *Fermat number* can be defined recursively by $F_0 = 3$ and

   $$F_n = 2 + \prod_{i=0}^{n-1} F_i$$

   for $n \geq 1$.

      i. Compute F_n for $0 \leq n \leq 5$.

      ii. Find a recursive definition for F_n that has order 1.

      iii. Find an explicit formula for F_n.

Remarks The great French mathematician Pierre Fermat (160?–1665) conjectured that F_n is prime for every nonnegative integer n. Indeed, F_n is prime for $0 \leq n \leq 4$. However, this conjecture of Fermat turned out to be false, as Leonhard Euler (1707–1783) discovered in 1732 that 641 is a factor of F_5 (checking this is easy, but finding factorizations of large numbers is a highly difficult problem). The only Fermat primes known to this day are the five values found above; today we know that F_n is definitely not prime for $n = 5, 6, \ldots, 32$. Fermat primes have an interesting connection with the constructibility of regular polygons (see Theorem B.3 in Appendix B).

7. Which of the following sequences are well defined? If a sequence is well defined, find its fifth term.

   (a) $a_1 = 1, a_n = a_{n-1} + a_{n-2}$ for $n \geq 2$.
   (b) $b_1 = 1, b_2 = 2, b_n = b_{n-1} + b_{n-2}$ for $n \geq 4$.
   (c) $c_1 = 1, c_2 = 2, c_{n+2} = c_{n+1} + c_n$ for $n \geq 1$.
   (d) $d_1 = 1, d_2 = 2, d_3 = 3, d_n = d_{n-1} + d_{n-2} + d_{n-3}$ for $n \geq 4$.
   (e) $e_1 = 1, e_2 = 2, e_n = e_{n-1} + e_{n-2} + e_{n-3}$ for $n \geq 3$.
   (f) $f_1 = 1, f_2 = 2, f_n = f_{n+1} + f_{n-1}$ for $n \geq 3$.
   (g) $g_1 - 1, g_2 = 2, g_n = g_{n-1} + g_{n-1}$ for $n \geq 2$.
   (h) $h_1 = 1, h_2 = 2, h_3 = 4, h_n = h_{n-1} + h_{n-2}$ for $n \geq 3$.

8. (a) Give a recursive definition for each of the following sequences. (Assume that the patterns continue indefinitely.)

      i. $41, 44, 47, 50, 53, 56, \ldots$
      ii. $41, 43, 47, 53, 61, 71, \ldots$
      iii. $41, 42, 44, 48, 56, 72, \ldots$

   (b) Give two recursive definitions for each of the following sequences, one of order 1 and one of order 2.

      i. $41, 83, 167, 335, 671, 1343, \ldots$
      ii. $41, 83, 165, 331, 661, 1323, \ldots$

Remark We will find explicit formulas for these sequences in Chapter 3.

9. Give a "recursive" solution to each of the following questions.

   (a) Andrew is standing at one end of a narrow bridge that is 12 feet long. He takes steps toward the other end of the bridge; each of his steps is either 1 foot or 2 feet long. In how many ways can he reach the other end of the bridge? (The order of his steps matters when counting the number of ways; for example, a 1-foot step followed by a 2-foot step is different from a 2-foot step followed by a 1-foot step.)
   (Hints: Let a_n denote the number of ways that Andrew can reach the end of a bridge that is n feet long. Find a_1, a_2, and a recursive formula for a_n.)

(b) Consider the following 2-by-12 board.

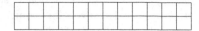

In how many ways can you cover the board with twelve 1-by-2 dominoes?
(Hints: Let b_n denote the number of ways that a 1-by-n board can be
covered by dominoes. Find b_1, b_2, and a recursive formula for b_n. Explain
carefully why your recursive formula for b_n must hold.)

(c) How does the answer to part (a) change if Andrew's steps can be 1 foot, 2
feet, or 3 feet long?

(d) How does the answer to part (b) change if a 3-by-12 board needs to be
covered with twelve 1-by-3 "trominoes"?

Remarks As a generalization of parts (b) and (d), one may ask for the number
of ways that an arbitrary board can be covered with dominoes, trominoes,
or even more general tiles. The general question remains largely unsolved,
but a stunning discovery from 1961 answers the case when an even-by-even
rectangular board is to be covered by dominoes. According to this result,
published in two separate physics journals by Fisher and Temperley (jointly)
and by Kastelyn (independently), a $2m$-by-$2n$ board has exactly

$$\prod_{j=1}^{m} \prod_{i=1}^{n} 4 \left(\cos^2 \frac{j\pi}{2m+1} + \cos^2 \frac{i\pi}{2n+1} \right)$$

different domino tilings. It is quite remarkable that this formula works—it is
not even clear why it should yield a positive integer value!

10. When discussing divisibility, we restrict our attention to the set of integers; for
example, we say that 2 is not divisible by 5 as there is no integer c for which
$2 = 5 \cdot c$ even though, of course, there is a rational number (namely, $c = 2/5$)
for which the equation holds. Similarly, we may limit ourselves to even integers
only. The concept of divisibility in the set of even integers is then different from
its meaning among all integers; for example, among the even integers 4 is not
a divisor of 12 since there is no even integer c (and we are ignoring odd ones!)
for which $4 \cdot c = 12$.

In this problem we examine what happens to Definitions 2.1, 2.1a, and 2.1b
among even integers. It turns out that, while the three definitions are equivalent
in the set of all integers, they describe different sets when considering even
integers only. For example, 12 is prime according to Definition 2.1 since it has
exactly two positive divisors: 2 and 6. However, 12 is not a prime according to
Definition 2.1a since it factors into the product $2 \cdot 6$. We can also see that 12
is not a prime according to Definition 2.1b either: 12 divides the product $4 \cdot 6$
(since for $c = 2$ we have $12 \cdot c = 24$), but 12 fails to divide either 4 or 6. On
the other hand, 10, for example, is a prime according to Definition 2.1a, but not
a prime according to Definitions 2.1 and 2.1b. So the three definitions are not
at all equivalent in the set of even integers!

(a) Characterize the even integers that are prime among even integers according to Definition 2.1. Justify your answer.

(b) Characterize the even integers that are prime among even integers according to Definition 2.1a. Justify your answer.

(c) Characterize the even integers that are prime among even integers according to Definition 2.1b. Justify your answer.

11. For a given positive integer n, we let $\pi(n)$ denote the number of positive primes up to, and including, n. Justify that the following formulas are correct. (The sums and the product below have no inherent meaning for the initial values of n and k; as is customary, we define the sum of no elements to be 0, and the product of no elements to be 1.)

(a)

$$\pi(n) = \sum_{k=2}^{n} \prod_{i=2}^{k-1} \left(\left\lceil \frac{k}{i} \right\rceil - \left\lfloor \frac{k}{i} \right\rfloor \right)$$

(Hints: Start by evaluating the function $f(k, i) = \left\lceil \frac{k}{i} \right\rceil - \left\lfloor \frac{k}{i} \right\rfloor$ for all integers $k \geq 2$ and $2 \leq i \leq k - 1$, then use that to compute $g(k) = \prod_{i=2}^{k-1} f(k, i)$.)

(b)

$$\pi(n) = \sum_{k=2}^{n} \left\lfloor \frac{1}{\sum_{i=1}^{k-1} \left\lfloor \frac{1}{1+k-i\lceil k/i \rceil} \right\rfloor} \right\rfloor$$

12. In this problem, we use the recursive method to define what is called *Nim addition* and *Nim multiplication*, denoted by $\oplus$ and $\otimes$, respectively.

(a) The *Nim sum* $a \oplus b$ of nonnegative integers a and b is defined as the smallest nonnegative integer that is not of the form $a \oplus j$ or $i \oplus b$ for any integers $0 \leq i < a$ and $0 \leq j < b$.
The *Nim* addition table begins as shown in Table 2.1.

Table 2.1 The *Nim* addition table

	0	1	2	3	4	5	6	...
0	0	1	2	3	4	5	6	
1	1	0	3	2	5	4	7	
2	2	3	0	1	6	7	4	
3	3	2	1	0	7	6	5	
4	4	5	6	7	0	1	2	
5	5	4	7	6	1	0	3	
6	6	7	4	5	2	3	0	
:								

Note that from the way we phrased our definition, no initial conditions are necessary; trivially, $0 \oplus 0 = 0$. To illustrate how one arrives at the values in the table, consider $2 \oplus 3$: It equals 1, as it is the smallest nonnegative integer that does not appear either to the left of or above this position in the table.

Verify each entry in the table above and extend the table to include the values $a \oplus b$ for all $0 \leq a \leq 8$ and $0 \leq b \leq 8$.

Remarks Not surprisingly, the name *Nim* addition refers to the game *Nim*, introduced in Problem 6 of Chapter 1. The connection between the game and the operation is the following: player Second has a winning strategy for the game $N(n_1, n_2, \ldots, n_m)$ exactly when the *Nim* sum

$$n_1 \oplus n_2 \oplus \cdots \oplus n_m = 0,$$

and player First is able to win exactly when this *Nim* sum is not 0. For example, $N(2, 3, 5)$ can be won by First, since $(2 \oplus 3) \oplus 5 = 1 \oplus 5 = 4 \neq 0$. We can, actually, see that if First starts the game by taking away 4 chips from the largest heap, then the resulting *Nim* sum is $2 \oplus 3 \oplus 1 = 0$, thus the second player of this game (i.e., First) will win.

In fact, *Nim* addition governs the game *Acrostic Twins* (cf. Problem 9 (a) of Chapter 1) also: Second has a winning strategy for the game played on an *m*-by-*n* board exactly when

$$\bigoplus_{i=0}^{m-1} \bigoplus_{j=0}^{m-1} (i \oplus j) = 0$$

(and First can win if the *Nim* sum is not 0). For example, First has a winning strategy on the 2-by-3 board as the *Nim* sum of the six relevant entries is 1. Furthermore, this also reveals what First's first move should be: leaving the board in such a way that the *Nim* sum of the coins showing heads is 0 guarantees that Second, who moves next, will lose. Therefore, First should turn over the two coins at the right of the board.

The explanations for these statements are beyond our scope for the moment; we return to a thorough analysis of these and other games in Chapter 22.

(b) The *Nim product* $a \otimes b$ of nonnegative integers a and b is defined as the smallest nonnegative integer that is not of the form

$$(i \otimes j) \oplus (a \otimes j) \oplus (i \otimes b)$$

for any integers i and j with $0 \leq i < a$ and $0 \leq j < b$.
The *Nim* multiplication table begins as shown in Table 2.2.

Table 2.2 The *Nim* multiplication table

	0	1	2	3	4	5	6	...
0	0	0	0	0	0	0	0	
1	0	1	2	3	4	5	6	
2	0	2	3	1	8	10	11	
3	0	3	1	2	12	15	13	
4	0	4	8	12	6	2	14	
5	0	5	10	15	2	7	8	
6	0	6	11	13	14	8	5	
$\vdots$								

To find, for example, the value of $2 \otimes 3$, we need to consider the rectangles with lower-right corners corresponding to $a = 2$ and $b = 3$. There are six such rectangles (with upper-left corners corresponding to $i = 0, 1$ and $j = 0, 1, 2$). For each of these rectangles, we need to compute the *Nim* sum of the entries of the other three corners:

$$
\begin{array}{c|ccc}
 & j = 0 & j = 1 & j = 2 \\
\hline
i = 0 & 0 \oplus 0 \oplus 0 & 0 \oplus 2 \oplus 0 & 0 \oplus 3 \oplus 0 \\
i = 1 & 0 \oplus 0 \oplus 3 & 1 \oplus 2 \oplus 3 & 2 \oplus 3 \oplus 3
\end{array}
$$

These values, using the table in part (a), come to $0, 2, 3, 3, 0$, and 2, respectively, and thus the smallest nonnegative integer that is not among them is 1, hence $2 \otimes 3 = 1$.

Verify each entry in the table above and extend the table to include the values $a \otimes b$ for all $0 \le a \le 8$ and $0 \le b \le 8$.

Remarks After our similar remarks above, it may not be surprising that *Nim* multiplication also has to do with games. Namely, Second is able to win *Turning Corners* (cf. Problem 9 (b) of Chapter 1) on an m-by-n board exactly when

$$
\bigoplus_{i=0}^{m-1} \bigoplus_{j=0}^{m-1} (i \otimes j) = 0,
$$

and First has a winning strategy when it is not 0. As above, this also reveals how the players should play for optimal outcome.

Chapter 3
How to Make a Statement

In the previous chapter, we learned how to introduce mathematical concepts with definitions or as primitives. Once we introduce a new concept, we are interested in its properties, usually stated as mathematical statements. *Statements* are sentences that are either true or false—but not both.

To understand the difference between statements and non-statements, let us consider some examples.

- $2^n - 1$.
- $2^n - 1$ is a prime number.
- This sentence is false.
- 6 is a nice number.
- The equation $x^2 + 1 = 0$ has no solutions.

None of these expressions is a statement. The first expression is not a statement since it is not even a sentence. The second expression is a sentence but is not a statement, as its truth cannot be determined until we know the value of n. For example, it will be a true statement if $n = 2$ and a false statement if $n = 4$. Such open sentences are called *predicates*. We could make this sentence into a statement by, for example, saying that "if $n = 5$, then $2^n - 1$ is a prime number"; this statement is clearly true as 31 is prime.

"This sentence is false" is not a statement because, as can quickly be verified, it can be neither true nor false. Such sentences are called *paradoxes*.

The sentence "6 is a nice number" is not a statement either, since its truth depends on the term "nice," which (fortunately!) does not have a universally agreed-upon definition. However, by replacing the word "nice" with "perfect," our sentence becomes a statement, as *perfect numbers* are mathematical concepts and are defined as follows:

Definition 3.1 *A positive integer n whose positive divisors other than n add up to exactly n is called* perfect.

B. Bajnok, *An Invitation to Abstract Mathematics*, Undergraduate Texts in Mathematics, https://doi.org/10.1007/978-3-030-56174-1_3

Having made this definition, we see that 6 is indeed a perfect number, as its positive divisors are 1, 2, 3, and 6, and $1 + 2 + 3 = 6$. We will see some other perfect numbers and examine their properties below.

Finally, the sentence "the equation $x^2 + 1 = 0$ has no solutions" becomes a statement only after we specify what kind of solutions we are looking for. The equation has no real number solutions, but it has two complex number solutions, namely $x = \pm i$, where i denotes the (imaginary) square root of -1.

Mathematicians are interested in developing new concepts, making statements about them, and deciding whether these statements are true or false. Coming up with statements that *seem* true may require a lot of experimentation, and to determine whether these statements indeed *are* true may require quite a bit of precise and thorough reasoning. Let us see an example.

Above we introduced the concept of perfect numbers and saw that 6 was a perfect number. Are there any others? It is not difficult to check that the only two-digit perfect number is 28, for which we have $1 + 2 + 4 + 7 + 14 = 28$. It takes more work (and perhaps a computer) to verify that the only three-digit perfect number is 496, the only four-digit perfect number is 8,128, and there are no perfect numbers with five, six, or seven digits. Perfect numbers are rather rare; to this day we only know of a few dozen (but it is widely assumed that there are infinitely many). The study of perfect numbers originated with the ancient Greeks who were interested in them for certain mystical beliefs. As we are about to see, perfect numbers are closely related to *Mersenne primes* (introduced in Problem 6 (a) of Chapter 2).

How can one make a true statement about perfect numbers? One can try to see if the four perfect numbers we have found follow a pattern. If we factor them into a product of primes, we get

$$6 = 2 \cdot 3,$$

$$28 = 2 \cdot 2 \cdot 7,$$

$$496 = 2 \cdot 2 \cdot 2 \cdot 2 \cdot 31,$$

and

$$8,128 = 2 \cdot 2 \cdot 2 \cdot 2 \cdot 2 \cdot 2 \cdot 127.$$

We see that each product contains a number of 2s and one relatively large prime. Let us be more precise. One can count the number of 2s and observe that the large prime is always one less than a power of 2. Namely, we find that

$$6 = 2^1 \cdot (2^2 - 1),$$

$$28 = 2^2 \cdot (2^3 - 1),$$

$$496 = 2^4 \cdot (2^5 - 1),$$

and

$$8, 128 = 2^6 \cdot (2^7 - 1).$$

Now it is easy to observe that all of these numbers have the form $2^{n-1}(2^n - 1)$; we get 6 for $n = 2$, we get 28 for $n = 3$, we get 496 for $n = 5$, and we get 8,128 for $n = 7$—thus we have found a pattern.

Our next task is to find out why only these n values appear in our pattern; in particular, why do we not get a perfect number for $n = 4$ or $n = 6$, for example. We see that for $n = 4$ the expression $2^{n-1}(2^n - 1)$ becomes $2^3 \cdot 15$ (which equals 120, not a perfect number) and for $n = 6$ we get $2^5 \cdot 63$ (another non-perfect number). It is apparent that the odd factors in these two products (15 and 63, respectively) are not primes. Thus, we are led to the following statement:

- The number $2^{n-1}(2^n - 1)$ is a perfect number for each positive integer n for which $2^n - 1$ is a prime number.

In order to claim that this statement is indeed true, one must verify that it holds in every case. That is, we need to show that there are no exceptional values of n for which $2^{n-1}(2^n - 1)$ is not a perfect number even though $2^n - 1$ is a prime number. We will see in the next chapter how this can be done. We also note in passing that it can be proved that every even perfect number is of the form $2^{n-1}(2^n - 1)$ but that only when $2^n - 1$ is prime does $2^{n-1}(2^n - 1)$ yield a perfect number. We do not know of any odd perfect numbers. It has been verified that there are no such numbers with fewer than $1, 500$ digits, and it is generally doubted that any exist at all.

Our discussion thus far leads us to wonder when the expression $2^n - 1$ gives a prime number. The number $2^n - 1$ is called the n-th *Mersenne number*. We have seen above that $2^2 - 1, 2^3 - 1, 2^5 - 1$, and $2^7 - 1$ were the primes $3, 7, 31$, and 127, respectively; however, $2^4 - 1$ and $2^6 - 1$ are not primes. We then might conjecture that the expression $2^n - 1$ is prime *if, and only if,* n itself is prime. In other words, we suspect that the following statements are true:

- If $2^n - 1$ is a prime number for some positive integer n, then n is a prime.
- If n is a positive prime number, then $2^n - 1$ is a prime.

How can we decide if these statements are true or false?

Let us first examine some examples. Consider Table 3.1.

Our table shows that $2^n - 1$ is prime for $n = 2, 3, 5$, and 7, but for no other value of n under thirteen. We then immediately see that the statement "If n is a positive prime number, then $2^n - 1$ is a prime" is false because we could find a prime number n, namely $n = 11$, for which $2^n - 1$ is not a prime. In this case we say that $n = 11$ is a *counterexample* for the statement. (The statement has other counterexamples as well, but it is enough to find one to conclude that the statement is false.)

Let us turn now to the statement "If $2^n - 1$ is a prime number for some positive integer n, then n is a prime." This statement seems true because in each of the four cases in which $2^n - 1$ was a prime number, n was a prime as well. This is not convincing, however, as perhaps a counterexample can be found by checking higher

Table 3.1 Prime
factorization of $2^n - 1$ for
$n = 1, 2, \ldots, 12$

n	$2^n - 1$	Factorization
1	1	—
2	3	Prime
3	7	Prime
4	15	$3 \cdot 5$
5	31	Prime
6	63	$3 \cdot 3 \cdot 7$
7	127	Prime
8	255	$3 \cdot 5 \cdot 17$
9	511	$7 \cdot 73$
10	1,023	$3 \cdot 11 \cdot 31$
11	2,047	$23 \cdot 89$
12	4,095	$3 \cdot 3 \cdot 5 \cdot 7 \cdot 13$

values of n. We can only claim that the statement is true if we can also explain why no such counterexample exists; and obviously, we need to do this without having to check every value of n since there are infinitely many cases to check. We will accomplish this in the next chapter.

While formulating statements may be easy, it can often be quite difficult to decide if they are true or false. As our example clearly shows, if our statement involves infinitely many cases, examining a small number of them—indeed, not examining all of them—may lead us to an incorrect or incomplete statement. Leonhard Euler, for example, pointed out in the eighteenth century that the quantity $n^2 - n + 41$ is prime for every integer value of n between -39 and 40, inclusive—a list of eighty consecutive values—but that the values of $n = -40$, $n = 41$, $n = 42$, and many others, yield composite numbers. (The number 41 is one of the *lucky numbers of Euler*—see Problem 3.) It is not known to this day whether the quantity $n^2 - n + 41$ yields primes infinitely often (though this is generally believed to be the case). An even more dramatic example was recently discovered by New York mathematician Kevin O'Bryant: The equation

$$\left\lfloor \frac{1}{\sqrt[n]{2} - 1} \right\rfloor = \left\lfloor \frac{n}{\ln 2} - \frac{1}{2} \right\rfloor$$

holds for every integer n that is greater than 1 and less than $777, 451, 915, 729, 368$, but is false for $n = 777, 451, 915, 729, 368$ (and certain other n values).

As these examples demonstrate, it is not hard to be misled by a pattern that seems true in many cases, only to discover later that it is false in others. Most mathematicians have at least once fallen for "capricious coincidences" to make "careless conjectures"; several such interesting false statements are discussed in Richard Guy's article, The Strong Law of Small Numbers, in the *American Mathematical Monthly*, Vol. 95, Issue 8 (Oct. 1988).

Problems

1. The first four perfect numbers are 6, 28, 496, and 8,128.

   (a) Verify that 496 and 8,128 are indeed perfect numbers.
   (b) Find the fifth perfect number.
      (Hints: Once you claim that a certain positive integer N is the fifth perfect number, you need to verify that N is perfect and explain why there are no other perfect numbers strictly between 8,128 and N. To verify that N is perfect, you may use Problem 5 (a) of Chapter 2. To prove that there are no others less than N, you need to first explain why only numbers of a certain form need to be considered.)

2. Consider the following sequences (cf. Problem 8 of Chapter 2). (Assume that the patterns continue indefinitely.) Find a pattern and then an explicit formula for each sequence. (You do not need to provide proofs; we will do so in Chapter 10.)

   (a) $41, 44, 47, 50, 53, 56, \ldots$
      (Hint: Since the difference sequence $a_{n+1} - a_n$ is constant, a_n is linear.)
   (b) $41, 43, 47, 53, 61, 71, \ldots$
      (Hint: Since the difference sequence $a_{n+1} - a_n$ is linear, a_n is quadratic.)
   (c) $41, 42, 44, 48, 56, 72, \ldots$
   (d) $41, 83, 167, 335, 671, 1343, \ldots$
      (Hint: Try a formula of the form $a \cdot 2^n + b$.)
   (e) $41, 83, 165, 331, 661, 1323, \ldots$
      (Hint: Try a formula of the form $a \cdot 2^n + b \cdot (-1)^n$.)

3. Let k be an arbitrary positive integer, and consider the function $f(n) = n^2 - n + k$.

   (a) Prove that there is no value of k for which $f(k)$ is prime.
   (b) Prove that there is no value of k for which $f(-k + 1)$ is prime.
   (c) A value of k that is more than 1 and for which $f(n)$ is prime for every n between $-k + 2$ and $k - 1$, inclusive, is called a *lucky number of Euler*. As we mentioned earlier, 41 is a lucky number of Euler; in fact, it is the largest one. Find all others.

   Remark The lucky numbers of Euler play a deep role in algebraic number theory (cf. Problem 12 in Chapter 21).

4. The following statements are all false. Provide a counterexample for each statement.

   (a) If we write down the positive integers starting with 41 in an infinite spiral, as partially shown in Table 3.2, then the entries in the Northwest–Southeast diagonal are all primes.

Table 3.2 A spiral of integers

83						
	61	62	63	64	65	66
	60	47	48	49	50	67
	59	46	41	42	51	68
	58	45	44	43	52	69
	57	56	55	54	53	70
77						71

(Hints: Extending the table will not yield a counterexample for quite a while, so a more abstract approach is needed. Start by using a bit of geometry to explain why the values in question are given by the recursive formula you developed in Problem 2 (b) above.)

(b) Let p_i denote the i-th positive prime ($i = 1, 2, 3, \dots$), and for a positive integer n, define

$$K_n = 1 + \prod_{i=1}^{n} p_i.$$

Then K_n is a prime number.

Remark The number K_n is called the *n-th Euclid number*; we will return to these numbers later.

(c) If n circles (n is a positive integer) are given in the plane in such a way that any two of them intersect in exactly two separate points but no three intersect in the same point, then the number of different regions they determine in the plane is 2^n.

(d) If n points (n is a positive integer) are given on a circle in general position (i.e., no three chords meet at a single point inside the circle), then the number of different regions that the chords determine inside the circle is 2^{n-1}.

5. Consider the following definition.

Definition 3.2 *Suppose that m is a positive integer and that r is a nonnegative integer that is less than m. We say that an integer n is* congruent to r mod m *if n leaves a remainder of r when divided by m, that is, if $n - r$ is divisible by m.*

(For example, positive integers with a last digit 3 are all congruent to 3 mod 10, as are negative integers with a last digit 7.)

For a given positive integer m and nonnegative integer r that is less than m, we define the predicate $C(m, r)$ as follows:

• Every positive integer other than 1 that is congruent to r mod m has a positive prime divisor that is also congruent to r mod m.

(For instance, $C(10, 3)$ is the statement that every positive integer that has last digit 3 has a positive prime divisor whose last digit is also 3. This statement is false; for example, the two positive prime divisors of 133 are 7 and 19, and neither has last digit 3.)

(a) Are there any positive integers below 133 that show that $C(10, 3)$ is false?
(b) For which values of m is $C(m, 0)$ true? Justify your answer.
(c) Explain why $C(2, 1)$ is true, but for every value of m between 3 and 10, inclusive, $C(m, 1)$ is false.
(d) Provide a thorough justification for the fact that $C(4, 3)$ is true.
(Hints: Suppose that n is an integer greater than 1 that is congruent to 3 mod 4. First explain why each positive prime divisor of n is congruent to either 1 or 3 mod 4, then explain why it cannot be that each of these prime divisors is congruent to 1 mod 4.)
(e) One can prove that there are exactly five positive integer values of m for which $C(m, m - 1)$ is true. Find these values. (No proofs are necessary.)

6. Consider the following four statements.

(a) Every positive integer can be written as the sum or the difference of two perfect squares. (Consider 0 to be a perfect square.)
(b) Every odd positive integer can be written as the sum or the difference of two perfect squares.
(c) Every positive integer can be written as the sum or the difference of two positive primes.
(d) Every even positive integer can be written as the sum or the difference of two positive primes.

Two of these statements are false, one is true, and one is an undecided open problem. Decide which is which. For each false statement, provide a counterexample; for the true statement, provide a justification. (The open problem is a famous *conjecture* that mathematicians have been unable to decide for hundreds of years—though most believe that it is true.)
(Hints: Note that to prove a statement of the form "N can be written as ...", it suffices to produce one appropriate expression that equals N, while to disprove it, one needs to rule out all (possibly infinitely many) possibilities. On the other hand, to prove that "all positive integers N can be written as ...", one needs a proof that handles every possible value of N, but a single counterexample suffices to disprove such a statement.)

7. (a) Observe that we can express the first few positive integers as follows:

$$1 = 2^0 \cdot 1$$
$$2 = 2^1 \cdot 1$$
$$3 = 2^0 \cdot 3$$
$$4 = 2^2 \cdot 1$$
$$5 = 2^0 \cdot 5$$
$$6 = 2^1 \cdot 3$$

Make a general statement that says that every positive integer can be (uniquely) expressed this way. Be as precise as necessary.

(b) Make a similar statement that involves the powers of 3.

8. The objective of this problem is to learn the lesson that diagrams (especially faulty ones like ours below) can lead us to false conjectures.

Recall that we defined the Fibonacci sequence recursively by $F_1 = 1$, $F_2 = 2$, and $F_n = F_{n-1} + F_{n-2}$ for $n \geq 3$ (cf. page 17). The following is a diagram of a square with side lengths F_n, divided into four regions (Figure 3.1).

Two of the regions are right triangles with side lengths F_{n-2} and F_n; the other two regions are trapezoids with two right angles and side lengths F_{n-1}, F_{n-1}, and F_{n-2}, as indicated. (Note that a segment of length F_n may be divided into a segment of length F_{n-1} and a segment of length F_{n-2}.) The area of the square equals F_n^2.

Our next diagram is a rearrangement of the four regions so that they form a rectangle of side lengths F_{n+1} and F_{n-1}, respectively. (Note that a segment of length F_{n+1} may be divided into a segment of length F_{n-1} and a segment of length F_n.) The area of the rectangle equals $F_{n-1} \cdot F_{n+1}$ (Figure 3.2).

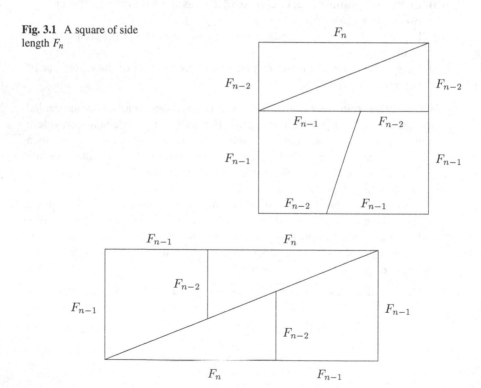

Fig. 3.1 A square of side length F_n

Fig. 3.2 A rectangle with side lengths F_{n+1} and F_{n-1}

Since the rectangle of the second diagram must have the same area as the square of the first diagram, we get the claim

$$F_{n-1} \cdot F_{n+1} = F_n^2.$$

(a) Find some counterexamples for this claim.
(b) Find the mistake(s) in the argument above. Be as specific as possible. (Hint: Draw accurate diagrams for some small values of n. Make sure to include both even and odd n values.)
(c) Although our claim is false, $F_{n-1} \cdot F_{n+1}$ and F_n^2 are quite close to one another. State a correct conjecture for an equation involving these quantities. (You do not need to prove your claim.)

9. For positive integers n, m, and k, let $P(n, m, k)$ denote the predicate that n can be written as the sum of m terms, each of which is a perfect k-th power (the k-th power of a nonnegative integer). For example, $P(100, 4, 2)$ is the statement that 100 can be written as the sum of four squares (true, since, for example, $100 = 10^2 + 0^2 + 0^2 + 0^2$ or $100 = 9^2 + 3^2 + 3^2 + 1^2$), and $P(100, 5, 3)$ is the statement that 100 can be written as the sum of five nonnegative cubes (also true, since $100 = 4^3 + 3^3 + 2^3 + 1^3 + 0^3$).
For each predicate below, find at least one example for n, m, and k for which the predicate becomes a true statement, and at least one example for which it becomes a false statement. If this is not possible, explain why.

(a) $P(100, 5, k)$
(b) $P(100, m, 4)$
(c) $P(n, 3, 2)$

Remark The predicate $P(n, 4, 2)$ is true for every n, as proved by Joseph Louis Lagrange in 1770.

(d) $P(n, 8, 3)$

Remark The predicate $P(n, 9, 3)$ is true for every n; this was proved by Arthur Wieferich and Aubrey Kempner around 1912.

(e) $P(n, k^2, k)$
(f) $P(2^k - 1, 2^k - 2, k)$
(g) $P(2^k \cdot \lfloor(\frac{3}{2})^k\rfloor - 1, 2^k + \lfloor(\frac{3}{2})^k\rfloor - 3, k)$

Remark These questions are related to a famous problem in additive number theory. For a given positive integer k, *Waring's Problem* asks for $g(k)$, the smallest integer m for which $P(n, m, k)$ is true for every n. Part (g) above yields the lower bound

$$g(k) \geq 2^k + \left\lfloor \left(\frac{3}{2}\right)^k \right\rfloor - 2.$$

At the present time we know that $g(k)$ indeed exists for every k (a 1909 result of David Hilbert) and agrees with this lower bound for every $k \leq 471,600,000$ and every "sufficiently large" k (thus leaving only finitely many cases open).

10. For each statement below, decide whether the statement is true or false. Provide as thorough a justification as you deem necessary. (Cf. Problem 8 in Chapter 1.)

   (a) In every group of five people, there are two people who know the same number of people in the group (assume that "knowing one another" is mutual).

   (b) In every group of five people, either there are three people who all know each other or there are three people so that no two of them know each other (or both).

   (c) In every group of six people, either there are three people who all know each other or there are three people so that no two of them know each other (or both).

Chapter 4
What's True in Mathematics?

In the last chapter we made the assertions that the statements

- the number $2^{n-1}(2^n-1)$ is a perfect number for each positive integer n for which $2^n - 1$ is a prime number, and
- if $2^n - 1$ is a prime number for some positive integer n, then n is a prime number

are true, and we promised arguments that demonstrate them without any doubt. We will provide these in this chapter.

Before addressing these two mathematical statements, let us discuss briefly what we mean by an argument that demonstrates a statement's truth without any doubt. We mean what we say: our argument should not leave *any* possible doubt to the truth of our statement. How does this notion compare to that in other fields? In criminal law, the highest level of proof requires the establishment of the claim to the extent that there is no reasonable doubt in the mind of a reasonable person. (Even lower levels of proof may be satisfactory in civil trials.)

The standards are higher in some branches of science, where a statement is considered true if it can be experimentally verified. While this assures that there is overwhelming evidence for the claim and there are no known counterexamples for it, the level of certainty in mathematics is absolute: mathematicians must verify that no one could ever possibly find any counterexamples for the statement in the future either. This is quite a demanding requirement!

The following puzzle demonstrates our point. Given a standard 8-by-8 chess board from which two diagonally opposite corner squares are removed, as shown in Figure 4.1, is it possible to tile the remaining 62 squares with 31 dominoes (2-by-1 rectangles)? Some experimentation will sooner or later lead everyone to conjecture that this cannot be done—but can we be sure?

Well, it is clear that the top row of the board, which only has seven squares, cannot be tiled without some dominoes also partially covering the next row. So the top row will fully house at most three dominoes, but perhaps none (maybe each of its seven squares will be covered by vertically placed dominoes). Soon we find that

© Springer Nature Switzerland AG 2020

B. Bajnok, *An Invitation to Abstract Mathematics*, Undergraduate Texts in Mathematics, https://doi.org/10.1007/978-3-030-56174-1_4

Fig. 4.1 A truncated chess
board

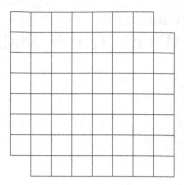

there are too many cases with too many subcases—so we cannot possibly expect to
explore every possible scenario. (In theory, one could delegate the verification of all
cases to an extensive computer program—something that not all mathematicians
view as satisfactory. In situations with infinitely many possibilities, a computer
cannot possibly suffice.)

Here is a beautifully simple argument that convinces us that the tiling cannot be
done. Recall that each square of the chess board is colored by one of two colors
(usually black or white) in an alternating pattern. This assures that each domino
will cover exactly one square of each color. Thus, 31 non-overlapping dominoes
will cover exactly 31 black and 31 white squares. But the two diagonally opposite
squares are of the same color, so, after removing them, we are left with 30 squares
of one color and 32 of the other; therefore, the required tiling is not possible.

The difficulty of this puzzle lies in the fact that we have to verify that something
is impossible; this requires evaluating *every* possibility. Nevertheless, our argument
achieves its task: after understanding this elegant argument, we can be absolutely
sure that no one in the future will find a tiling! In fact, it achieves more; it shows that
the board cannot be tiled whenever any two squares of the *same* color are removed.
In Problem 1, we examine the case when two squares of *opposite* color are removed.

In mathematics, a rigorous logical argument that will convince everyone (that
is, everyone with sufficient mathematical background) of the truth of the statement
is called a *proof*. A mathematical statement that is supported by a proof is called
a *theorem*. (The term *proposition* is usually used instead when we regard our
statement as less important.) The idea of proofs was originated by Thales of Miletus
more than twenty-five hundred years ago and is the cornerstone of mathematics. We
will see a large number and variety of proofs in this book.

Let us now return to the two statements at the beginning of our chapter and see
how we can provide proofs for them so that they become theorems. These particular
theorems were chosen because their statements involve very few terms that require
definition and explanation, yet their proofs are complex enough to illustrate various
components and techniques that often appear in proofs.

We start with the following:

Theorem 4.1 *Let n be a positive integer for which $2^n - 1$ is a prime number. Then the number $2^{n-1}(2^n - 1)$ is a perfect number.*

Before proving Theorem 4.1, let us make some comments. We have already seen in Chapter 3 that our theorem holds for $n = 2$ and $n = 3$, and in Problem 1 (a) of Chapter 3 we also verified the cases $n = 5$ and $n = 7$. Note, however, that, as we pointed out earlier, we need to prove that our statement has no counterexamples, and we know nothing about the positive integer n other than the fact that $2^n - 1$ is a prime number. We have infinitely many n values to consider, so we cannot possibly evaluate every case individually. Our proof below will address all cases.

Let us make another observation. The definition of a perfect number (cf. Definition 3.1) refers to the sum of the positive divisors of the number other than the number itself. It will be convenient for us to use the notation $\sigma(n)$ for the sum of *all* of the positive divisors of a given integer n (that is, the number n itself is included). In terms of this notation, we can say that the positive integer n is perfect exactly when $\sigma(n) = 2n$. Now let us see the proof of Theorem 4.1.

Proof We are given that n is a positive integer for which $2^n - 1$ is a prime number, and our goal is to prove that $2^{n-1}(2^n - 1)$ is a perfect number. According to the notation we just introduced, for $2^{n-1}(2^n - 1)$ to be perfect, we would need to verify that

$$\sigma(2^{n-1}(2^n - 1)) = 2 \cdot 2^{n-1}(2^n - 1) = 2^n(2^n - 1).$$

We first list the positive divisors of $2^{n-1}(2^n - 1)$. Since we are assuming that the number $2^n - 1$ is prime, the divisors of $2^{n-1}(2^n - 1)$ are as follows:

$$1, 2, 2^2, 2^3, \ldots, 2^{n-1},$$

and

$$2^n - 1, 2(2^n - 1), 2^2(2^n - 1), 2^3(2^n - 1), \ldots, 2^{n-1}(2^n - 1).$$

The sum of the divisors in the first row is

$$1 + 2 + 2^2 + 2^3 + \cdots + 2^{n-1} = 2^n - 1,$$

and the sum in the second row is

$$(2^n - 1)(1 + 2 + 2^2 + 2^3 + \cdots + 2^{n-1}) = (2^n - 1)(2^n - 1).$$

Therefore, we get

$$\sigma(2^{n-1}(2^n - 1)) = (2^n - 1) + (2^n - 1)(2^n - 1) = 2^n(2^n - 1),$$

and this is what we intended to prove. Therefore, $2^{n-1}(2^n - 1)$ is a perfect number. $\square$

Our proof consists of the recollection of the relevant definitions (e.g., divisor, perfect number) and a sequence of true statements, and we mark the end of a proof by the symbol □. When reading a proof, one needs to carefully verify that each statement in the proof is true. How can one be sure that each statement is true?

Consider, for example, the statement that

$$1 + 2 + 2^2 + 2^3 + \cdots + 2^{n-1} = 2^n - 1.$$

(This is, perhaps, the least obvious of the statements involved in our proof.) This statement is clearly true for $n = 1$ (we get $1 = 1$); we can also easily verify our statement for $n = 2$ $(1 + 2 = 2^2 - 1)$, for $n = 3$ $(1 + 2 + 4 = 2^3 - 1)$, etc. As n gets larger, however, these identities get more complicated, and there are infinitely many n values to check! To claim that our equation holds for *every* n, we must provide an argument. We will indeed provide such an argument in Chapter 10 using the proof technique of *mathematical induction*.

If the proof of a theorem, say Theorem A, uses the statement of Theorem B, then we say that Theorem A is a *corollary* of Theorem B and that Theorem B is a *lemma* for Theorem A. For example, our theorem above is a corollary of the stated identity for $1 + 2 + 2^2 + 2^3 + \cdots + 2^{n-1}$, and this identity is a lemma for our theorem. In turn, this identity will itself be the corollary (indeed, a special case) of the following more general lemma.

Lemma 4.2 *If a and b are arbitrary real numbers and n is a positive integer, then*

$$(a - b) \cdot \left(a^{n-1} + a^{n-2}b + a^{n-3}b^2 + \cdots + ab^{n-2} + b^{n-1} \right) = a^n - b^n.$$

In our proof above, we used Lemma 4.2 with $a = 2$ and $b = 1$.

Let us now consider another statement involved in our proof, the statement that

$$(2^n - 1) + (2^n - 1)(2^n - 1) = 2^n(2^n - 1).$$

How do we know that this is true? This statement seems quite obvious for everyone who is familiar with basic algebra: we can factor out $2^n - 1$, then combine the terms 1 and $2^n - 1$ to write $2^n(2^n - 1)$. We could, in fact, state these statements as lemmas, but one feels that they are pretty obvious and, therefore, the need for a formal proof is not as keen. How can we decide if a particular statement requires a proof or not?

Clearly, one cannot reduce all statements to previously proved lemmas, as these lemmas would have to rely on earlier ones as well—the process cannot be traced back indefinitely. So, just as we had to build our definitions on a collection of undefined concepts (primitives), we have to build our proofs on a collection of statements that we regard as true. Statements whose truth we accept without proofs are called *axioms*.

Each branch of mathematics has its own axioms; however, as with primitives, the choice of axioms is somewhat flexible. For example, when we study geometry, we use Euclid's five axioms, two of which can be stated as follows:

Axiom 4.3 *Given two points P and Q in the plane, there is a unique line l that contains both P and Q.*

Axiom 4.4 *Given two points P and Q on a line l, and given an arbitrary distance d, there is a unique point R on l such that Q is between P and R and the points Q and R have distance d.*

Our proof above uses the axioms of algebra, such as the associativity, commutativity, and distributivity of addition and multiplication of numbers. These axioms of algebra allow us, for example, to write

$$(2^n - 1) + (2^n - 1)(2^n - 1) = 2^n(2^n - 1).$$

Our goal in this book is *not* to reduce every statement to the axioms—this would take too long and would not be very interesting. We will, however, see later how this *could* be done.

The truth of our mathematical system depends on the set of axioms we choose. If we start with a different set of axioms, then we develop a different theory as its consequence. Thus, what is true in mathematics (the question posed in the title of this chapter) depends very much on what axiomatic system we are working with. It is its perceived importance and applicability that determines which set of axioms is most useful.

We call a collection of axioms

- *consistent* if they do not lead to a contradiction and
- *independent* if no axiom can be proved using the others.

We require that any system of axioms be consistent; independence, while desirable, is not essential.

It is not always easy—and, as we will see, it is sometimes impossible—to decide whether a particular axiomatic system is independent and consistent. The usual way to prove that a system is consistent is to create a *model* for the system. For example, it is possible to construct models for the various number systems (e.g., the integers or the real numbers), and thus their axiomatic systems are consistent (we will carry this out in Chapter 21).

The question of independence is also difficult. A famous historical example is the case of the fifth of Euclid's axioms, referred to as the *Parallel Postulate*:

Axiom 4.5 (The Parallel Postulate) *If P is a point and l is a line so that P does not lie on l, then there is a unique line l' that contains P and is parallel to l.*

It was unknown for more than two thousand years whether the Parallel Postulate could be proved using Euclid's other four axioms. Finally, in the 1830s, three mathematicians—the Hungarian János Bolyai (1802–1860), the German Carl Friedrich Gauss (1777–1855), and the Russian Nicolai Ivanovitch Lobachevsky (1792–1856)—proved (independently of each other, although, according to some accounts, Gauss was aware of Bolyai's work) that this is not the case:

Theorem 4.6 *The Parallel Postulate is independent of Euclid's other four axioms.*

Sometimes two contradicting axioms may both give rise to interesting and applicable theories. For example, assuming that the line l' that contains P and is parallel to l is not unique, we can develop the theory of a non-Euclidean geometry (in particular, *elliptic geometry* assumes that there is no parallel line, and *hyperbolic geometry* assumes that there are infinitely many). It is for physicists and astronomers to decide which of these models describes the geometry of the universe—mathematicians, staying out of such arguments, are only concerned with the consequences of the particular choice of axioms.

As we have mentioned above, the proof of a mathematical statement consists of other true statements, and each statement, in turn, needs to follow from the axioms. The concept of one statement following from another can be made more precise, but we will not explain this here. We will, instead, develop a heuristic understanding of the logical structure of our proofs. In some cases, this structure will be apparent and easy to follow. Our second theorem includes an example of a logical structure that is not necessarily evident, yet is typical for many proofs.

Theorem 4.7 *Let n be a positive integer. If $2^n - 1$ is a prime number, then n is a prime number as well.*

Proof We are given that n is a positive integer for which $2^n - 1$ is a prime number, and our goal is to prove that n is a prime; that is, n has exactly two positive divisors. We will accomplish this by showing that n has at least two, but not more than two, positive divisors.

First we note that n cannot be 1 if $2^n - 1$ is prime, since for $n = 1$ we have $2^n - 1 = 1$, which is not a prime. Since n is certainly divisible by 1 and n and these two divisors are different (as $n \neq 1$), n has to have at least two positive divisors.

To prove that n has no divisors other than 1 and n, we assume that c is a positive divisor of n, and we will show that then either $c = 1$ or $c = n$. Because c is a positive divisor of n, by definition, there is a positive integer k for which $n = c \cdot k$, and, therefore, $2^n - 1 = 2^{c \cdot k} - 1$. We can rewrite this latter quantity as $(2^c)^k - 1$.

With $a = 2^c$, $b = 1$, and $n = k$, Lemma 4.2 implies that $(2^c)^k - 1$ is divisible by $2^c - 1$. But, according to our assumption, $2^n - 1$ is a prime, so it can only have $2^c - 1$ as a divisor if $2^c - 1 = 1$ or $2^c - 1 = 2^n - 1$. From these equations we get that $c = 1$ or $c = n$, as claimed.

Thus we have proved that n has at least two, but not more than two, positive divisors. Therefore, n must be a prime number. □

Theorem 4.7 gives us the opportunity to point out a logical structure that is rather peculiar. When proving that n had to be a prime, we needed to establish that it had exactly two positive divisors. We did not prove this directly. Instead, we proved that n could not have less than two or more than two positive divisors—in other words, n could not be 1 or a composite number. This only left the case of n having exactly two positive divisors, so n had to be a prime by Definition 2.1. Proofs that use this kind of an argument are called *indirect proofs*; they are based on the notion that if a statement cannot be false then it has to be true. We will study indirect proofs in detail later.

There are many different kinds of proofs—mathematical induction and indirect proofs, mentioned above, are two of the most important techniques—and one can write a proof in a variety of different ways. In fact, there is no single correct way to write a proof; it depends on who it is intended for (beginner students, experts in the field, etc.) and what we wish to emphasize (why our statement is true, how we discovered it, etc.). But proofs, above all, must be clear and convincing.

Needless to say, constructing and writing (and sometimes even understanding) a proof can be quite challenging. We will practice these skills extensively (but gradually) in this book.

Problems

1. Above we proved that when two diagonally opposite corner squares are removed from a standard 8-by-8 chess board, it is not possible to tile the remaining 62 squares with 31 dominoes. The argument that we presented actually proves more: the covering is impossible whenever *any* two squares of the same color are removed from the board. Prove that, on the other hand, if any two squares of opposite color are removed, then it is always possible to tile the remaining squares with dominoes.
   (Hints: Consider Figure 4.2, featuring a closed path through the 64 squares on the board; that is, a path that always takes us from a square to a vertically or horizontally neighboring square, going through each square exactly once, with the final square in the path the same as the initial square. There are, of course, many other such closed paths; any one of them will do. Use the diagram to generate a proof that handles all of the cases simultaneously, regardless of which two squares of opposite color are removed.)
2. Prove that there is no collection of points in the Euclidean plane that intersects every line exactly once.
   (Hints: Rely on two of Euclid's axioms mentioned above to prove that neither a collection of at least two points nor one consisting of a single point works.)

Fig. 4.2 A walk through the squares of a chess board

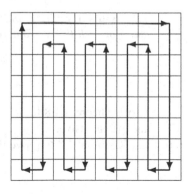

Remark It is known that a point set that intersects every line in the plane exactly twice does exist; however, one cannot visualize such a set (cf. page 272).

3. Suppose that c is an arbitrary real number and m is a positive integer. Use Lemma 4.2 to find a direct formula (one without summation) for each of the following.

  (a) $1 + c + c^2 + c^3 + \cdots + c^m$
  (b) $(1) + (1 + c) + (1 + c + c^2) + \cdots + (1 + c + c^2 + c^3 + \cdots + c^m)$

  (Hint: Note that the cases when $c = 1$ and when $c \neq 1$ need to be considered separately.)
4. Alvin has a New Year's resolution: he wants to start saving money toward his ultimate goal of $100,000. His bank accounts earn a 0.3 % daily interest rate, compounded each day; the bank is open for business every day of the year. (Why can't we all have access to this bank?)

  (a) How long would it take him to achieve his goal if he opened a new account every day with a $1 investment?
      (Hint: Use Problem 3 (a) from above.)
  (b) Not being satisfied with the answer above, Alvin decides to adopt the following strategy: starting on January 1, and each day of the year, he (i) opens a new account with a $1 investment and (ii) deposits $1 into each of his other existing accounts. (So this will cost him $1 on January 1, $2 on January 2, etc.) Will Alvin achieve his goal by December 31? (Note: A year contains either 365 or 366 days.)
      (Hint: Use Problem 3 (b) from above.)

5. (a) Let a, b, k, and n be positive integers, and suppose that $a \neq b$ (and therefore $a^n \neq b^n$). Prove that Lemma 4.2 implies that

$$\frac{a^{kn} - b^{kn}}{a^n - b^n} \quad \text{and} \quad \frac{a^k - b^k}{a - b}$$

  are integers.
  (b) Consider the following theorem.

  Theorem 4.8 *Suppose that a and b are distinct positive integers and that k and n are relatively prime positive integers. Then*

$$\frac{a^{kn} - b^{kn}}{a^n - b^n} \quad \text{is divisible by} \quad \frac{a^k - b^k}{a - b}.$$

  Verify Theorem 4.8 for $a = 2, b = 1, n = 3$, and each value of k between 1 and 10, inclusive.
  (c) Use Theorem 4.8 to prove the following:

Lemma 4.9 *If a and b are positive integers and n is an odd positive integer, then $a^n + b^n$ is divisible by $a + b$.*

(Hint: Since Theorem 4.8 assumes that a and b are distinct, you need to treat the case when $a = b$ separately.)

(d) Prove Lemma 4.9 using Lemma 4.2 but without relying on Theorem 4.8.
(e) Use Theorem 4.8 to prove the following:

Lemma 4.10 *If a and b are positive integers and n is a positive integer that is not divisible by 3, then $a^{2n} + a^n b^n + b^{2n}$ is divisible by $a^2 + ab + b^2$.*

(In Chapter 11 we will use induction to prove Lemma 4.10.)

6. (a) Prove that the only positive integer n for which $4^n - 1$ is a prime number is $n = 1$.
   (b) Prove that the only positive integer n for which $n^7 - 1$ is a prime number is $n = 2$.
   (c) Prove that the only positive integer n for which $n^7 + 1$ is a prime number is $n = 1$.
   (d) Prove that the only positive integer n for which $4n^4 + 1$ is a prime number is $n = 1$.
   (Hint: Use *Sophie Germain's Identity* that for all real numbers a and b we have

$$a^4 + 4b^4 = (a^2 + 2ab + 2b^2)(a^2 - 2ab + 2b^2).$$

The identity is named after the French mathematician Sophie Germain (1776–1831).)

Remark It is generally very hard to see how many prime values a given expression yields. For example, it is still unknown whether n^2+1 assumes finitely many or infinitely many prime values as n ranges through the positive integers.

7. Recall that a positive integer n is perfect whenever $\sigma(n)$, the sum of all its positive divisors, equals $2n$. It is customary to call n *deficient* when $\sigma(n) < 2n$, and *abundant* when $\sigma(n) > 2n$. For example, among the first thirty positive integers, two (6 and 28) are perfect, five (12, 18, 20, 24, and 30) are abundant, and the rest are deficient. Below we investigate deficient and abundant numbers; in particular, we show that there are infinitely many deficient numbers and infinitely many abundant numbers.

Remark While the number of deficient numbers and the number of abundant numbers are both infinite, one can prove that, in a certain sense, there are more than three times as many deficient ones than abundant ones.

(a) Prove that the number 2^n is deficient for every positive integer n.
(b) Prove that the number 3^n is deficient for every positive integer n.
(c) Prove that the number $2^n \cdot 3$ is perfect for $n = 1$ and abundant for every integer $n \geq 2$.

(d) Prove that the number $2 \cdot 3^n$ is perfect for $n = 1$ and abundant for every integer $n \geq 2$.

(e) Prove that $2^{n-1}(2^n - 1)$ is abundant whenever n is a positive integer for which the number $2^n - 1$ is composite.

(f) Let m and n be positive integers and assume that $2^n - 1$ is a prime number. Theorem 4.1 says that $2^m(2^n - 1)$ is a perfect number when $m = n - 1$. Prove that $2^m(2^n - 1)$ is a deficient number when $m < n - 1$ and an abundant number when $m > n - 1$.

8. We say that a positive integer n is *super-perfect* whenever $\sigma(\sigma(n)) = 2n$. For example, 16 is a super-perfect number since

$$\sigma(\sigma(16)) = \sigma(1 + 2 + 4 + 8 + 16) = \sigma(31) = 1 + 31 = 32.$$

Find, with proof, a statement similar to Theorem 4.1 about super-perfect numbers.

9. (a) Prove the following result about Fermat numbers.

Theorem 4.11 *Let n be a positive integer. If $2^n + 1$ is a prime number, then n is a power of 2, that is, $n = 2^k$ for some nonnegative integer k.*

(Hints: Adapt the proof of Theorem 4.7. Use Problem 7 (a) of Chapter 3 and Lemma 4.9 above.)

Remarks As we mentioned in Problem 6 (b) in Chapter 2, the only prime Fermat numbers known to this day are $2^1 + 1 = 3, 2^2 + 1 = 5, 2^4 + 1 = 17, 2^8 + 1 = 257$, and $2^{16} + 1 = 65, 537$.

(b) Prove the following.

Theorem 4.12 *Let n be a positive integer. If $4^n + 2^n + 1$ is a prime number, then n is a power of 3, that is, $n = 3^k$ for some nonnegative integer k.*

(Hints: Again, adapt the proof of Theorem 4.7. Use Problem 7 (b) of Chapter 3 and Lemma 4.10 above.)

Remarks One can check that $N_k = 4^{3^k} + 2^{3^k} + 1$ is a prime number for $k = 0$ (when $N_0 = 7$), $k = 1$ (when $N_1 = 73$), and $k = 2$ (when $N_2 = 262, 657$); however, N_k is composite for all other values of $k \leq 10$. It is not known how many values of k there are for which $4^{3^k} + 2^{3^k} + 1$ is a prime number.

10. In the Plutonian alphabet, used on planet Pluto (more precisely, since 2006 Pluto is officially only a "dwarf planet"), there are only four letters: A, B, C, and D. Therefore, any word in Plutonian is a finite string of these four letters (where each letter can appear any number of times or not at all). Suppose that the following rules hold.

- (W1) A and BCD are words in the language.
- (W2) Whenever a word contains the letter B, it also remains a word if the B is deleted (and the remaining space is closed up). Similarly, strings obtained from Plutonian words by deleting two consecutive Cs or three consecutive Ds are also Plutonian words.
- (W3) If the letter A appears in a word, then it can be replaced by the string DCB and the result is another word.
- (W4) The letter B can be replaced by the string CDA and we get another word.
- (W5) If any two words are written consecutively, then we can delete the space between them and the resulting string is another word in the language.

   Prove that every finite string of the four letters is a Plutonian word.
11. Policies at a certain college require that the following "axioms" hold.

- (C1) There must be at least two classes offered each semester.
- (C2) Each class must have at least three students in it.
- (C3) Every student must take at least three classes each semester.
- (C4) For each two students, there must be exactly one class that both students take (during the same semester).
- (C5) For each two classes, there must be exactly one student who is in both classes.

(a) Suppose that the college offers seven classes in a certain semester. Suppose that the enrollments in these classes are as follows (students are numbered 1, 2, 3, etc.):

   Class A: 1, 2, 4
   Class B: 2, 3, 5
   Class C: 3, 4, 6
   Class D: 4, 5, 7
   Class E: 5, 6, 1
   Class F: 6, 7, 2
   Class G: 7, 1, 3

   Decide which of the five axioms are satisfied.
(b) Design an example where all five axioms are satisfied and at least one class has four students in it.
   (Hints: Suppose that thirteen classes are offered and start with students 1, 2, 4, and 10 taking Class A.)
(c) Explain why the five axioms are consistent.
(d) Explain why axiom (C1) is independent from the other four axioms.
(e) Explain why axiom (C2) is independent from the other four axioms.
(f) Explain why the five axioms are not independent.
   (Hints: Prove that axioms (C1), (C2), (C4), and (C5) imply axiom (C3), as follows. Suppose that the students are $X_1, X_2, \ldots, X_n$ and that the classes are $A_1, A_2, \ldots, A_m$. To show that X_1 is in at least three classes, separate

two cases: when X_1 is in every class and when X_1 is not in some class A_i ($1 \leq i \leq m$). In the first case, prove that $m \geq 3$; in the second case, consider three students who take A_i.)

Remarks This problem is modeled after an important and well-studied mathematical structure. A system of points and lines satisfying the five axioms above (with "students" playing the role of points and "classes" interpreted as lines) is called a *projective plane*. Note that the Parallel Postulate does not hold in a projective plane; in fact, by axiom (C5), any two lines meet at a point. We return to projective planes in Problem 11 of Chapter 13.

Appendix A
Ten Famous Conjectures

There are many perplexing questions in mathematics that have not been resolved. Researchers have been able to guess the answers for most of these open questions—particularly since computing power has increased; the question is not considered settled, however, until a rigorous proof has been established. A *conjecture* is a statement, believed to be true, for which there is no proof yet known.

While conjectures turn into theorems on a daily basis, one suspects that, as mathematicians are introducing and investigating new concepts in an ever-growing variety of fields, the number of unsolved questions in mathematics is actually increasing. Even about such familiar concepts as the positive integers, there is a lot that we do not yet know. Here we present our selection for the top ten most famous conjectures about the integers, together with some information on how much we currently know about them.

Our first conjecture is about perfect numbers. Recall from Definition 3.1 that a positive integer n is called perfect when it equals the sum of its positive divisors other than itself, equivalently, if $\sigma(n)$—denoting the sum of all of its positive divisors—equals $2n$. The first four perfect numbers are 6, 28, 496, and 8, 128 (see Problem 1 in Chapter 3). Perfect numbers are rare: the next perfect number has eight digits, and we only know of a few dozen at this time, though we believe that there are infinitely many. As we mentioned in Chapter 3, one can show that all even perfect numbers are of the form $2^{n-1}(2^n - 1)$ for some positive integer n, and a number of this form is indeed perfect whenever $2^n - 1$ is prime (see Theorem 4.1 in Chapter 4). Even perfect numbers are related to Mersenne primes, which we study below.

But how about odd perfect numbers? At the present time, we are not aware of any, and we believe that there are none:

Conjecture A.1 (Odd Perfect Number Conjecture) *There are no odd perfect numbers.*

© Springer Nature Switzerland AG 2020

B. Bajnok, *An Invitation to Abstract Mathematics*, Undergraduate Texts in Mathematics, https://doi.org/10.1007/978-3-030-56174-1

We know that, if there are any odd perfect numbers, then they must be huge: it has been verified that no odd positive integer with fewer than 1500 digits is perfect. There are numerous other facts we know—for example, odd perfect numbers cannot be divisible by 3, 5, or 7—but the conjecture remains open. We should also point out that, while there seem to be no odd perfect numbers, there are infinitely many odd deficient numbers (where $\sigma(n)$ is less than $2n$) and also infinitely many odd abundant numbers (where $\sigma(n)$ is greater than $2n$). As we remarked in Problem 7 of Chapter 4, it is more likely that a number is deficient than that it is abundant, and this is true about odd numbers as well. In fact, all odd positive integers under one thousand are deficient with one exception: this exception is 945, as $945 = 3^3 \cdot 5 \cdot 7$ implies that

$$\sigma(945) = (1 + 3 + 3^2 + 3^3) \cdot (1 + 5) \cdot (1 + 7) = 1920,$$

which is more than $2 \cdot 945$. In Problem 1 below we generalize this example to construct infinitely many odd abundant numbers.

Next, we consider prime values of polynomial expressions. Recall from Chapter 3 that, as Leonard Euler observed, the polynomial $n^2 - n + 41$ is prime for every positive integer n up to $n = 40$, but is, obviously, not prime for $n = 41$. We in fact do not know whether there are infinitely many values of n for which this polynomial is prime. In fact, we do not know of any quadratic (or higher degree) polynomial that returns infinitely many prime values, though we believe that there are some:

Conjecture A.2 (Conjecture on Prime Values of Polynomials) *There are polynomials of one variable and of degree two or more that assume infinitely many prime values.*

In particular, it is believed that there are infinitely many values of n for which $n^2 - n + 41$ yields prime numbers, and the same was conjectured about $n^2 + 1$. As a more modest result, it has been known about both of these polynomials that there are infinitely many values of n for which they either yield primes or a product of two primes. Note that we have polynomials of degree one that are prime infinitely many times; for example, n, $n + 1$, and $2n + 1$ are obviously prime infinitely many times. In fact, according to Dirichlet's Theorem (see Theorem 12.13), every linear polynomial $an + b$ is prime for infinitely many n as long as the coefficients a and b satisfy the obvious requirement that they are relatively prime.

Rather similarly, we may look for prime values of exponential expressions. In particular, our next two conjectures are about the number of n values for which $2^n - 1$ and $2^n + 1$ are primes, respectively.

Conjecture A.3 (Mersenne Prime Conjecture) *There are infinitely positive integers n for which $2^n - 1$ is a prime number.*

Conjecture A.4 (Fermat Prime Conjecture) *There are only five nonnegative integers n for which $2^n + 1$ is a prime number.*

The conjectures, and the corresponding prime number values, are named after the French mathematicians who first studied them: Marin Mersenne (1588–1648) and Pierre Fermat (160?–1665). It may be surprising that Conjectures A.3 and A.4 are so dissonant, but we gain more insight by recalling from Theorem 4.7 that if $2^n - 1$ is prime, then n itself must be prime, and from Theorem 4.11 that if $2^n + 1$ is prime, then n must be a nonnegative integer power of 2. Since prime numbers are a lot more frequent than powers of 2, it is reasonable to expect more Mersenne primes than Fermat primes. To illuminate this further, in Problem 3 we investigate a common generalization of Mersenne primes and Fermat primes.

At the time of writing, there are only fifty-one Mersenne primes known. The one most recently found is the largest of them, having close to twenty-five million decimal digits. For more information on the *Great Internet Mersenne Prime Search (GIMPS)*, see www.mersenne.org. As we pointed out above, Mersenne primes are aligned with even perfect numbers.

As for Fermat primes, we know that $2^{2^k} + 1$ is prime for $k = 0, 1, 2, 3, 4$. Fermat—one of the greatest mathematicians of all time—believed that there are infinitely many Fermat primes; indeed, he conjectured that $2^{2^k} + 1$ is prime for every nonnegative integer k. The Swiss mathematician Leonhard Euler discovered in 1732 that $2^{2^6} + 1$ is not prime as it is divisible by 641; we now also know that $2^{2^k} + 1$ is definitely not prime for $k = 7, 8, \ldots, 32$. Fermat primes have an interesting connection with the constructibility of regular polygons (see Theorem B.3 in Appendix B).

Next, we revisit Waring's Problem, named after the British mathematician Edward Waring (1736–1798) and discussed in Chapter 3: How many terms does it take to write every positive integer as a sum of perfect squares, as a sum of perfect cubes, and so on? Letting $g(k)$ denote the smallest integer m for which every positive integer can be expressed as the sum of m perfect k-th powers, we see that obviously $g(1) = 1$ (every positive integer n can be written as the "1-term sum" n^1), but values of $g(k)$ are difficult to find for $k > 1$.

With a remarkable breakthrough in 1770, Joseph-Louis Lagrange (1736–1813) settled a fifteen-hundred-year-old conjecture of Diophantus that every positive integer can be written as the sum of four perfect squares. (The fact that four squares may be necessary can be seen by observing that 7—indeed, any positive integer congruent to 7 mod 8—requires four squares. Note that the k-th powers need not be distinct and, while they have to be nonnegative, they may be zero; for example, we may write 6 as the sum of four squares as $2^2 + 1^2 + 1^2 + 0^2$.) Thus we have $g(2) = 4$. It took almost another century and a half to prove that $g(3) = 9$; the next value to be found was $g(6) = 73$ in 1940, followed by $g(5) = 37$ in 1964, then $g(4) = 19$ in 1986. The fact that these values are lower bounds can be explained by exhibiting numbers that cannot be expressed with fewer terms; indeed, as we saw in Problem 9 (g) of Chapter 3, the number

$$n = 2^k \cdot \left\lfloor (3/2)^k \right\rfloor - 1$$

requires at least

$$2^k + \left\lfloor (3/2)^k \right\rfloor - 2$$

terms. It is widely believed that this yields the value of $g(k)$:

Conjecture A.5 (Waring's Conjecture) *The smallest integer m for which every positive integer n can be expressed as the sum of m terms, each of which is the k-th power of a nonnegative integer, is*

$$g(k) = 2^k + \left\lfloor (3/2)^k \right\rfloor - 2.$$

We can check that the formula in Conjecture A.5 yields the above-stated values for $k = 1, 2, 3, 4, 5, 6$. In 1909, David Hilbert proved that $g(k)$ exists for every k, and we now also know that Conjecture A.5 holds for every $k \leq 471, 600, 000$ and every "sufficiently large" k (thus leaving only finitely many cases open).

Rather than wanting to express integers as sums of squares, cubes, and other powers, it is natural to wonder how we can express them as sums of positive primes. Since primes are more frequent than squares, one may guess that fewer than $g(2) = 4$ terms may suffice. Among positive integers under 20—ignoring 1—each is either a prime or the sum of two primes: indeed, $4 = 2+2, 6 = 3+3, 8 = 5+3, 9 = 7+2$, $10 = 7 + 3, 12 = 7 + 5, 14 = 11 + 3, 15 = 13 + 2, 16 = 13 + 3, 18 = 11 + 7$. But two primes will not always suffice: 27 is not prime and, since 2 is the only even prime and 25 is not prime, 27 cannot be written as the sum of two primes. However, it can be written as the sum of three primes: for example, $27 = 19 + 5 + 3$.

At this point, it seems appropriate to consider even and odd values separately. With a bit more experimentation, we can then come to the belief that all positive even integers can be written as the sum of at most two primes, and all odd integers greater than 1 can be written as the sum of at most three primes.

Let us consider even integers first. Since 2 is the only even prime, the statement that all positive even integers can be written as the sum of at most two primes is clearly equivalent to the following:

Conjecture A.6 (Goldbach's Conjecture) *Every even integer that is greater than 2 can be expressed as the sum of two positive prime numbers.*

The conjecture is named after the German mathematician Christian Goldbach (1690–1764) who first stated this in a conversation with Leonhard Euler in 1742. At the present time, Goldbach's Conjecture has been verified for integers up to four quintillion, but a proof remains elusive.

We can easily see that Conjecture A.6 implies that every odd integer greater than 5 can be expressed as the sum of three positive prime numbers: Indeed, if n is odd and greater than 5, then $n - 3$ is even and greater than 2, so by Conjecture A.6, we can find primes p_1 and p_2 so that $n - 3$ is the sum of p_1 and p_2, and thus

$$n = p_1 + p_2 + 3.$$

Therefore, the statement that every odd integer greater than 5 is the sum of three primes is called the *Weak Goldbach Conjecture*. A few years ago, the Peruvian mathematician Harald Helfgott released a proof of the Weak Goldbach Conjecture; while the proof is generally believed to be correct, it has yet to appear in a refereed journal. We should point out that the Weak Goldbach Conjecture implies that every even integer greater than 8 can be expressed as the sum of four prime numbers: Indeed, suppose that n is even and $n > 8$, then $n - 3$ is odd and greater than 5, so it is the sum of primes p_1, p_2, and p_3, with which we then have

$$n = p_1 + p_2 + p_3 + 3.$$

With our next conjecture, we return to Euclid numbers, introduced in Problem 4 (b) of Chapter 3. The n-th Euclid number is defined as the product of the first n positive prime numbers, plus 1; we thus have

$$K_1 = 2 + 1 = 3,$$

$$K_2 = 2 \cdot 3 + 1 = 7,$$

$$K_3 = 2 \cdot 3 \cdot 5 + 1 = 31,$$

$$K_4 = 2 \cdot 3 \cdot 5 \cdot 7 + 1 = 211,$$

$$K_5 = 2 \cdot 3 \cdot 5 \cdot 7 \cdot 11 + 1 = 2311,$$

$$K_6 = 2 \cdot 3 \cdot 5 \cdot 7 \cdot 11 \cdot 13 + 1 = 30031,$$

and so on. The numbers received their name after Euclid's famous proof for the infinitude of primes in which they were introduced (see Theorem 12.2). In that proof, the claim is made that K_n is either prime or has a prime divisor different from the first n positive primes—we thus may wonder when we are in the case that K_n itself is prime. As we saw in Problem 4 (b) of Chapter 3, K_n is prime for $n = 1, 2, 3, 4, 5$, but K_6 is composite as it factors as $59 \cdot 509$. In fact, it seems that most Euclid numbers are composite: It has been shown that the only values of n under 200 for which K_n is prime are $n = 1, 2, 3, 4, 5, 11, 75, 171$, and 172. Nevertheless, we believe that there are infinitely many such n:

Conjecture A.7 (Euclid Number Conjecture) *Let* $p_1, p_2, p_3, \ldots$ *be the sequence of positive primes in increasing order. There are infinitely many positive integers n for which Euclid's number*

$$K_n = p_1 \cdot p_2 \cdots \cdot p_n + 1$$

is a prime number.

While we believe that there are infinitely many Euclid numbers that are prime, at the present time we only know of about two dozen.

Our final conjecture here involving primes is perhaps the most famous one of all: there are infinitely many pairs of consecutive odd integers that are both prime. For example, 3 and 5 are both prime, and so are 5 and 7, 11 and 13, 17 and 19, 29 and 31, and so on.

Conjecture A.8 (Twin Prime Conjecture)   *There are infinitely many pairs of prime numbers that differ by 2.*

An equivalent way to state the Twin Prime Conjecture is to say that 2 can be written as the difference of two primes in infinitely many ways; more generally, the French mathematician Alphonse de Polignac (1826–1863) conjectured in 1846 that every even integer can be written as the difference of two primes in infinitely many ways.

An interesting though weaker result was achieved by the Chinese mathematician Chen Jingrun when he proved in 1966 that there are infinitely many primes p for which $p + 2$ is either prime or the product of two primes. A breakthrough in a different direction occurred in 2013 when the American mathematician Yitang Zhang proved that there are infinitely many positive integers n for which the n-th and $(n + 1)$-st positive primes differ by at most 70,000,000; this bound was reduced by Terence Tao and others to 246. While we thus see that the gap between consecutive primes is small infinitely often, we note that it is not hard to see that the gap can also be arbitrarily large—see Problem 5.

The sequence of prime numbers offers many intriguing open questions, but we round out our top ten list with two conjectures that don't involve them, at least not directly.

The so-called $3x + 1$ Conjecture or Collatz Conjecture is as follows. On the positive integers, define the function $f(x) = 3x + 1$ if x is odd and $f(x) = x/2$ if x is even; then consider the iteration of f starting with any initial value a, that is, the sequence

$$(a, \ f(a), \ f(f(a)), \ f(f(f(a))), \ \ldots),$$

called the *orbit* of a. For example, the orbit of $a = 5$ is

$$(5, \ 16, \ 8, \ 4, \ 2, \ 1, \ 4, \ 2, \ 1, \ \ldots),$$

with $a = 6$ we get

$$(6, \ 3, \ 10, \ 5, \ 16, \ 8, \ 4, \ 2, \ 1, \ 4, \ 2, \ 1, \ \ldots),$$

and with $a = 7$ we have

$$(7, \ 22, \ 11, \ 34, \ 17, \ 52, \ 26, \ 13, \ 40, \ 20, \ 10, \ 5, \ 16, \ 8, \ 4, \ 2, \ 1, \ 4, \ 2, \ 1, \ \ldots).$$

In 1937, the German mathematician Lothar Collatz (1910–1990) made the following conjecture:

Conjecture A.9 (Collatz Conjecture)   *Define the function f on the positive integers by setting $f(x) = 3x + 1$ when x is odd, and $f(x) = x/2$ when x is even. For every positive integer a, the orbit of a under the iterations by f always includes 1.*

Of course, once we have reached 1, the iteration assumes the infinite cycle

$$(1, 4, 2, 1, 4, 2, 1, 4, 2, 1, \ldots).$$

The Collatz function has been studied from a variety of perspectives. For example, one can try to characterize values of a for which it takes a relatively large number of iterations to reach 1: We find, for instance, that there are 15 values of a under one hundred for which it takes over 100 iterations (with $a = 97$ taking the most: 118 iterations). Or, one can examine the maximum value reached in the orbit: It turns out, for example, that starting with $a = 703$ or $a = 937$, the orbit contains the integer $250, 504$ (this is the highest for any a under one thousand).

The Collatz Conjecture is considered to be one of the most difficult conjectures in mathematics. Paul Erdős, one of the greatest mathematicians of all time, commented that "mathematics is just not ready to solve problems like this." Nevertheless, the conjecture attracts many mathematicians and computer scientists and, as is the case with most conjectures that are simple to state, a large number of amateurs as well.

Our final conjecture is also strikingly easy to present. Consider powers of the form a^n with positive integer values of a and integer values of n that are at least 3. How can the sum of two of these powers equal a third? It is not difficult to find examples, for instance,

$$2^5 + 2^5 = 2^6,$$

$$9^3 + 18^3 = 9^5,$$

and

$$26^3 + 26^4 = 78^3.$$

(Each of these examples generalizes as well.) But note that in each case, the three bases were quite special: all equal to 2 in the first case, all multiples of 9 in the second equation, and all multiples of 26 in the last case. In 1993, the banker and amateur mathematician Andrew Beal (1952–) made the following conjecture:

Conjecture A.10 (Beal's Conjecture)   *If a, b, and c are relatively prime positive integers, then the equation*

$$a^k + b^m = c^n$$

has no positive integer solutions with k, m, and n all greater than 2.

Mr. Beal offers $1,000,000 for a proof of his conjecture.

Note that the requirement that each exponent be greater than 2 is essential, as the examples

$$2^5 + 7^2 = 3^4$$

and

$$13^2 + 7^3 = 2^9$$

show.

We should mention two famous special cases. First, Beal's Conjecture holds when the exponents k, m, and n are equal, as we get Fermat's Last Theorem in that case (see Theorem B.9). Second, the conjecture holds when a or b equals 1: As Preda Mihailescu (1955–) proved in 2002, the only ways for $1 + b^m$ to equal c^n for positive integers b, c, m, and n are when m or n equals 1 or when $b = 2$, $m = 3$, $c = 3$, and $n = 2$:

$$1 + 2^3 = 3^2.$$

We have to admit that, in our selection for our top ten conjectures, we had a preference for those questions that could be stated without complicated technicalities. The questions considered most important, however, often require background that would take more time to explain. In the year 1900, the German mathematician David Hilbert (1862–1943), at the second International Congress of Mathematicians held in Paris, stated twenty-three open questions ranging over most branches of mathematics. Many of *Hilbert's Problems*, as they became known, are still open. (As it turns out, Hilbert was mistaken in thinking that every well-phrased problem has a solution: The first problem on his list, the so-called *Continuum Hypothesis*, is now known to be independent of the usual axioms of mathematics; see Chapter 20.)

One hundred years later, at the turn of the twenty-first century, several mathematicians collected what they believed were the most important open questions. Seven of these questions are known as the *Millennium Problems* (e.g., the Riemann Hypothesis, the Poincaré Conjecture, and "$P = NP$"); the Clay Mathematics Institute offers one million dollar prizes for the solution of any of them—see www. claymath.org for more information. As of now, six of the Millennium Problems remain open; the Poincaré Conjecture, however, was solved in 2003 in a fantastic achievement by Grigori Perelman of the Steklov Institute of Mathematics in St. Petersburg. (Perelman chose not to accept the Millennium Prize, as he also declined the most prestigious award in mathematics, the Fields Medal. As he told the President of the International Mathematical Union, the prize "was completely irrelevant for me. Everybody understood that if the proof is correct then no other recognition is needed.") We shall discuss the Poincaré Conjecture (now theorem) in Appendix B.

The problems and assignments below offer you a wide variety of possibilities for further investigation of these conjectures and results, as well as additional questions that await mathematicians of the future.

Problems and Assignments

1. In Problem 7 (b) of Chapter 4 we proved that 3^n is deficient for every nonnegative integer n, and thus an odd perfect number must have a prime divisor that is at least 5.

   (a) Prove that $3^n \cdot 5^m$ is deficient for all nonnegative integers n and m, and thus an odd perfect number must have a prime divisor that is at least 7.

   (b) As the example of 945 shows, a number of the form $3^n \cdot 5^m \cdot 7^k$ may be abundant. Prove that, in fact, all numbers of this form are abundant, as long as $n \geq 3$, $m \geq 1$, and $k \geq 1$.
   (Hints: Let $N = 3^n \cdot 5^m \cdot 7^k$ with $n \geq 3$, $m \geq 1$, and $k \geq 1$. Verify that

   $$\sigma(N) = \frac{(3 \cdot 3^n - 1)(5 \cdot 5^m - 1)(7 \cdot 7^k - 1)}{2 \cdot 4 \cdot 6},$$

   and thus

   $$48\sigma(N) = (3 \cdot 3^n - 1)(5 \cdot 5^m - 1)(7 \cdot 7^k - 1)$$
   $$\geq 105N - (15 \cdot 3^n \cdot 5^m + 21 \cdot 3^n \cdot 7^k + 35 \cdot 5^m \cdot 7^k)$$
   $$\geq 105N - (15/7 + 21/5 + 35/27)N$$
   $$> 96N,$$

   as needed.)

2. Recall that we call a positive integer n perfect if its positive divisors other than itself add up to n. Related to perfect numbers, we define a pair of positive integers m and n amicable if the positive divisors of m other than m add up to n, and the positive divisors of n other than n add up to m. At the time of writing, over ten million pairs of amicable numbers are known, and it is a famous old conjecture that there are infinitely many.

   (a) Verify that 220 and 284 form an amicable pair.

   (b) There have been numerous rules discovered that, under certain conditions, yield amicable numbers. One such rule, known since the tenth century, states that if $p_1 = 3 \cdot 2^k - 1$, $p_2 = 3 \cdot 2^{k+1} - 1$, and $p_3 = 9 \cdot 2^{2k+1} - 1$ are primes for some positive integer k, then $m = 2^{k+1} \cdot p_1 \cdot p_2$ and $n = 2^{k+1} \cdot p_3$ are amicable. (For $k = 1$ we get $m = 220$ and $n = 284$.) Verify this result.

3. In this problem, we introduce a common generalization of Mersenne numbers and Fermat numbers, and investigate cases in which they are primes. Namely, for positive integers m and n, let us define

$$F(m, n) = 1^n + 2^n + 4^n + \cdots + \left(2^{m-1}\right)^n.$$

(a) Prove that

$$F(m, n) = \frac{2^{mn} - 1}{2^n - 1},$$

in particular, that $F(2, n)$ is the Fermat number $2^n + 1$ and $F(m, 1)$ is the Mersenne number $2^m - 1$. (Note also that $F(1, n) = 1$ for all n.)

(b) Prove that if $F(m, n)$ is a prime number, then m must be prime and n must be a (nonnegative integer) power of m.

Remarks Observe that our claim generalizes Theorem 4.7 (which says that $F(m, 1)$ can only be prime when m is prime) and Theorem 4.11 (which says that $F(2, n)$ can only be prime when n is a nonnegative integer power of 2).

(c) In Table A.1, entries (m, n) with $2 \leq m \leq 10$ and $1 \leq n \leq 9$ are marked by X if they are ruled out to be prime by part (b); we also labeled spots with **P** for which we have already seen that the number in question is prime, namely:

- $F(m, 1)$ with $m = 2, 3, 5, 7$ (these are the first four Mersenne primes; note that, as we pointed out on page 29, $F(11, 1)$ is composite);
- $F(2, n)$ with $n = 1, 2, 4, 8$ (these are the first four Fermat primes); and
- $F(3, n)$ with $n = 1, 3, 9$ (see Problem 9 (b) in Chapter 4).

This leaves only two table entries open: $F(5, 5)$ and $F(7, 7)$. Decide if these values are prime.

Table A.1 Some prime and non-prime values of $F(m, n)$

	$n = 1$	$n = 2$	$n = 3$	$n = 4$	$n = 5$	$n = 6$	$n = 7$	$n = 8$	$n = 9$
$m = 2$	P	P	X	P	X	X	X	P	X
$m = 3$	P	X	P	X	X	X	X	X	P
$m = 4$	X	X	X	X	X	X	X	X	X
$m = 5$	P	X	X	X		X	X	X	X
$m = 6$	X	X	X	X	X	X	X	X	X
$m = 7$	P	X	X	X	X	X		X	X
$m = 8$	X	X	X	X	X	X	X	X	X
$m = 9$	X	X	X	X	X	X	X	X	X
$m = 10$	X	X	X	X	X	X	X	X	X

Remarks As we see, $F(m, n)$ is rarely prime; in particular, by part (b) above, for each $n \geq 2$, there is at most one value of m for which $F(m, n)$ is prime (and definitely none if n is not a prime power). In particular, the only other cases with $2 \leq m \leq 100$ and $2 \leq n \leq 100$ that yield prime values for $F(m, n)$ are $F(2, 16) = 65, 537$ (the largest known Fermat prime) and $F(59, 59)$ (a number with 1031 decimal digits).

4. For a positive integer k, we define $G(k)$ as the smallest integer m for which *all but finitely many* positive integers can be expressed as the sum of m perfect k-th powers.

   (a) Prove that $G(2) = 4$.
       (Hints: Since $G(k) \leq g(k)$ holds for all positive integers k, we have $G(2) \leq 4$ by Lagrange's theorem on the sum of four squares. To prove that $G(2) \geq 4$, use the fact that squares can only be congruent to 0, 1, or 4 mod 8.)
   (b) Prove that $G(3) \geq 4$.
       (Hints: Use the fact that cubes can only be congruent to 0, 1, or 8 mod 9.)

Remarks Besides $G(1) = 1$ (which is obvious) and $G(2) = 4$ (which we proved in part (a)), the only other value of $G(k)$ that we know at the present time is $G(4) = 16$ (proved by the British mathematician Harold Davenport in 1939). Unlike for $g(k)$, we do not even have a conjecture for the exact value of $G(k)$ in general. We do know, however, that $G(k)$ grows substantially slower than the exponentially-growing $g(k)$; for example, while $g(10) = 1079$ and $g(20) = 1, 051, 899$, we have the upper bounds $G(10) \leq 59$ and $G(20) \leq 142$.

5. Prove that the gap between two consecutive prime numbers can be arbitrarily large.
   (Hint: For any given positive integer N, prove that there are no prime numbers between $N! + 2$ and $N! + N$, inclusive.)

Remark This statement provides a contrast with the Twin Prime Conjecture, which says that the gap between two consecutive primes is 2 infinitely often.

6. In this problem, we consider variations of the Collatz Conjecture.

   (a) Investigate what happens when the Collatz function is defined on the set of negative integers. Make a conjecture.
       (Hints: Try various initial values a and observe the orbits. Make sure to include $a = -1, a = -5$, and $a - -17$.)
   (b) Investigate what happens when we alter the Collatz function so that it maps the positive integer x to $x/2$ when x is even to $3x - 1$ when x is odd. Make a conjecture.
       (Hints: Try various initial values a and observe the orbits. Make sure to include $a = 1, a = 5$, and $a = 17$.)
   (c) Explain why your conjectures in parts (a) and (b) above are, in a certain sense, equivalent.

7. Here we investigate prime numbers in arithmetic progressions; that is, a sequence of prime numbers of the form

$$a, a + d, a + 2d, \ldots, a + kd$$

for some positive integers a, d, and k.

(a) Find the longest arithmetic progression consisting of positive primes under 100.

(b) A recent breakthrough in this area was achieved in 2004 when Ben Green and Terence Tao proved the long-standing conjecture that there are arbitrarily long arithmetic progressions made up of primes. Their proof was non-constructive, so the question of finding explicit examples remains open. Review the current records known.

(c) The question of primes in arithmetic progressions becomes even more difficult if we insist on the primes being consecutive. Find four consecutive primes in arithmetic progression. Review the current records in this area.

8. Prime numbers seem to wonderfully combine regularity with unpredictability; the list of primes has been studied by professional mathematicians and interested amateurs alike. They seem to reveal just enough of their mystery to keep everyone interested. Investigate some of the unsolved problems regarding primes that we have not mentioned yet. A great place to start your research is Richard Guy's book *Unsolved Problems in Number Theory*, 3rd ed. (Springer-Verlag, 2004) that has thousands of results and open questions and provides relevant references. There is also a wealth of information on prime number records on the website http://primes.utm.edu.

9. (a) Describe some of Hilbert's Problems and review their status quo.

(b) Describe some of the six unsolved Millennium Problems.

10. The late Hungarian mathematician Paul Erdős, the "Prince of Problem-Solvers and the Monarch of Problem-Posers," was famous for offering cash prizes for certain of his unsolved problems. The prizes ranged from $25 to $10,000. (Since Erdős's 1996 death, the rewards can be collected from Ronald Graham, a former president of the American Mathematical Society. Though many mathematicians have received checks for their solutions, most have decided to frame the checks rather than to cash them.) Describe some of Erdős's "cash problems."

Appendix B
Ten Famous Theorems

There are many famous theorems in mathematics. Some are known for their importance, others for their depth, usefulness, or sheer beauty. In this chapter we discuss ten of the most remarkable theorems of all time. Our choices for this top ten list were motivated primarily by how influential they were; we also avoided theorems that are presented elsewhere in this book. (A more representative top forty list can be found in Appendix G at the end of the book.) Here we include theorems that are considered to be the oldest, the most well-known, the most surprising, and the most unsettling. Some of the theorems on our list were disappointing to mathematicians of the time, others were celebrated instantly by most. We will not be presenting any proofs here: for most of our theorems this would be far beyond the scope of this book. (We will outline proofs for our first two results in Problems 1 and 2, respectively.)

We start with what historians of mathematics regard as the oldest theorem in mathematics, oldest in the sense that it was the first statement for which a rigorous proof was given. This is the following theorem discovered in the sixth century BCE by the Greek mathematician and scientist Thales of Miletus.

Theorem B.1 (Thales's Theorem) *If a triangle is inscribed in a circle so that one of its sides goes through the center of the circle, then the angle of the triangle that is opposite to this side is a right angle.*

A proof to Thales's Theorem, using basic properties of triangles, can be established easily—we leave this as Problem 1.

Our next theorem might be the most well-known theorem in mathematics. While once thought to have been discovered by Pythagoras and his circle of friends at the end of the sixth century BCE, we now know that the Babylonians as well as the Chinese knew of this result about a thousand years earlier.

Theorem B.2 (The Pythagorean Theorem) *If a and b are the lengths of the two legs of a right triangle and c is the length of its hypotenuse, then*

© Springer Nature Switzerland AG 2020

B. Bajnok, *An Invitation to Abstract Mathematics*, Undergraduate Texts in Mathematics, https://doi.org/10.1007/978-3-030-56174-1

$$a^2 + b^2 = c^2.$$

There are many nice proofs of this theorem (a collection of 370 proofs, published by Elisha Scott Loomis, appeared in 1927); one such proof is suggested in Problem 2.

Our next theorem also originates in antiquity, but its proof was attained only in the nineteenth century. The ancient Greeks had a keen interest in constructing various geometric shapes, and by construction they meant a (finite) sequence of markings where each step relied either on drawing straight lines or on drawing circles. One shape they had a special interest in was the regular polygon. (A regular polygon is one whose sides are all equal and whose angles are all equal.) The ability to construct certain regular polygons, such as triangles, squares, hexagons, and octagons, was known early on, and Euclid described a method for constructing regular pentagons. The Greeks also knew how to get a regular $2n$-gon from a regular n-gon, and how to get a regular $(n_1 n_2)$-gon from a regular n_1-gon and a regular n_2-gon when n_1 and n_2 are relatively prime. However, they were unable to achieve more; in particular, the only odd values of n for which they knew constructions were 3, 5, and 15.

There were no new developments for over two thousand years, until the nineteen-year-old Carl Friedrich Gauss discovered a construction for the regular 17-gon in 1796. His construction relied on the fact that 17 is a Fermat prime: a prime of the form $2^{2^k} + 1$ for some nonnegative integer k (in the case of 17, $k = 2$). Note also that 3 and 5 are Fermat primes: $3 = 2^{2^0} + 1$ and $5 = 2^{2^1} + 1$, and that 15 is the product of two distinct Fermat primes. About four decades later, the French mathematician Pierre Wantzel (1814–1848) completed Gauss' work, and proved the following result on the constructibility of regular polygons.

Theorem B.3 (The Gauss–Wantzel Theorem) *For an integer $n \geq 3$, the regular n-gon is constructible if, and only if, n is the product of a (nonnegative integer) power of 2 and some (possibly zero) distinct Fermat primes.*

Recall that, at the moment, we are only aware of five Fermat primes— $3, 5, 17, 257$, and $65, 537$—and we believe that there are no others, which means that there are likely to be only 31 odd values of n for which a regular n-gon is constructible, with the largest being

$$3 \cdot 5 \cdot 17 \cdot 257 \cdot 65,537 = (2^1 + 1)(2^2 + 1)(2^4 + 1)(2^8 + 1)(2^{16} + 1) = 2^{32} - 1.$$

We now turn to what may be the most astonishing result in all of mathematics, one that was entirely unexpected at the time. First some background.

For a given nonnegative integer n and for real numbers $c_n, \ldots, c_2, c_1, c_0$, with $c_n \neq 0$, a *polynomial* of *degree* n is a function f defined on the set of real numbers $\mathbb{R}$ with

$$f(x) = c_n x^n + \cdots + c_2 x^2 + c_1 x + c_0.$$

Here $c_n, \ldots, c_0$ are called the *coefficients* of f. If $n = 0$, $n = 1$, $n = 2$, or $n = 3$, then the polynomial is called *constant*, *linear*, *quadratic*, or *cubic*, respectively. (Also, so that the set of polynomials forms a closed set for addition and subtraction—for example, x^2 can be added to $-x^2$—we define the *zero-polynomial* with $f(x) = 0$; it is customary to say that the zero-polynomial has degree -1.)

Clearly, a nonzero constant polynomial has no roots; a linear polynomial $c_1 x + c_0$ has one root: $-c_0/c_1$. We are all familiar with the Quadratic Formula; it determines the roots of a polynomial of degree 2, and it involves square roots as well as the four arithmetic operations (addition, subtraction, multiplication, and division). The prominent question of whether there are similar formulas for polynomials of higher degree wasn't settled until the sixteenth century, when Niccolo Tartaglia developed a formula for the roots of the general cubic (degree 3) polynomial, a formula that uses only the arithmetic operations, square roots, and cube roots. Soon after, a formula for quartic (degree 4) polynomials was developed by Tartaglia's nemesis Ludovico Ferrari. Both formulas were published by Gerolamo Cardano in 1545 in his book *Ars Magna*; in fact, Tartaglia's solution for the cubic equation is known to this day as Cardano's Formula. (The question of who really deserves full credit for discovering these formulas was hotly debated at the time; the vicious fight among Tartaglia, Ferrari, Cardano, and several other Renaissance Italians was a low point of sixteenth-century mathematics.)

The search eventually turned to finding an algebraic formula (one that only uses arithmetic operations and roots) for the solution of the general quintic (degree 5) equation. In one of the most surprising moments in the history of mathematics, in 1824 the Norwegian mathematician Niels Abel (1802–1829) announced that no such formula can exist. Later it was discovered that the Italian mathematician Paolo Ruffini (1765–1822) had published a proof of the same result twenty-five years earlier. Although Ruffini's proof was not complete, the theorem is now known under the names of both Abel and Ruffini.

Theorem B.4 (The Abel–Ruffini Theorem) *There is no algebraic formula for the roots of the general polynomial of degree 5 or higher.*

We should point out that there are (infinitely many) quintic polynomials whose roots can be expressed algebraically, but this does not contradict the Abel–Ruffini Theorem. For example, the roots of $x^5 - 4x^3 + 3x$ can be easily determined: factoring it as

$$x(x^2 - 1)(x^2 - 3)$$

shows that the five roots are 0, ± 1, and $\pm\sqrt{3}$. It may take more work, but the roots of

$$x^5 - 15x^4 + 85x^3 - 225x^2 + 274x - 120$$

can be determined as well: they are $1, 2, 3, 4$, and 5. But, as it turns out, the roots of $x^5 - 6x + 3$ or $x^5 + 20x + 16$ (for example) cannot be written algebraically, so there cannot be a general algebraic formula for the solution of the quintic. The question of which equations have algebraic solutions was settled a few years after Abel's work by the French mathematician Évariste Galois (1811–1832). (It is a very sad fact that neither Abel nor Galois received much recognition for his work in his lifetime; both young men died under tragic circumstances in their twenties.) Galois theory, and the proof of the Abel–Ruffini Theorem, in particular, are discussed in most textbooks on abstract algebra.

We may ask a variety of other questions regarding the roots of polynomials. One such question is as follows: Is every real number the root of a nonzero polynomial with integer coefficients? The answer is clearly 'yes' for rational numbers (fractions of integers): a/b (with a and b integers, $b \neq 0$) is the root of the (linear) polynomial $bx - a$. Similarly, square roots of rational numbers are roots of integer polynomials too: $\sqrt{a/b}$ is the root of $bx^2 - a$; this clearly generalizes to cube roots, fourth roots, etc.

We say that a real number (or, more generally, a complex number) is *algebraic* if it is the root of some nonzero polynomial with integer coefficients; if that is not the case, we call the number *transcendental*. As we just saw, rational numbers and their various roots are all algebraic. How about sums of algebraic numbers? For example, is $\sqrt{2} + \sqrt[3]{3}$ algebraic? We can find out with a bit of work: If $x = \sqrt{2} + \sqrt[3]{3}$, then

$$x - \sqrt{2} = \sqrt[3]{3};$$

cubing both sides and simplifying yields

$$x^3 + 6x - 3 = (3x^2 + 2)\sqrt{2}.$$

Squaring both sides and simplifying again gives

$$x^6 - 6x^4 - 6x^3 + 12x^2 - 36x + 1 = 0,$$

which means that $\sqrt{2} + \sqrt[3]{3}$ is algebraic. Indeed, there is nothing atypical about this example: It can be proven that applying arithmetic operations on algebraic numbers or taking their square roots, cube roots, etc., always yields algebraic numbers. So are all real numbers algebraic, or are there some that are transcendental?

While it was long suspected that numbers such as e and π are transcendental, this was not known until the second half of the nineteenth century. (Charles Hermite proved in 1873 that e was transcendental, and Ferdinand von Lindemann established in 1882 that π was transcendental.) But the most important and influential result on this topic was given by the German mathematician Georg Cantor (1845–1918), who laid the foundations of what is now called modern set theory, and introduced the idea of different magnitudes of infinity. In particular, he distinguished between *countably infinite* and *uncountably infinite* sets, and proved that the sets of natural numbers, integers, and rational numbers were all countably infinite, while the set of real numbers was uncountably infinite. Moreover, he proved the following:

Theorem B.5 (Cantor's Theorem)   *The set of algebraic numbers is countable, but the set of transcendental numbers is uncountable.*

Later in the book we will study carefully what is meant by the terms "countable" and "uncountable"—and we will prove Theorem B.5—but for now we just surmise that they represent two different levels of infinity. In particular, we interpret Theorem B.5 to say that there are a lot more transcendental numbers than algebraic ones. This may somewhat defy our imagination as we can readily exhibit algebraic numbers, but it is not easy to come up with examples for transcendental numbers. For example, to this day we still don't know if numbers such as $e + \pi$, $e \cdot \pi$, and π^e are transcendental (but we do know that e^π is).

Our next theorem is the famous Prime Number Theorem. We chose this theorem for two reasons. First, it answers a long-standing question of the eighteenth century that is of fundamental importance. Second, its first proof, given by Charles-Jean de la Vallée Poussin (1866–1962) and Jacques Hadamard (1865–1963) in 1896, was greatly simplified by a different approach a half century later by both Paul Erdős (1913–1996) and Atle Selberg (1917–2007) in 1949. Paul Erdős, who published more papers than any mathematician in history, used to refer—somewhat jokingly— to an imaginary *Book* that contains the best proof for each theorem in mathematics. While one considers a theorem proven as long as it has *any*—correct and complete— proof, it is its "Book-proof" (often given many years after the first proof) that is most beautiful and insightful. The proofs given by Erdős and Selberg for the Prime Number Theorem were such Book-proofs.

We mention, in passing, another phenomenon inspired by Erdős: the so-called Erdős number, defined recursively as follows. The only one with Erdős number 0 was Paul Erdős himself. Then, for any natural number n, the family of people with Erdős number n is made up of those who have a joint publication with someone who has Erdős number $n - 1$ but not with anyone whose Erdős number is less than $n - 1$. (Those without a finite Erdős number are said to have Erdős number equal to infinity.) Thus, the people who themselves collaborated with Erdős have Erdős number 1 (there are currently 511 such people, and this is unlikely to increase further as Erdős died in 1996); those who collaborated with anyone with Erdős number 1 but not with Erdős himself have Erdős number 2 (currently, there are 11,002 such individuals); and so on. It has been estimated that more than 90 percent of the approximately four hundred thousand authors with mathematical publications have an Erdős number of 8 or less; the largest finite Erdős number currently known is 13.

Let us return to the Prime Number Theorem. After spending time with the (infinite) sequence of positive prime numbers, one discovers that they behave rather randomly. In particular, we see that the sequence of primes is occasionally quite rare (there is only one prime between 90 and 100), but at other times, quite dense (the odd numbers between 100 and 110 are all primes except for 105). (See Problem 5 of Appendix A on long gaps between consecutive primes.) Therefore, there are no good practical formulas that tell us exactly how many primes we have in a given interval (but cf. Problem 11 below). In addition, while we have reasonably simple tests for

Table B.1 Comparing values of $\pi(N)$ and $N/\ln N$

N	$\pi(N)$	$N/\ln N4$	$\pi(N)/(N/\ln N)$
10	4	4.3	0.921
100	25	21.7	1.151
1000	168	144.8	1.161
10^6	78,498	72,382.4	1.084
10^9	50,847,534	48,254,942.4	1.054
10^{12}	37,607,912,018	36,191,206,825.3	1.039
10^{15}	29,844,570,422,669	28,952,965,460,216.8	1.031
10^{18}	24,739,954,287,740,860	24,127,471,216,847,323.8	1.025
10^{21}	21,127,269,486,018,731,928	20,680,689,614,440,563,221.5	1.022

deciding if a given positive integer is prime or not, for a given large composite number, it may be a difficult task to find its prime factorization. (Cryptography, the field of encryption and decryption of secret transactions, takes advantage of this discrepancy.)

However, when one looks at the density of the primes on a large interval, the behavior is much more regular: we have a pretty good way to *estimate* how many positive primes we have up to a given value. Namely, if we choose a large positive integer N, then the number of primes between 1 and N, denoted by $\pi(N)$, will be quite close to $\frac{N}{\ln N}$, as Table B.1 demonstrates.

This observation is stated more precisely in the following theorem.

Theorem B.6 (The Prime Number Theorem) *Let N be an integer greater than 1, and let $\pi(N)$ be the number of primes between 1 and N. Then $\pi(N) \sim \frac{N}{\ln N}$, that is, we have*

$$\lim_{N \to \infty} \frac{\pi(N)}{N/\ln N} = 1.$$

The Prime Number Theorem can be summarized by saying that $\pi(N)$ and $\frac{N}{\ln N}$ are *asymptotically equal*. Note that the statement is about the ratio of these two functions and not their difference (as the table suggests, the difference $\pi(N) - \frac{N}{\ln N}$ actually goes to infinity!).

Our next theorem is perhaps the theorem that generated the greatest interest outside mathematics. As we have discussed in Chapter 4, all branches of mathematics strive to build their theory on the basis of a set of axioms; these axioms are desired to be both independent (no axiom can be proved using the others) and consistent (they do not lead to contradictions). In the 1930s, the Austrian logician Kurt Gödel (1906–1978) proved a two-fold result, which we state in a simplified fashion.

Theorem B.7 (Gödel's Incompleteness Theorems) *No system of axioms (which is rich enough to contain the axioms of arithmetic) is complete; that is, there will be statements that can be stated using the concepts of the system, but will remain independent from the axioms and thus neither their truth nor their falsehood can be* *proved using the axioms. Furthermore, it will not be possible to decide (within the system) if the system of axioms is consistent; that is, the statement that the system is consistent will itself remain an independent statement in the system.*

Recall that we have already learned (cf. Chapter 4) that the Parallel Postulate is independent from Euclid's other axioms. According to Gödel's result, adding more and more axioms to our system will not suffice: there will be new statements whose truth cannot be established within this larger system. Later in the book we learn another famous case in which a statement was proved to be independent from a system, namely the so-called Continuum Hypothesis. There is still considerable mystery to understanding the full scope of Gödel's results.

Our next theorem has a proof that, at present, relies heavily on computers. As the proof cannot be read and verified manually, some mathematicians do not accept it as a theorem that has been proven. We present this result as follows.

In 1852 the South African botanist and mathematician Francis Guthrie, while trying to color the map of the counties of England, asked the following question: How many different colors are needed if one wants to color a map of connected regions in the plane (or on the surface of the globe) in such a way that regions with a common boundary segment (of positive length) receive different colors? The minimum number of colors needed for such a coloring is called the *chromatic number* of the map. Guthrie noticed that his map had chromatic number 4, and he wondered if four colors would suffice for all maps. In 1976, more than one hundred and twenty years after Guthrie's investigations, Kenneth Appel and Wolfgang Haken proved that this is indeed the case for any planar map.

Theorem B.8 (The Four-Color Theorem) *Any planar map has chromatic number at most 4.*

The proof given by Appel and Haken consists of two parts. First they prove that every map can be reduced to one of 1,482 configurations, and then they use a computer program to verify that these 1,482 maps can indeed be colored with four colors. Since the computer program is too long for a human to check, an argument can be made that the Four-Color Theorem has not been proven in the traditional sense. However, from a practical point of view, the chance of a computer error on all of the many successful runs of the program is small, even smaller than the likelihood of a human error during the same amount of case-checking. Thus most mathematicians, but not all, agree that we indeed have a proof of the Four-Color Theorem. Efforts to reduce the number of configurations to be checked are under way.

While the Four-Color Theorem has a rather playful nature, the theory of *graph coloring* has a wide range of applications; in fact, there is an entire branch of mathematics that deals with chromatic numbers. Consider, for example, the following *scheduling problem*. A set of tasks has to be performed under the conditions that certain pairs of tasks cannot be done at the same time. We want to know how long it will take to perform all tasks (for simplicity, assume that every task alone takes a unit amount of time). Assign a color to every task so that conflicting tasks will get different colors. The minimum amount of time needed to perform all tasks will then be the least number of colors that can be used. Note that this number might be larger than four, as we may have a situation that does not correspond to a planar map. Problem 8 provides some examples for the scheduling problem.

The next theorem that we will discuss here is another example for a statement that is very easy to state but whose proof eluded mathematicians for a long time.

It is not hard to find positive integers x, y, and z for which

$$x^2 + y^2 = z^2.$$

For example, we can choose $x = 3$, $y = 4$, and $z = 5$ or $x = 5$, $y = 12$, and $z = 13$. In fact, all (infinitely many) integer solutions to the equation $x^2 + y^2 = z^2$ can be determined using elementary number theory—see Problem 9. In about 1637, the French mathematician Pierre Fermat, while reading Diophantus's *Arithmetica* (the 1621 edition), wrote the following (in Latin) in the margin of the book.

It is impossible to write a cube as the sum of two cubes, a fourth power as the sum of two fourth powers, and, in general, any power beyond the second as the sum of two similar powers. For this, I have discovered a truly wonderful proof, but the margin is too small to contain it.

Fermat often left his results unpublished, and while all his other statements have been proven, this proof resisted both professionals and numerous amateurs for more than 350 years. The statement came to be known as "Fermat's Last Theorem," and can be stated as follows:

Theorem B.9 (Fermat's Last Theorem) *If n is an integer and $n \geq 3$, then the equation*

$$x^n + y^n = z^n$$

has no positive integer solutions x, y, and z.

A very large number of mathematicians (as well as amateurs, cf. Problem 13) worked on Fermat's Last Theorem with little progress. Even in the 1970s, the statement was only proven for some special types of exponents (Fermat himself knew a proof for the case $n = 4$, the only proof he ever cared to write down). The news traveled almost instantaneously around the world when the English mathematician Andrew Wiles (1953–) announced in 1993 that he finally succeeded in proving Fermat's Last Theorem after several years of solitary work. As Wiles's work became the subject of scrupulous investigation, a substantial gap was found in the proof. Wiles managed, however, to fill this gap soon after with the help of Richard Taylor (1962–). The proof was finally published in 1995. As is the case with the Four-Color Theorem, to this day only a few people have checked the proof; it involves a large amount of and a wide variety of very difficult mathematics. While the significance of proving Fermat's Last Theorem may only be symbolic, the mathematics that was discovered in the process has far-reaching applications.

Our final theorem to complete our top ten list is the Poincaré Theorem, which takes us to the twenty-first century. The precise statement of the theorem would require an extensive discussion; here we only provide a simplified and heuristic treatment.

The theorem deals with the characterization of n-dimensional spheres, a fundamental topic in topology. We attempt to explain the question as follows. In topology, one considers two shapes equivalent if each one can be continuously transformed into the other—breaking and punching holes are not allowed. For example, a topologist would say that a doughnut and a coffee cup are equivalent, but an apple is different. One way to see the difference is by considering various loops on these surfaces. Every loop on an apple can be continuously moved to any other loop—this, however, cannot be said about the coffee cup where, for example, a loop around the ear of the cup cannot be moved to a loop on the side (without breaking up the loop). In 1904, the French mathematician Henri Poincaré conjectured that this simple loop test is enough to identify S^n, the n-dimensional sphere.

The 2-dimensional case was quickly solved by Poincaré himself, but several decades passed before the next breakthrough occurred. In 1961, Stephen Smale proved the conjecture for every $n \geq 5$; then, in 1982, Michael Freedman resolved the case $n = 4$. This left only the 3-dimensional case open, which became the *non plus ultra* question in topology, until the Russian mathematician Grigori Perelman (1966–) solved it in 2003. Using more precise (yet here undefined) terminology, the result can be stated as follows:

Theorem B.10 (The Poincaré Theorem) *Every simply-connected closed 3-dimensional manifold is homeomorphic to the sphere S^3.*

Perelman's breakthrough came after deep and extensive work done by many mathematicians. The proof, published in 2006, introduced new ideas and techniques that can now be used in the solution of other problems. As we mentioned in Appendix A, the Poincaré Theorem is one of the seven *Millennium problems*, the only one which has been solved thus far.

The ten theorems mentioned in this appendix are some of the most well-known theorems in mathematics. However, there are more new results every day as research in abstract mathematics is more active now than ever. It is a common misconception that the questions of mathematics have all been answered—this is far from the truth. (In Appendix A we discussed a top ten list of open questions.) The history of mathematics is a fascinating saga spanning several thousand years and taking place in every corner of the world inhabited by humans. It is a story full of hopes, successes, and disappointments; its greatest achievements are interspersed with perplexing questions unresolved to this day.

Problems and Assignments

1. Use Figure B.1 below to prove Thales's Theorem. You may use, without proof, the following two lemmas.

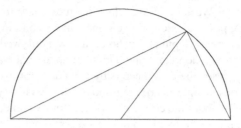

Fig. B.1 Thales's Theorem

> **Lemma B.11** *If in a triangle the lengths of the sides opposite to two vertices are equal, then the measures of the angles at these two vertices are also equal.*
>
> **Lemma B.12** *The sum of the radian measures of the three angles in any triangle equals π.*

2. Use the two diagrams in Figure B.2 to prove the Pythagorean Theorem. You may use, without proof, Lemma B.12 and the following four lemmas.

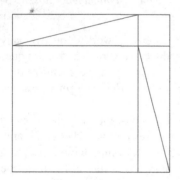 

Fig. B.2 The Pythagorean Theorem

> **Lemma B.13** *Given a triangle ABC, let us denote the angles at vertices A, B, and C by α, β, and γ, respectively, and let us denote the lengths of the sides opposite to A, B, and C by a, b, and c, respectively. We introduce notations for the angles and side lengths in triangle A'B'C' similarly. If*
>
> - $a = a', b = b', c = c'$, *or*
> - $a = a', b = b', \gamma = \gamma'$, *or*
> - $a = a', \beta = \beta', \gamma = \gamma'$,

then triangles ABC and A'B'C' are congruent, that is, we have

$$a = a', b = b', c = c', \alpha = \alpha', \beta = \beta', \gamma = \gamma'.$$

Lemma B.14 *The area of a square of side length d equals d^2.*

Lemma B.15 *Congruent triangles have equal areas.*

Lemma B.16 *If a polygon is divided into a finite number of polygonal parts, then the sum of the areas of the parts equals the area of the whole polygon.*

3. We say that a positive real number a is *constructible* when, given a segment of length 1, a segment of length a can be constructed with a straightedge and compass.

   (a) Prove that if a and b are constructible numbers, then so are $a + b$, $|a - b|$, $a \cdot b$ and a/b.
   (Hints: To construct $a \cdot b$, consider Figure B.3, where the distances of points I, A, and B from O are 1, a, and b, respectively, and the lines IA and BC are parallel. How far is C from O?)

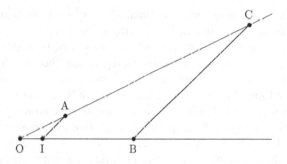

Fig. B.3 A construction of $a \cdot b$

   (b) Prove that if a is a constructible number, then so is $\sqrt{a}$.
   (Hints: Consider Figure B.4, where ABC is a right triangle, CM is perpendicular to AB, and the distance of M from A and B is a and 1, respectively. How far is C from M?)

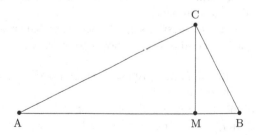

Fig. B.4 A construction of $\sqrt{a}$

Remarks According to our claims, a positive real number a is constructible when it can be expressed from the number 1 using additions, subtractions, multiplications, divisions, and taking square roots. For example,

$$\cos\left(\frac{2\pi}{5}\right) = \frac{\sqrt{5}-1}{4}$$

is constructible, and so is

$$\cos\left(\frac{2\pi}{17}\right) = \frac{-1+\sqrt{17}+\sqrt{34-2\sqrt{17}}+2\sqrt{17+3\sqrt{17}-\sqrt{170+38\sqrt{17}}}}{16},$$

explaining why the regular pentagon and the regular heptadecagon are constructible. By the results on Wantzel, the converse is true as well: if a number is constructible, then it must be expressible using additions, subtractions, multiplications, divisions, and taking square roots. While the proof of this is beyond our scope, we can shed some light on the result by noting that the equations of straight lines and circles are linear and quadratic, respectively, hence their intersections can be expressed by the algebraic operations above.

4. Write all five real roots of the polynomial $x^5 - 7x^3 + 2x$ algebraically. Explain how this does *not* contradict the Abel–Ruffini Theorem.

5. Two players, Nate and Ria, play a game where one of them thinks of a secret sequence (s_1, s_2, s_3, s_4) of four positive integers, and the other has to find out, using as few questions as possible, what s_1, s_2, s_3, s_4 are. Each question must also be of the form of a sequence of length four; in Nate's case, the terms of the sequence must be natural numbers, for Ria, they can be arbitrary real numbers. (The sequences may vary from question to question.) The answer, in each case, must be the linear combination of s_1, s_2, s_3, and s_4 with the four terms of the sequence of the question; that is, if the question is the sequence (a_1, a_2, a_3, a_4), then the answer must be

$$a_1 s_1 + a_2 s_2 + a_3 s_3 + a_4 s_4.$$

(This expression is also called the *dot product* of (a_1, a_2, a_3, a_4) and (s_1, s_2, s_3, s_4).)

(a) What is the minimum number of questions Nate will need to always be able to discover the four numbers?
(Hints: Prove that two questions may be necessary but are always sufficient. Note that Nate's second question may depend on Ria's answer to his first question.)

(b) What is the minimum number of questions Ria will need to always be able to discover the four numbers?
(Hint: Prove that the single question $(1, t, t^2, t^3)$ always reveals the secret sequence when t is any transcendental real number.)

6. Our goal here is to illuminate the Prime Number Theorem by comparing the relative density of primes, perfect squares, and integers that are divisible by a million. Before answering the questions below, you may want to guess which is most common: prime numbers, square numbers, or integers that are divisible by a million? Which is least common?

For an integer N greater than 1, we define $S(N)$ to be the number of perfect square numbers between 1 and N (inclusive), and we define $M(N)$ to be the number of integers between 1 and N (inclusive) that are divisible by a million (Table B.2).

(a) Find each entry in the following table.

(b) (Note: This problem requires a basic understanding of limits.) Rank the three magnitudes $\pi(N)$, $S(N)$, and $M(N)$ as N approaches infinity.

Table B.2 Comparing densities of primes, perfect squares, and integers that are divisible by a million

N	$\pi(N)$	$S(N)$	$M(N)$
10	4	3	0
100	25		
1000	168		
10^6	78,498		
10^9	50,847,534		
10^{12}	37,607,912,018		
10^{15}	29,844,570,422,669		
10^{18}	24,739,954,287,740,860		
10^{21}	21,127,269,486,018,731,928		

7. (a) For an integer $n \geq 3$, the map W_n, called the *wheel graph of order n*, is a configuration of $n + 1$ regions where one region is at the center and the other n regions each have exactly three neighbors, of which one is the center region. (Two regions are said to be neighbors whenever they share a boundary of positive length.) An illustration of W_5, for example, is given in Figure B.5.

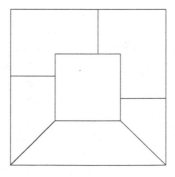

Fig. B.5 An illustration of W_5

Find the chromatic number of W_n for each value of n.

(b) What is the chromatic number of a map of the forty-eight contiguous states of the United States? Justify your answer.

(Hint: Use part (a).)

8. (a) A certain (very small) college offers eight different mathematics courses each semester. The Xs in Table B.3 indicate which two distinct classes have students in common.

Table B.3 Scheduling conflicts in a mathematics department

	1	2	3	4	5	6	7	8
1		X		X	X	X		
2	X		X	X				
3		X		X		X		X
4	X	X	X		X		X	
5	X			X		X		X
6	X		X		X			X
7				X				X
8		X			X	X	X	

Find a planar map containing eight regions representing the eight courses, with two regions sharing a common boundary precisely when the two corresponding courses have students in common. Then color your map so that neighboring regions receive different colors. How many colors are needed? How many exam periods does the college have to offer for these classes during finals week, at the minimum?

(Hints: Since there is a planar map corresponding to the courses, by the Four-Color Theorem the chromatic number—and thus the minimum number of exam periods needed—is at most four. To establish that the chromatic number is in fact four, one needs to prove that the map cannot be colored using three or fewer colors.)

(b) Suppose now that the time conflicts are as in Table B.4.

Table B.4 More scheduling conflicts

	1	2	3	4	5	6	7	8
1		X		X	X	X		
2	X		X	X				
3		X		X		X		X
4	X	X	X		X		X	
5	X			X		X		X
6	X		X		X		X	X
7				X		X		X
8		X			X	X	X	

This time, there is no planar map representing the situation like in the previous part (so don't even try—we will prove this in Problem 8 (b) of Appendix E). Nevertheless, find the minimum number of exam periods needed. Prove your answer.

(Hints: Note that, with one exception, the conflicts are exactly the same as they were in part (a) above. What does this say about the corresponding chromatic numbers?)

9. (a) Verify that $x = k \cdot (m^2 - n^2)$, $y = k \cdot (2mn)$, $z = k \cdot (m^2 + n^2)$ is a solution of the equation $x^2 + y^2 = z^2$ for all choices of the integers k, m, and n.

   Remark It can also be shown that every solution is of this form.

   (b) Use the remark above to find all positive integers A under 100 for which there is a right triangle with area A and with all three sides having integer lengths.

   Remarks A positive integer A for which there is a right triangle with area A and with all three sides having rational number lengths is called a *congruent number*. (Congruent numbers are not to be confused with two integers being congruent to each other or two triangles being congruent to each other!) For example, 6 is a congruent number as it is the area of a right triangle with side lengths 3, 4, and 5. It takes a bit more effort to see that 5 is also a congruent number; it is the area of a right triangle with side lengths 3/2, 20/3, and 41/6. Leonhard Euler established that 7 is also a congruent number; the next congruent number is 13. At the present time, we know that exactly twenty-three numbers under 50 are congruent numbers, but, to this day, we do not have a good understanding of congruent numbers. Indeed, deciding whether a given positive integer is a congruent number or not touches on some of the deepest and most advanced areas of mathematics.

10. According to Fermat's Last Theorem, there are no positive integers x, y, z, and $n \geq 3$ for which

$$x^n + y^n = z^n;$$

but there is considerable interest in finding examples where the two sides of the equation are "very close." One such example,

$$1782^{12} + 1841^{12} = 1922^{12},$$

was featured in 1995 (after Andrew Wiles's announcement) on the animated television series *The Simpsons* (the longest running sitcom on American television). When David Cohen (a writer for the show responsible for the program that generated this near miss) was told that his example can be too easily seen to be wrong by observing that the left-hand side is odd while the right-hand side is even, he adapted his program to check for parity. His subsequent example,

$$3987^{12} + 4365^{12} = 4472^{12},$$

was exhibited on the show three years later. Prove (without relying on Fermat's Last Theorem) that this second equation cannot hold either.

Remark Most calculators will—incorrectly—evaluate $(1782^{12} + 1841^{12})^{1/12}$ to 1922 and $(3987^{12} + 4365^{12})^{1/12}$ to 4472.

11. Show that it is enough to prove Fermat's Last Theorem when the exponent is 4 or an odd prime.
    (Hints: The problem is asking you to prove that the statements

    (1): Fermat's Last Theorem holds when the exponent is 4
    (2): Fermat's Last Theorem holds when the exponent is an odd prime

    imply that Fermat's Last Theorem holds whenever the exponent is any integer greater than or equal to 3. Use Problem 7 (a) of Chapter 3.)
12. Prove the following statement:

    Theorem B.17 *If n is an integer and $n \geq 3$, then the equation*

    $$n^x + n^y = n^z$$

    has no positive integer solutions x, y, and z.

    Remarks This statement superficially resembles Fermat's Last Theorem but is considerably easier to prove. Note that the assumption that $n \geq 3$ is necessary here as well: $2^x + 2^y = 2^z$ holds whenever $x = y = z - 1$.

    (Hints: Start by first ruling out the cases when $x \geq z$ or $y \geq z$. How much can the left-hand side be at the most when x and y are both less than z?)
13. It is believed that Fermat's Last Theorem is the mathematical statement for which the greatest number of false proofs has been published. (For example, during the four years after the Academy of Science at Göttingen offered a 100,000 mark prize for a correct solution in 1908, more than a thousand alleged proofs appeared, mostly printed as private pamphlets.)
    What is right and what is wrong with the following "proofs" of Fermat's Last Theorem?

    (a) Let $n \geq 3$ be an integer and suppose, indirectly, that the equation

    $$x^n + y^n = z^n$$

    has a positive integer solution (x, y, z). By Problem 11 above, we can assume that $n = 4$ or n is an odd prime. The case $n = 4$ was already proved by Fermat himself, so we only have the case when n is an odd prime left. Since n is odd, we can use Lemma 4.9 to conclude that $x^n + y^n$ is divisible by $x + y$. Therefore, the right-hand side must also be divisible. But if z is divisible by $x + y$, then $z \geq x + y$, so

$$z^n \geq (x + y)^n.$$

Since for positive x and y we have

$$(x + y)^n > x^n + y^n,$$

we get

$$z^n > x^n + y^n,$$

a contradiction.

(b) Let $n \geq 3$ be an integer and suppose, indirectly, that the equation

$$x^n + y^n = z^n$$

has a positive integer solution (x, y, z). Substituting $2n$ for n into our equation yields

$$x^{2n} + y^{2n} = z^{2n}.$$

On the other hand, squaring both sides of the original equation results in

$$x^{2n} + 2x^n y^n + y^{2n} = z^{2n}.$$

Comparing our two derived equations yields $2x^n y^n - 0$, but that is impossible if x and y are positive.

(c) Let $n \geq 3$ be an integer and suppose, indirectly, that the equation

$$x^n + y^n = z^n$$

has a positive integer solution (x, y, z). Squaring both sides yields

$$(x^n + y^n)^2 = z^{2n}.$$

By Problem 9 above, we know that the equation

$$x^2 + y^2 = z^2$$

has some (actually, infinitely many) solutions; raising this equation to the n-th power gives

$$(x^2 + y^2)^n = z^{2n}.$$

Comparing our two equations, we can write

$$(x^n + y^n)^2 = (x^2 + y^2)^n.$$

We will now prove that this can never happen.

Assume first that $x \leq y$ (the case $x \geq y$ can be done similarly). Dividing both sides of our equation by y^{2n} gives

$$\left(\left(\tfrac{x}{y}\right)^n + 1\right)^2 = \left(\left(\tfrac{x}{y}\right)^2 + 1\right)^n.$$

Since $n > 2$, for the right-hand side we have

$$\left(\left(\tfrac{x}{y}\right)^2 + 1\right)^n > \left(\left(\tfrac{x}{y}\right)^2 + 1\right)^2.$$

Furthermore, since $n > 2$ and $\tfrac{x}{y} \leq 1$, for the left-hand side above we have

$$\left(\left(\tfrac{x}{y}\right)^n + 1\right)^2 \leq \left(\left(\tfrac{x}{y}\right)^2 + 1\right)^2.$$

Therefore,

$$\left(\left(\tfrac{x}{y}\right)^n + 1\right)^2 < \left(\left(\tfrac{x}{y}\right)^2 + 1\right)^n,$$

a contradiction with the two sides being equal.

(d) Let $n \geq 3$ be an integer and suppose, indirectly, that the equation

$$x^n + y^n = z^n$$

has a positive integer solution (x, y, z). Rewriting our equation, we get

$$\left(\frac{x}{z}\right)^n + \left(\frac{y}{z}\right)^n = 1.$$

This implies that $\tfrac{x}{z}$ and $\tfrac{y}{z}$ are both less than 1. Therefore, if n is a large enough positive integer, then

$$\left(\frac{x}{z}\right)^n < \frac{1}{2}$$

and

$$\left(\frac{y}{z}\right)^n < \frac{1}{2},$$

which makes

$$\left(\frac{x}{z}\right)^n + \left(\frac{y}{z}\right)^n = 1$$

impossible. This proves that Fermat's Last Theorem holds for every value of n that is large enough, leaving only finitely many cases. These cases then can be verified individually.

14. The assignments below ask you to bring some of the great moments in the history of mathematics to light. Your research should probably start by skimming some of the literature on the history of mathematics. A particularly useful place to start is the MacTutor History of Mathematics archive page, http://turnbull.mcs.st-and.ac.uk/history, which is maintained by the University of St. Andrews in Scotland.

   (a) What were some of the most important mathematical discoveries of antiquity?
   (b) Who were the mathematicians involved in the discovery of the formula for the roots of general cubic and quartic polynomials? Who do you think deserves most of the credit?
   (c) How and when did the concept of *limits* develop?
   (d) How has our concept of *numbers* evolved over history?
   (e) What were the greatest contributions of the following mathematicians: Benjamin Bannekcr, René Descartes, Euclid, Evariste Galois, Carl Friedrich Gauss, David Hilbert, Emmy Noether, Srinivasa Ramanujan? What were some of the challenges they had to face in their work or in their personal lives?
   (f) Where were the most influential mathematical centers of antiquity? Where were the greatest centers in the nineteenth century? How about today?

15. (a) What are the major branches of mathematics today?
   (b) Where and on what subjects are some of the biggest mathematical conferences this year?
   (c) Which mathematician has had the greatest number of works published to this day? What is the median number of publications of a mathematician today?
   (d) Find some mathematical results that were achieved by undergraduate students.
   (e) Your professor and other mathematicians at your institution are undoubtedly working on their own unsolved questions. Find out what some of these questions are.

Part II
The Foundations of Mathematics

Chapter 5
Let's Be Logical!

In arithmetic we learned how to perform arithmetical operations (addition, multiplication, taking negatives, etc.) on numbers; then, in algebra, we generalized arithmetic using variables instead of numbers. Similarly, we can build compounded statements from simple statements and we can study their general structures. The branch of mathematics dealing with the structure of statements is called *logic*. A study of the rules of logic is essential when one studies correct reasoning.

We start with the "arithmetic" of logic. The operations on statements are called *logical connectives*. We are about to define the *unary* (one-variable) operation of negation and the four *binary* (two-variable) operations of disjunction, conjunction, conditional, and biconditional. The variables we use for general statements will be P, Q, R, etc.; recall that their *truth values* can either be T (true) or F (false), but that no statement can be both true and false.

Definition 5.1 *The* negation *of a statement P is the statement $\neg P$ (read as "not P"), which is true when P is false and false when P is true.*

For example, consider the statement P that "$3 > 4$." Then the negation $\neg P$ of P is the statement "$3 \not> 4$," which can be rewritten in the more customary form "$3 \leq 4$." In this example P is false and $\neg P$ is true; clearly, we always have exactly one of P or $\neg P$ true.

Definition 5.2 *The* disjunction *of statements P and Q is the statement $P \vee Q$ (read as "P or Q"), which is true when P or Q (or both) is true and false only when both P and Q are false.*

For example, if P is the statement "345 is even" and Q stands for "345 is divisible by 5," then $P \vee Q$ is the statement that "345 is even or divisible by 5." Since Q is a true statement, the disjunction $P \vee Q$ is true as well (regardless of the fact that P is false).

© Springer Nature Switzerland AG 2020

B. Bajnok, *An Invitation to Abstract Mathematics*, Undergraduate Texts in Mathematics, https://doi.org/10.1007/978-3-030-56174-1_5

Note that the definition of disjunction differs somewhat from its everyday usage. Our "or" is an *inclusive or*: it is true when either or both components are true. For example, the statement "$3^2 = 9$ or $4^2 = 16$" is a true statement. (In everyday life "or" is usually used as an *exclusive or*. When we say "I will call you or send you a text," we normally mean that we do not intend to both call and text.)

Definition 5.3 *The* conjunction *of statements P and Q is the statement $P \wedge Q$ (read as "P and Q"), which is true only when both P and Q are true, and is false when P or Q (or both) is false.*

If again P is the statement "345 is even" and Q is "345 is divisible by 5," then $P \wedge Q$ is the statement that "345 is even and divisible by 5." Since P is a false statement, the conjunction $P \wedge Q$ is false as well (regardless of the fact that Q is true).

It is also possible to define the disjunction and conjunction of more than two statements: The disjunction of $P_1, \ldots, P_n$ (here n is a natural number), denoted by

$$\bigvee_{i=1}^{n} P_i = P_1 \vee \cdots \vee P_n,$$

is true when at least one of the statements $P_1, \ldots, P_n$ is true and false if they are all false; similarly, the conjunction of $P_1, \ldots, P_n$, denoted by

$$\bigwedge_{i=1}^{n} P_i = P_1 \wedge \cdots \wedge P_n,$$

is true when all of the statements $P_1, \ldots, P_n$ are true and false if at least one of them is false. Note that, as we will see in Chapter 13, disjunctions and conjunctions are associative, so we do not need to write parentheses in conjunctions and disjunctions of more than two variables. However, parentheses are necessary when both conjunctions and disjunctions are involved; for example, $P \wedge (Q \vee R)$ and $(P \wedge Q) \vee R$ have different truth values when P is false and R is true.

We have two more logical connectives to define: conditionals and biconditionals. We start with the latter.

Definition 5.4 *The* biconditional *of statements P and Q is the statement $P \Leftrightarrow Q$ (read as "P if, and only if, Q" or "P iff Q" for short), which is true when P and Q have the same truth value and false otherwise.*

Thus, $P \Leftrightarrow Q$ means that either P and Q are both true or they are both false. We can rephrase our definition, therefore, to say that $P \Leftrightarrow Q$ holds exactly when the logical expression

$$(P \wedge Q) \vee (\neg P \wedge \neg Q)$$

is true. We should also point out that, according to our definition, the biconditional of two statements may hold even when there is no evident connection between the two statements. For example, the biconditional that "345 is odd if, and only if, 345 is divisible by 5" is true in spite of the fact that divisibility by 2 and by 5 are unrelated.

Definition 5.5 *The* conditional *of statements P and Q is the statement* $P \Rightarrow Q$ *(read as "if P then Q" or "P only if Q"), which is true when P is false or Q is true, and false when P is true and Q is false.*

Note that, according to our definition, $P \Rightarrow Q$ holds whenever statement P is false, regardless of the validity of statement Q. For example, the statements "if $3 = 4$ then $3^2 = 4^2$" and "if $3 = 4$ then $3^2 \neq 4^2$" are both true statements! Furthermore, we should point out that, like with biconditionals, any hidden meaning of causation is removed from our definition; for example, the statement "if $3^2 = 9$ then $5 + 5 = 10$" is true!

We can also observe that conditionals are not commutative: $P \Rightarrow Q$ and its *converse*, $Q \Rightarrow P$, may have different truth values. Indeed, $P \Rightarrow Q$ is true exactly when the logical expression $\neg P \vee Q$ is true, while $Q \Rightarrow P$ holds exactly when $P \vee \neg Q$ does. Thus, when, for example, P is false and Q is true, the conditional $P \Rightarrow Q$ is true but its converse is false. For example, "if 345 is even then 345 is divisible by 5" is true, but the statement that "if 345 is divisible by 5 then 345 is even" is false.

While we consider biconditionals and conditionals logical connectives with truth values as defined above, it is often also helpful to think of them as possible relationships that two statements might have. In this context, we say that statements P and Q are *equivalent* (or that P is *necessary and sufficient* for Q) if their biconditional, $P \Leftrightarrow Q$, is true, and we say that P *implies* Q (or that P is *sufficient* for Q, or that Q is *necessary* for P) when the conditional $P \Rightarrow Q$ holds.

We will study relations in general in Chapters 16 and 17, but we can already point out some of their properties. Regarding equivalence, we see that the relation is

- *reflexive*: for every statement P, $P \Leftrightarrow P$;
- *symmetric*: if $P \Leftrightarrow Q$ holds for some statements P and Q, then $Q \Leftrightarrow P$ holds too; and
- *transitive*: if $P \Leftrightarrow Q$ and $Q \Leftrightarrow R$ for statements P, Q, and R, then $P \Leftrightarrow R$ holds as well.

Note that the same relations hold for equality of numbers: Equality is reflexive (any number equals itself), symmetric (if one number equals another, then that number also equals the original number), and transitive (if the first number equals the second number and the second number equals the third, then the first number also equals the third number). We intend to use these properties without mention. Analogous properties of implications are examined in Problem 3.

Let us now turn to the "algebra" of logic where, instead of dealing with operations between specific statements, we examine the same logical connectives between *logical variables*. The computations of truth values of a logical expression

Table 5.1 The truth tables of negation, disjunction, and conjunction

P	$\neg P$
T	F
F	T

P	Q	$P \vee Q$	$P \wedge Q$	$P \Leftrightarrow Q$	$P \Rightarrow Q$
T	T	T	T	T	T
T	F	T	F	F	F
F	T	T	F	F	T
F	F	F	F	T	T

Table 5.2 The truth table of the logical formula $(P \vee \neg Q \vee R) \wedge (\neg(P \vee R))$

P	Q	R	$\neg Q \vee R$	$P \vee \neg Q \vee R$	$\neg(P \vee R)$	$(P \vee \neg Q \vee R) \wedge (\neg(P \vee R))$
T	T	T	T	T	F	F
T	T	F	F	T	F	F
T	F	T	T	T	F	F
T	F	F	T	T	F	F
F	T	T	T	T	F	F
F	T	F	F	F	T	F
F	F	T	T	T	F	F
F	F	F	T	T	T	T

depending on its logical variables can be conveniently done via *truth tables*. The above definitions, for example, can be summarized by the truth tables in Table 5.1.

Consider now a more complicated expression such as

$$(P \vee \neg Q \vee R) \wedge (\neg(P \vee R)).$$

We call such expressions *logical formulas*: they are predicates that become statements once we know what the truth values of P, Q, and R are. The truth table of the logical formula above is shown in Table 5.2. (Only the first three columns and the last column are considered to form the truth table; we included the other three columns only as auxiliaries.)

It may happen that two different logical formulas have identical truth tables; after all, we can have only finitely many different truth tables on a given finite number of variables, but we can build up infinitely many different formulas. For example, from the truth table of

$$(P \vee \neg Q \vee R) \wedge (\neg(P \vee R))$$

above one can see that this expression has the same truth table as the less complicated

$$\neg P \wedge \neg Q \wedge \neg R.$$

Indeed, both predicates are true exactly when P, Q, and R are all false. In this case, we say (somewhat sloppily) that the two logical formulas are *equivalent*; more precisely, the two formulas become equivalent statements for every choice of their variables. Similarly, we say that one logical formula *implies* another if the implication holds whatever the truth values of their variables.

The truth table of a logical expression might indicate that it is always true or that it is always false. We have special terms for such situations.

Definition 5.6 *A logical formula is called a* tautology *whenever it is true for every assignment of truth values to its variables.*

Definition 5.7 *A logical formula is called a* contradiction *whenever it is false for every assignment of truth values to its variables.*

It is easy to verify, for example, that $P \vee (\neg P)$ is a tautology and that $P \wedge (\neg P)$ is a contradiction. We will see further examples for tautologies and contradictions later; in fact, we study the algebra of logic in more depth in Chapters 13 and 14.

It is not an exaggeration to say that logic is the subsistence of mathematics. In fact, every definition is, technically, an equivalence. For example, by Definition 2.1 we meant that "an integer is a prime if, and only if, it has exactly two positive divisors," even though, as customary, the "only if" part was skipped. Similarly, we often use words such as "when" or "whenever" in definitions (like in Definitions 5.6 and 5.7 above) although we really have equivalence in mind.

Moreover, most mathematical theorems can be put in the form of an implication or an equivalence. For example, Theorem 4.1 can be stated as "The implication

$$2^n - 1 \text{ is a prime number} \implies 2^{n-1}(2^n - 1) \text{ is a perfect number}$$

is true for every positive integer n." As we saw in Problem 7 (e) of Chapter 4, the converse implication that

$$2^{n-1}(2^n - 1) \text{ is a perfect number} \implies 2^n - 1 \text{ is a prime number}$$

is true for every positive integer n as well, so the predicates

$$A(n) : 2^n - 1 \text{ is a prime number}$$

and

$$B(n) : 2^{n-1}(2^n - 1) \text{ is a perfect number}$$

are equivalent for every positive integer n. For example, if $n = 3$, then $A(n)$ is true (7 is prime) and $B(n)$ is true (28 is perfect), so $A(3) \Leftrightarrow B(3)$; if $n = 4$, then $A(n)$ is false (15 is not prime) and $B(n)$ is false (120 is not perfect), so $A(4) \Leftrightarrow B(4)$ holds again. (Theorem 4.1 and Problem 7 (e) of Chapter 4 prove all cases.)

Similarly, using the language of logic more explicitly, Theorem 4.7 can be stated as "The implication

$$2^n - 1 \text{ is a prime number } \Rightarrow n \text{ is a prime number}$$

is true for every positive integer n." However, the converse implication

$$n \text{ is a prime number } \Rightarrow 2^n - 1 \text{ is a prime number}$$

is not true for every positive integer n: as we pointed out in Chapter 4, 11 is a prime number but $2^{11} - 1$ is not. As we mentioned before, care has to be taken so that we don't confuse our implication $P \Rightarrow Q$ with its converse $Q \Rightarrow P$—these two implications are generally not equivalent! It is true, however (cf. Problem 1), that $P \Rightarrow Q$ is always equivalent to $\neg Q \Rightarrow \neg P$; this latter statement is called the *contrapositive* of the implication. We will return to the usage of implications and their converses and contrapositives in proofs in Chapter 9. For now, we only discuss the two most fundamental logical schemes used in proofs: the Law of Modus Ponens and the Law of Modus Tollens.

Perhaps the most often used logical structure in a proof is what is referred to as the

- *Law of Modus Ponens*: For arbitrary statements P and Q, the statement

$$[P \wedge (P \Rightarrow Q)] \Rightarrow Q$$

is a true statement.

One can easily verify the Law of Modus Ponens by considering the appropriate truth table, but most people with a bit of experience in logic would find this law quite convincing without a proof. It is perhaps a bit less obvious to see the following variation, called the

- *Law of Modus Tollens*: For arbitrary statements P and Q, the statement

$$[\neg Q \wedge (P \Rightarrow Q)] \Rightarrow \neg P$$

is a true statement.

A closer look at the Law of Modus Tollens reveals that it is essentially the contrapositive of the Law of Modus Ponens. Indeed, applying the Law of Modus Ponens to statements $\neg Q$ and $\neg P$ rather than to P and Q gives

$$[\neg Q \wedge (\neg Q \Rightarrow \neg P)] \Rightarrow \neg P,$$

from which the Law of Modus Tollens follows by the fact that $P \Rightarrow Q$ and its contrapositive, $\neg Q \Rightarrow \neg P$, are equivalent.

The laws of Modus Ponens and Modus Tollens are frequently applied not only in mathematics but in everyday situations as well. For example, if we assume that the statements

- P: "it is raining" and
- $P \Rightarrow Q$: "If it is raining then there are clouds in the sky"

are both true, then using Modus Ponens we can safely conclude that

- Q: "There are clouds in the sky."

Similarly, assuming that

- $\neg Q$: "There are no clouds in the sky" and
- $P \Rightarrow Q$: "If it is raining then there are clouds in the sky"

both hold lets us conclude, using Modus Tollens, that

- $\neg P$: "It is not raining."

Care needs to be taken, however, that we use these laws correctly. For example, the statements

- $\neg P$: "It is not raining" and
- $P \Rightarrow Q$: "if it is raining then there are clouds in the sky"

do not imply that

- $\neg Q$: "There are no clouds in the sky."

This last argument has the logical form

$$[\neg P \wedge (P \Rightarrow Q)] \Rightarrow \neg Q,$$

which is not a tautology! (Modus Pollens, while a perfectly plausible name for an allergy medication, is not a legitimate combination of Modus Ponens and Modus Tollens!)

The laws of logic in this chapter provide the bases for a variety of amusing (and often quite tricky) puzzles. We close this chapter by examining one such puzzle.

Suppose that every inhabitant of a certain distant planet has one of two possible occupations: educator or politician. The educators always tell the truth and the politicians always lie. After arriving at this planet, we meet three inhabitants: A, B, and C. A turns to us and says "B and C have the same occupation." Someone then asks C: "Do A and B have the same occupation?" What will C answer?

We provide two different solutions for this puzzle.

For the first solution, let us construct a table that lists the possible occupations for A, B, and C, as well as the truth values of the statements that "A and B have the same occupation" and "B and C have the same occupation," denoted by statement R and S, respectively.

We see that the second and third lines in Table 5.3 are impossible: if A is an educator, then statement S must be true. Similarly, the fifth and the eighth lines are

Table 5.3 The truth values
of statements R and S

	A	B	C	R	S
1	E	E	E	T	T
2	E	E	P	T	F
3	E	P	E	F	F
4	E	P	P	F	T
5	P	E	E	F	T
6	P	E	P	F	F
7	P	P	E	T	F
8	P	P	P	T	T

impossible: if A is a politician, then S must be false. This leaves us with lines 1 and 7, where C is an educator, and lines 4 and 6, where C is a politician. In lines 1 and 7, R is true and, since C is an educator, he or she will say that R is true. R is false in lines 4 and 6 but, since C is a politician there, he or she will say that R is true. Thus we see that C will say that R is true in every case.

A different solution can be presented as follows. We know that A is either an educator or a politician. We examine the two cases separately.

If A is an educator, then B and C have the same occupation: either both are educators or both are politicians. In the first case C will say the truth, namely that A and B have the same occupation; in the second case C will lie and, therefore, he or she will say that A and B have the same occupation.

If A is a politician, then either B is an educator and C is a politician, or B is a politician and C is an educator. Again, in both cases C will have to say that A and B have the same occupation.

Mathematical logic is of primary importance in the foundations of mathematics. It came into prominence in the nineteenth century as the need to formalize the rules by which we build proofs for our statements became more urgent. Eventually, mathematical logic developed into an independent field of study; beyond mathematics itself, it has contributed to other fields such as computer science and philosophy.

Problems

1. Suppose that P and Q are logical variables. Use truth tables to prove each of the following facts about logical formulas.

   (a) $(P \Rightarrow Q) \Leftrightarrow (\neg Q \Rightarrow \neg P)$
   (This says that an implication is equivalent to its contrapositive.)
   (b) $(P \Rightarrow Q) \not\Leftrightarrow (Q \Rightarrow P)$
   (This says that an implication is not equivalent to its converse.)
   (c) $\neg(P \wedge Q) \Leftrightarrow (\neg P \vee \neg Q)$ and $\neg(P \vee Q) \Leftrightarrow (\neg P \wedge \neg Q)$
   (These equivalences are called *De Morgan's Laws*; cf. Theorem 14.10.)

2. Use truth tables to decide which of the following implications are tautologies.

   (a) If $P \Leftrightarrow Q$, then $(P \vee R) \Leftrightarrow (Q \vee R)$.
   (b) If $P \Leftrightarrow Q$, then $(P \wedge R) \Leftrightarrow (Q \wedge R)$.
   (c) If $(P \vee R) \Leftrightarrow (Q \vee R)$, then $P \Leftrightarrow Q$.
   (d) If $(P \wedge R) \Leftrightarrow (Q \wedge R)$, then $P \Leftrightarrow Q$.

   Remark This problem is asking you to analyze how similar equivalences are to equalities. In particular, parts (a) and (b) ask whether you can apply a disjunction/conjunction to the two sides of an equivalence and still preserve the equivalence; parts (c) and (d) ask whether you can cancel a disjunction/conjunction from both sides.

3. (a) Formulate definitions for the relation of implication to be

      i. reflexive,
      ii. symmetric,
      iii. transitive.

   (b) Using truth tables, decide if the relation of implication is reflexive, symmetric, or transitive.

4. Suppose that Γ is an irrational number given by its infinite (non-periodic) decimal representation. State the negation of the following statements without using the word "not."

   (a) The 100th decimal digit of Γ is odd.
   (b) The 100th decimal digit of Γ is at most 4.
   (c) The 100th decimal digit of Γ is odd or it is at most four.
   (d) The 100th decimal digit of Γ is odd and at most 4.
   (e) If the 100th decimal digit of Γ is odd, then it is at most 4.
   (f) If the 100th decimal digit of Γ is at most 4, then it is odd
   (g) The 100th decimal digit of Γ is odd iff it is at most 4.
   (h) The 100th decimal digit of Γ is even iff it is at most 4.

5. Consider the following pairs of statements. Which pairs are equivalent?

   (a) P: I will not be accepted at law school if I don't do well on the LSAT.
       Q: If I do well on the LSAT, I will be accepted at law school.
       (Hints: Let R denote the statement that "I will be accepted at law school" and let S denote that "I do well on the LSAT." Write the *logical form* of P and Q using R and S; that is, write a logical expression capturing the meaning of the two sentences using R, S, and appropriate logical operations.)
   (b) P: In order for it to rain, there must be clouds.
       Q: If it does not rain, then there are no clouds.
   (c) P: If it is sunny tomorrow, then I'll go hiking.
       Q: I'll go hiking only if it is sunny tomorrow.
   (d) P: Only completely justified answers will receive full credit.

Q: If complete justification is not given, the answer will not receive full credit.

6. State the contrapositive and the converse of each of the following statements.

   (a) In order for it to rain, there must be clouds.
   (b) In order for it to rain, it is sufficient that there are clouds.
   (c) What is good for the goose is good for the gander.
   (d) If wishes were horses then beggars would ride.

7. Consider the predicates

$$a = 0; \quad a > 0; \quad a < 0; \quad b = 0; \quad b > 0; \quad b < 0;$$
$$a = b; \quad -a = b; \quad a = b^2; \quad a > b; \quad -a > b; \quad a > b^2.$$

Many other predicates that are frequently used in algebra are equivalent to logical expressions built from these predicates. For example, for arbitrary real numbers a and b, we have

$$a \cdot b = 0 \Leftrightarrow (a = 0 \vee b = 0)$$

(cf. Corollary 14.7 for a proof of this statement).

For each of the predicates below, combine some of the twelve given predicates above into a logical expression using conjunctions and/or disjunctions, so that your expression is equivalent to the given predicate for all real number values of a and b. (You do not need to provide proofs for your statements.)

   (a) $a \cdot b > 0$
   (b) $|a| = b$
   (c) $|a| > b$
   (d) $\sqrt{a} = b$
   (e) $\sqrt{a} > b$

(Hints: Recall that the absolute value of a real number x is defined to be x when $x \geq 0$ and $-x$ when $x < 0$; also, for a nonnegative real number y, $\sqrt{y}$ is defined as the nonnegative real number whose square is y—cf. Problem 1 of Chapter 2.)

8. There are four cards on a desk, each containing a letter on one side and a whole number on the other. We see the top of the four cards showing the characters $R, U, 4$, and 1. Which cards do we have to turn over if we want to verify the following statements?

   (a) Each card contains a vowel or an even number.
   (b) If the letter on one side of the card is a vowel, then the number on the other side is even.
   (c) The letter on one side of the card is a vowel if, and only if, the number on the other side is even.

9. (a) How many pairwise non-equivalent logical formulas (of arbitrary length) are there on two variables?

   (b) For each expression in part (a), find a representation that only involves negation and disjunction. (You do not need to use both.)

10. Every citizen of the town of Logicville is either a knight who always tells the truth or a knave who always lies (but they can't be both, of course).

    (a) A and B are citizens of Logicville, and A says "I am a knave or B is a knight." What are A and B?

    (b) One day you meet three of the citizens of Logicville, C, D, and E, and they tell you the following:

    C says: "All three of us are knaves."
    D says: "Exactly two of us are knaves."

    What can you determine about who is a knight and who is a knave?

    (c) The next day you meet F, G, and H, and the following conversation takes place.

    F says: "All three of us are knights."
    G says: "Exactly two of us are knights."

    What can you determine now about who is a knight and who is a knave?

11. Every inhabitant of a certain distant planet has one of three possible occupations: educator, lawyer, or politician. The educators always tell the truth, the politicians always lie, and the lawyers sometimes say the truth and sometimes lie—whatever they happen to feel like. Educators are considered to be members of the upper class, lawyers are members of the middle class, and politicians belong to the lower class.

    (a) A and B, who live on this planet, say the following:

    A says: "B is in a higher class than I am."
    B says: "That's not true."

    Can we determine the occupations of A and B?

    (b) One day we meet three inhabitants of the planet: C, D, and E. We know that all three have different occupations. They tell us the following:

    C says: "I am a lawyer."
    D says: "That's true."
    E says: "I am not a lawyer."

    Can we determine the occupations of these three people?

12. At one time or another, we have all heard the saying that "not everything is black or white." Fuzzy logic, an extension of the classical two-valued logic discussed above, takes this sentiment to heart. In *fuzzy logic*, a statement P may have any real number truth value $\tau(P)$ between 0 and 1, with $\tau(P) = 0$ and $\tau(P) = 1$ corresponding to P being completely false and completely true, respectively.

For example, the statement that "This classroom is warm" may be given a truth value of 0.9 if its temperature is 80 degrees Fahrenheit, but only 0.1 if it cools to 60 degrees. While fuzzy logic has been discussed for a very long time, its formal development started with a 1965 paper of Lofti A. Zadeh (1921–2017).

The definitions of negation, disjunction, conjunction, equivalence, and implication can be generalized as follows:

Definition 5.8 *Let P and Q be fuzzy statements, that is, statements with truth values $\tau(P)$ and $\tau(Q)$ between 0 and 1, inclusive. Then,*

- *the negation of P, ¬P, has truth value*

$$\tau(\neg P) = 1 - \tau(P);$$

- *the disjunction of P and Q, P $\vee$ Q, has truth value*

$$\tau(P \vee Q) = \max\{\tau(P), \tau(Q)\};$$

and
- *the conjunction of P and Q, P $\wedge$ Q, has truth value*

$$\tau(P \wedge Q) = \min\{\tau(P), \tau(Q)\}.$$

Furthermore, we say that

- *P and Q are equivalent if $\tau(P) = \tau(Q)$ and*
- *P implies Q if $\tau(P) \leq \tau(Q)$.*

(a) Verify that Definition 5.8 indeed generalizes our corresponding definitions for (non-fuzzy) statements.
(b) Generalize De Morgan's Laws (cf. Problem 1 (c) above) for fuzzy statements, and prove that the identities still hold.

Remarks A somewhat different way to generalize two-valued logic is to assign one of several pre-determined truth values to each statement—this modification of fuzzy logic is referred to as *multi-valued logic*. In the simplest case, we may restrict the possible truth values to 0 (false), 0.5 (undecided), or 1 (true). An even more general idea is the development of *set-valued logic*, which assigns not a single truth value but an entire collection of truth values to each statement. For example, one might choose to say that, for P being the statement that "this classroom is warm," $\tau(P)$ is the entire interval $[0.6, 0.75]$.

There is considerable discussion among mathematicians, computer scientists, engineers, and other scientists as to whether fuzzy logic or any of its variations are valuable. Some claim that fuzzy logic is nothing more than imprecise logic; others argue that there is no need for fuzzy logic as one has probability theory available to discuss statements whose truth values are not exactly known. In response, fans of fuzzy logic point to the ever-growing number of applications (e.g., *fuzzy control systems, artificial intelligence, pattern recognition*); there are even attempts to modify probability theory to serve the needs of fuzzy logic (this not-fully accepted field is referred to as *possibility theory*). In any case, it seems that fuzzy logic has many fans and has indeed developed into a vast research area.

Chapter 6
Setting Examples

When we are interested in studying several objects at the same time, we may put these objects into a *set*. Some of the most commonly used sets in mathematics are the following sets of numbers.

- $\mathbb{N}$ is the set of natural numbers (positive integers);
- $\mathbb{Z}$ is the set of integers (natural numbers, their negatives, and zero);
- $\mathbb{Q}$ is the set of rational numbers (positive and negative fractions and zero);
- $\mathbb{R}$ is the set of real numbers (finite and infinite decimals); and
- $\mathbb{C}$ is the set of complex numbers (real numbers, imaginary numbers, and their sums)

Other standard notations for sets of numbers include *intervals*. For example, the intervals $(-2, 3)$, $[-2, 3]$, and $(-\infty, 3)$ include all real numbers between -2 and 3 exclusive, between -2 and 3 inclusive, and below 3, respectively.

Sets may also contain objects other than numbers. We may talk about, for example, the set of students in a class, the set of days in a week, or the set of letters in an alphabet. As we will see shortly, however, not every collection of objects can be considered a set!

In this chapter we develop a heuristic understanding of sets; our study will be carried out through examples (hence the title). We do not give a definition for sets; as we mentioned in Chapter 2, we treat the concepts of "set" and "element" as primitives. Appendix C discusses a more advanced approach to set theory.

We usually denote sets by capital letters and their elements by lower case letters. If x is an element of the set S, we denote this by $x \in S$ and say that S *contains* x or x is a *member* of S. If x is not an element of S, we write $x \notin S$.

Let us see how we can describe sets and the various notations used for this purpose. We may identify sets by listing their elements: For example, $A = \{1, 4, 9\}$ denotes the set with the three elements 1, 4, and 9, while $B = \{1, 4, 9, \dots\}$ has infinitely many elements in that it contains all positive perfect squares. (When using "$\dots$" we must implicitly understand what all the other elements of the set are!) This

© Springer Nature Switzerland AG 2020
B. Bajnok, *An Invitation to Abstract Mathematics*, Undergraduate Texts in Mathematics, https://doi.org/10.1007/978-3-030-56174-1_6

description of sets is called the *list notation*. We should note that a proper list cannot contain more than one "....". Even though, in some cases, descriptions such as

$$C = \{\ldots, \ -7\pi/2, \ -3\pi/2, \ \pi/2, \ 5\pi/2, \ 9\pi/2, \ \ldots\}$$

or even

$$D = \{1, \ 1.1, \ 1.11, \ 1.111, \ \ldots, \ 2, \ 2.1, \ 2.11, \ 2.111, \ \ldots, \ 3, \ 3.1, \ 3.11, \ 3.111, \ \ldots\}$$

may identify a set clearly enough, these notations are not considered to be proper lists!

Can every set be written as a list? Obviously, any finite set can be given as a list. The situation is not so clear for infinite sets. Listing the natural numbers causes no concern:

$$\mathbb{N} = \{1, 2, 3, 4, 5, \ldots\}.$$

The set of all integers is not hard either:

$$\mathbb{Z} = \{0, 1, -1, 2, -2, 3, -3, 4, -4, \ldots\}.$$

(Other lists for these sets are, of course, possible.) The sets C and D of the previous paragraph can also be put in lists:

$$C = \{\pi/2, \ -3\pi/2, \ 5\pi/2, \ -7\pi/2, \ 9\pi/2, \ -11\pi/2, \ 13\pi/2, \ \ldots\}$$

and

$$D = \{1, \ 2, \ 1.1, \ 3, \ 2.1, \ 1.11, \ 4, \ 3.1, \ 2.11, \ 1.111, \ 5, \ 4.1, \ 3.11, \ 2.111, \ 1.1111, \ \ldots\}.$$

(The meanings of "..." in these descriptions are clear enough.) Even the set of all rational numbers can be listed (cf. Problem 4).

It might come as a surprise that the elements of some infinite sets cannot be listed! We will see, for example, that the set of all real numbers cannot be listed; as we will explain in Chapter 20, there are just too many of them to be put in a single list. It is quite an interesting question to decide if a particular set can be listed or not.

There are some more concise ways to describe sets. One such description follows the pattern

$$\{\text{formula} \mid \text{variable(s)}\}.$$

For example, $B = \{n^2 \mid n \in \mathbb{N}\}$ is the set of positive perfect squares, $E = \{2k \mid k \in \mathbb{Z}\}$ is the set of even integers, and $O = \{2k + 1 \mid k \in \mathbb{Z}\}$ is the set of odd integers. The symbol "$\mid$" is read as *such that* or *for which*. This description of sets is called the *formula notation*.

It is not always easy—or even possible—to describe a set using the formula notation. A famous example in which the list notation is more practical than the formula notation is the case of *Fibonacci numbers* listed as

$$\{1, 2, 3, 5, 8, 13, 21, 34, 55, 89, 144, 233, \dots \}$$

(cf. Chapter 2). It might be quite surprising that the precise formula notation for this set is

$$\left\{ \frac{1}{\sqrt{5}} \left(\left(\frac{1 + \sqrt{5}}{2} \right)^{n+1} - \left(\frac{1 - \sqrt{5}}{2} \right)^{n+1} \right) \,\middle|\, n \in \mathbb{N} \right\};$$

even the fact that this formula, which involves irrational numbers such as the *golden ratio* $\frac{1+\sqrt{5}}{2}$, yields integer values is less than obvious (we will prove this in Chapter 11). For other such amazing examples visit the On-Line Encyclopedia of Integer Sequences at https://oeis.org.

Alternatively, we may describe our set with one or more condition(s) that its elements must satisfy; this description has the structure

$$\{\text{variable(s)} \mid \text{condition(s)}\}.$$

For example,

$$\{n \in \mathbb{Z} \mid 0 \leq n \leq 5\} = \{0, 1, 2, 3, 4, 5\},$$

while

$$\{x \in \mathbb{R} \mid 0 \leq x \leq 5\} = [0, 5].$$

This description of sets is called the *conditional notation*.

The conditional notation allows us to describe a large variety of finite or infinite sets. For example,

$$\{n \in \mathbb{N} \mid n \text{ has exactly two positive divisors}\}$$

is the set of positive prime numbers, while

$$\{n \in \mathbb{N} \mid n \text{ can be written as the sum of four perfect squares}\}$$

is, according to Lagrange's Theorem (see the remark after Problem 9 (c) in Chapter 3) the set $\mathbb{N}$ of all positive integers. The fact that one is able to define sets by restricting a given set to the collection of those elements that satisfy a certain property (or properties) is an axiom that can be stated as follows:

Axiom 6.1 (The Axiom of Separation) *If U is an arbitrary set and P is an arbitrary predicate defined on the elements of U, then*

$$\{x \in U \mid P(x) \text{ is a true statement}\}$$

is a set.

We now address the question of whether every collection of objects is a set. The fact that this is not the case was first observed in 1902 by the Welsh philosopher and mathematician Bertrand Russell (1872–1970). Before we examine *Russell's Paradox*, let us note that some collections contain themselves and some do not. For example, $T = \{2\}$ only contains the number 2 and not the set $\{2\}$, so $T \notin T$. Similarly, the set P of all people in the world only contains people and not the set of all people P, so $P \notin P$. However, we can say that the collection I of all abstract ideas is certainly an abstract idea, so $I \in I$; similarly, if J is the collection of all infinite collections, then J itself is infinite, hence $J \in J$.

Now define Z to be the collection of all sets that do not contain themselves, that is

$$Z = \{A \mid A \text{ is a set and } A \notin A\}.$$

By our considerations above, $T \in Z$ and $P \in Z$, but $I \notin Z$ and $J \notin Z$. Now we can ask whether Z is an element of Z or not. It is easily seen that if Z were a set, then both $Z \in Z$ and $Z \notin Z$ would be false (check!). Thus, Z cannot be a well-defined set; this is referred to as Russell's Paradox.

Note that the Axiom of Separation does not contradict Russell's Paradox. In general, one cannot assume that $\{x \mid P(x) \text{ is true}\}$ is a set; instead, we only require that inside a given set (whose existence we already assume) the elements satisfying predicate P form a set.

For this reason, when describing a set S, we often have an underlying *universal set U* given, and we specify which elements of U belong to S and which do not. (We may omit the designation of the universal set when it is already clear from the context.)

At the other end of the spectrum from the universal set we have the *empty-set*, denoted by $\emptyset$: it is the unique set with no elements. For example,

$$\{x \in \mathbb{R} \mid x + 1/2 \in \mathbb{R}\} = \mathbb{R},$$

but

$$\{x \in \mathbb{Z} \mid x + 1/2 \in \mathbb{Z}\} = \emptyset.$$

Let us now turn to set operations. We introduce three operations on sets: the unary (one-variable) operation of complementation and the two binary (two-variable) operations of union and intersection. In the definitions below we assume that the elements of sets A and B come from a given universal set U. To emphasize the strong similarity between operations on sets and statements (which we will investigate further in Chapters 13 and 14), we state these definitions using the terminology of logic from Chapter 5.

Definition 6.2 *The* complement *of a set A (with respect to the universal set U) is the set*

$$\overline{A} = \{x \in U \mid \neg(x \in A)\}.$$

Definition 6.3 *The* union *of sets A and B is the set*

$$A \cup B = \{x \in U \mid (x \in A) \vee (x \in B)\}.$$

Definition 6.4 *The* intersection *of sets A and B is the set*

$$A \cap B = \{x \in U \mid (x \in A) \wedge (x \in B)\}.$$

Furthermore, we define two relations that may hold between two sets.

Definition 6.5 *We say that sets A and B are* equal *and write A = B if, and only if,*

$$(x \in A) \Leftrightarrow (x \in B).$$

Definition 6.6 *We say that A is a* subset *of B and write A $\subseteq$ B if, and only if,*

$$(x \in A) \Rightarrow (x \in B).$$

We also say that A is a *proper subset* of B and write $A \subset B$, whenever $A \subseteq B$ but $A \neq B$.

It is easy to see that the relation of equality between two sets is

- reflexive: $A = A$ for every set A,
- symmetric: if $A = B$ then $B = A$, and
- transitive: if $A = B$ and $B = C$, then $A = C$;

while the relation of being a subset is

- reflexive: $A \subseteq A$ for every set A,
- *antisymmetric*: $A \subseteq B$ and $B \subseteq A$ cannot both be true, unless $A = B$, and
- transitive: if $A \subseteq B$ and $B \subseteq C$, then $A \subseteq C$.

We can also define the union and intersection of more than two sets. The union of sets $A_1, \ldots, A_n$ ($n \in \mathbb{N}$), denoted by

$$\bigcup_{i=1}^{n} A_i = A_1 \cup \cdots \cup A_n,$$

consists of all elements of the universal set that are in at least one of them; similarly, the intersection of $A_1, \ldots, A_n$, denoted by

$$\bigcap_{i=1}^{n} A_i = A_1 \cap \cdots \cap A_n,$$

contains those elements of the universal set that are in all of them. (As was the case with the disjunction and conjunction of statements, we may freely omit parentheses in expressions involving only unions or only intersections.) We can define the union and intersection of infinitely many sets analogously; if sets A_i are "indexed" by the elements of any set I, then their union and intersection are denoted by

$$\bigcup_{i \in I} A_i \quad \text{and} \quad \bigcap_{i \in I} A_i,$$

respectively. (If I is the set of positive integers, then the notations

$$\bigcup_{i=1}^{\infty} A_i \quad \text{and} \quad \bigcap_{i=1}^{\infty} A_i$$

are also used.)

Comparing Definitions 6.2 through 6.6 above to Definitions 5.1 through 5.5 of Chapter 5, we see a strong parallel between sets and statements. In particular, complements, unions, and intersections of sets seem to correspond to negations, disjunctions, and conjunctions of statements, respectively. Furthermore, the relations of implication and equivalence between two logical expressions seem to correspond to a set being a subset of another set and to two sets being equal, respectively.

To further demonstrate the similarity between the algebra of statements and the algebra of sets, consider the example in Chapter 5 that the logical formulas

$$(P \vee \neg Q \vee R) \wedge (\neg(P \vee R))$$

and

$$\neg P \wedge \neg Q \wedge \neg R$$

are equivalent; that is

$$(P \vee \neg Q \vee R) \wedge (\neg(P \vee R)) \Leftrightarrow \neg P \wedge \neg Q \wedge \neg R$$

is a tautology. Making the appropriate changes described above—and replacing the letters P, Q, and R by A, B, and C, respectively (as it is more customary to denote sets by letters chosen from the beginning of the alphabet)—one gets that the identity

$$(A \cup \overline{B} \cup C) \cap \overline{(A \cup C)} = \overline{A} \cap \overline{B} \cap \overline{C}$$

Table 6.1 The truth table verification of the identity $(A \cup \bar{B} \cup C) \cap \overline{(A \cup C)} = \bar{A} \cap \bar{B} \cap \bar{C}$

$x \in A$	$x \in B$	$x \in C$	$x \in A \cup \bar{B} \cup C$	$x \in \overline{(A \cup C)}$	$x \in (A \cup \bar{B} \cup C) \cap \overline{(A \cup C)}$	$x \in \bar{A} \cap \bar{B} \cap \bar{C}$
T	T	T	T	F	F	F
T	T	F	T	F	F	F
T	F	T	T	F	F	F
T	F	F	T	F	F	F
F	T	T	T	F	F	F
F	T	F	F	T	F	F
F	F	T	T	F	F	F
F	F	F	T	T	T	T

holds for all sets A, B, and C. To see this, we can use a truth table very similar to Table 5.2.

We will provide a deeper analysis of the similarities between the algebra of statements and the algebra of sets in Chapters 13 and 14.

In addition to truth tables, combinations of sets can sometimes be conveniently visualized using *Venn diagrams*. The Venn diagram of an expression involving sets consists of an outer box representing the universal set U and simple closed curves inside the box corresponding to the individual sets involved. Regions within the box should correspond to the rows of the truth table. For example, the Venn diagram for the general position of the set A consists of two regions (A and A); the Venn diagram for the general position of the sets A and B consists of four regions ($A \cap B$, $A \cap \bar{B}$, $\bar{A} \cap B$, and $\bar{A} \cap \bar{B}$); and the Venn diagram for the general position of the sets A, B, and C consists of eight regions (Figure 6.1).

The Venn diagram of more than three sets cannot be drawn using circles only (cf. Problem 9 (c) in Chapter 10). However, one can find other simple closed curves for the Venn diagram of an arbitrary (finite) number of sets (cf. Problem 12).

If we have additional information about our sets (so they will not be in "general position"), then we wish to reflect that in our diagram. Suppose, for example, that we are given sets A, B, and C for which we have $C \subseteq A$. In this case, we may draw a diagram like the one in Figure 6.2.

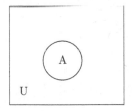 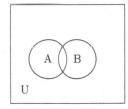 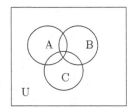

Fig. 6.1 The Venn diagrams of one, two, and three sets in general position

Fig. 6.2 The diagram of sets
A, B, and C for which $C \subseteq A$

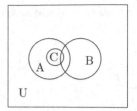

Fig. 6.3 Two diagrams of
sets A, B, and C for which
$A \cap B \subseteq C$

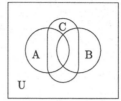

Note that this time we only have six regions and not eight; there are no regions corresponding to $\overline{A} \cap B \cap C$ and $\overline{A} \cap \overline{B} \cap C$. Such diagrams are often convenient for illustrations; to distinguish them from Venn diagrams, we refer to them as *Euler diagrams*.

We need to make additional concessions for some other situations. For example, suppose that we have an expression involving the sets A, B, and C, and that we know that $A \cap B \subseteq C$. Then our desired diagram should consist of only seven regions (there will be no region for $A \cap B \cap \overline{C}$). The first diagram in Figure 6.3 has two separate regions corresponding to $\overline{A} \cap \overline{B} \cap C$, while the second diagram has the correct seven regions, but it has three curves intersecting in a single point.

Venn diagrams can be used to verify statements involving sets. For example, to see that

$$(A \cup \overline{B} \cup C) \cap \overline{(A \cup C)} = \overline{A} \cap \overline{B} \cap \overline{C}$$

holds for arbitrary sets A, B, and C, one can check that the region of the Venn diagram corresponding to $(A \cup \overline{B} \cup C) \cap \overline{(A \cup C)}$ matches the region corresponding to $\overline{A} \cap \overline{B} \cap \overline{C}$. Indeed, we find that the regions (marked by $\checkmark$) corresponding to $A \cup \overline{B} \cup C$ and $\overline{(A \cup C)}$ are respectively, as shown in Figure 6.4, with the diagrams

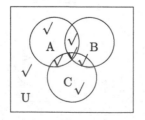   and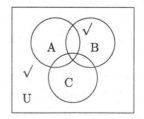

Fig. 6.4 Venn diagram verification of the identity $(A \cup \overline{B} \cup C) \cap \overline{(A \cup C)} = \overline{A} \cap \overline{B} \cap \overline{C}$

corresponding to the columns of $A \cup \overline{B} \cup C$ and $\overline{(A \cup C)}$ in the truth table in Figure 6.1. It is now easy to see that their intersection is $\overline{A} \cap \overline{B} \cap \overline{C}$, as claimed.

While Venn diagrams and Euler diagrams often help us with visualizing or illustrating such statements, mathematicians prefer not to rely on them in written proofs.

Let us now define an important set whose elements are themselves sets.

Definition 6.7 *Suppose that S is a set. The* power set $P(S)$ *of S is the set of all subsets of S; that is, $P(S) = \{A \mid A \subseteq S\}$.*

For example, if $S = \{1, 2\}$, then

$$P(S) = \{\emptyset, \{1\}, \{2\}, \{1, 2\}\};$$

similarly, $P(\mathbb{N})$ consists of all (finite and infinite) subsets of $\mathbb{N}$ (we will see in Chapter 20 that the subsets of $\mathbb{N}$ cannot be listed!). The reason behind the name "power set" lies in the fact that, as we will see in Chapter 19, if S has n elements, then $P(S)$ has 2^n elements; for this reason, the power set of S is sometimes denoted by 2^S. Note that our definition above does not automatically tell us that $P(S)$ is a set (there is no universal set given); thus we need to rely on this fact as an axiom.

Finally, we turn to the Cartesian product (sometimes called cross product or direct product) of sets. Note that, by our definition, the sets $\{x, y\}$ and $\{y, x\}$ are equal. However, we often need to distinguish between the *ordered pairs* (x, y) and (y, x). For this purpose, we make the following definition.

Definition 6.8 *The Cartesian product or direct product of sets A and B is the set*

$$A \times B = \{(x, y) \mid (x \in A) \wedge (y \in B)\}.$$

The elements of $A \times B$ are called *ordered pairs*. The Cartesian product of sets $A_1, A_2, \ldots, A_n$ ($n \in \mathbb{N}$) is defined similarly:

$$\prod_{i=1}^{n} A_i = A_1 \times A_2 \times \cdots \times A_n$$

$$= \{(x_1, x_2, \ldots, x_n) \mid (x_1 \in A_1) \wedge (x_2 \in A_2) \wedge \cdots \wedge (x_n \in A_n)\}.$$

(The direct product of an infinite number of sets will be defined in Chapter 16.) We should mention that it can be proven that Cartesian products are sets and the proof, although skipped here, is necessary as the definition above does not rely on a universal set.

For example, if $A = \{1, 2, 3\}$ and $B = \{2, 4\}$, then

$$A \times B = \{(1, 2), (1, 4), (2, 2), (2, 4), (3, 2), (3, 4)\}$$

and

$$B \times B \times B = \{(2, 2, 2), (2, 2, 4), (2, 4, 2), (2, 4, 4), (4, 2, 2), (4, 2, 4), (4, 4, 2), (4, 4, 4)\}.$$

When $A_1 = A_2 = \cdots = A_n = A$, we write

$$A_1 \times A_2 \times \cdots \times A_n = A^n.$$

For example, $\mathbb{R}^2$ is the set of ordered pairs of real numbers (this set can be identified with the set of points in the plane), and $\mathbb{R}^3$ is the set of ordered triples of real numbers (this set can be identified with the set of points in three-dimensional space).

Sets are used in all branches of mathematics. In fact, it is possible (though rarely worthwhile) to translate every mathematical statement to the language of sets. The branch of mathematics dealing with the precise treatment of sets is *set theory*. We will use and further study sets in every subsequent chapter in this book.

Problems

1. (a) Which of the following expressions describe subsets of the set of real numbers?

   i. $\{n \in \mathbb{N} \mid n^2 + 9\}$
   ii. $\{n^2 + 9 \mid n \in \mathbb{N}\}$
   iii. $\{n \in \mathbb{N} \mid n^2 + 9 > 100\}$
   iv. $\{n^2 + 9 > 100 \mid n \in \mathbb{N}\}$
   v. $\{n \in \mathbb{N} \mid n \in \mathbb{R}\}$
   vi. $\{n \in \mathbb{R} \mid n \in \mathbb{N}\}$

   (Hint: It may be helpful to read out loud the expressions as written.)
   (b) Find all pairs of sets above that are equal.

2. Describe the following sets using the formula notation (cf. Problem 2 in Chapter 3).

   (a) $\{41, 44, 47, 50, 53, 56, \ldots\}$
   (b) $\{41, 43, 47, 53, 61, 71, \ldots\}$
   (c) $\{41, 42, 44, 48, 56, 72, \ldots\}$
   (d) $\{41, 83, 167, 335, 671, 1343, \ldots\}$
   (e) $\{41, 83, 165, 331, 661, 1323, \ldots\}$

3. Describe the following sets using list notation. (Recall that a list may not contain more than one "$\ldots$".)

   (a) $\{n \in \mathbb{N} \mid (-1)^n = -1\}$
   (b) $\{x \in \mathbb{R} \mid \sqrt{x} + 6 = x\}$
   (c) $\{x \in \mathbb{R} \mid x^2 = 2\}$
   (d) $\{x \in \mathbb{Z} \mid x^2 = 2\}$
   (e) $\{(x, y) \in \mathbb{R}^2 \mid (x + 2)^2 + (y - 3)^2 = 0\}$
   (f) $\{(x, y) \in \mathbb{N}^2 \mid x = y\}$
   (g) $\{(x, y) \in \mathbb{Z}^2 \mid x = y\}$

(h) $\{(x, y) \in \mathbb{Z}^2 \mid x^2 = y^2\}$

(i) $\{x \in \mathbb{R} \mid \sin x = 1\}$

(j) $\{(x, y) \in \mathbb{Z}^2 \mid 2x + 3y = 1\}$

4. (a) List *all* elements of the Cartesian product $\mathbb{N} \times \mathbb{N}$.

    (Hint: Placing the elements of $\mathbb{N} \times \mathbb{N}$ in the usual coordinate system may inspire some geometric ideas that provide such a list. Of course, there are infinitely many possible lists.)

    (b) List *all* elements of the set of rational numbers $\mathbb{Q}$.

    (Hint: Use part (a).)

Remarks The method we employed here to list the elements of $\mathbb{Q}$ does not yield a convenient formula. Such a formula, however, does exist, as was recently discovered by Neil Calkin and Herbert Wilf.

Theorem 6.9 *The infinite sequence* $q_1, q_2, q_3, \ldots$, *defined recursively by* $q_1 = 1$ *and*

$$q_{n+1} = \frac{1}{2\lfloor q_n \rfloor - q_n + 1}$$

for $n \geq 1$, *provides a listing of all positive rational numbers; that is, each positive rational number appears exactly once in the sequence.*

The list generated by the Calkin–Wilf sequence starts as

$$1, \tfrac{1}{2}, 2, \tfrac{1}{3}, \tfrac{3}{2}, \tfrac{2}{3}, 3, \tfrac{1}{4}, \tfrac{4}{3}, \tfrac{3}{5}, \tfrac{5}{2}, \tfrac{2}{5}, \tfrac{5}{3}, \tfrac{3}{4}, 4, \ldots$$

The proof of Theorem 6.9 is quite elementary, though it is a bit too lengthy to be presented here.

5. Consider the following sets: $\emptyset$, $\{\emptyset\}$, $\{\{\emptyset\}\}$, $\{\emptyset, \emptyset\}$, $\{\{\emptyset\}, \{\emptyset\}\}$, $\{\{\{\emptyset\}\}, \{\{\emptyset\}\}\}$, $\{\emptyset, \{\emptyset\}\}$, $\{\emptyset, \{\{\emptyset\}\}\}$, $\{\{\emptyset\}, \{\{\emptyset\}\}\}$, $P(\emptyset)$, $P(P(\emptyset))$, $\emptyset \times \emptyset$, and $\emptyset \times \{\emptyset\}$.

    (a) How many different sets are listed?

    (b) Find the smallest possible set that has all of the given sets as elements.

    (c) Find the smallest possible set that contains all the given sets as subsets.

6. Assume that U is a set, and let $Z = \{A \in U \mid A$ is a set and $A \notin A\}$.

    (a) Explain why Z is a set.

    (b) Explain why $Z \subseteq U$.

    (c) Prove that $Z \notin U$.

    (Hint: Show that $Z \in U$ would imply the contradiction that neither $Z \in Z$ nor $Z \notin Z$ can hold.)

    (d) Prove that $Z \notin Z$.

    (Hint: Use part (c).)

7. For sets A and B, we define the difference $A \setminus B$ (read as "A minus B") of A and B as

$$A \setminus B = A \cap \overline{B} = \{a \in A \mid a \notin B\}.$$

Use Venn diagrams to verify each of the following:

(a) $A \cap (B \setminus C) = (A \cap B) \setminus (A \cap C)$
(b) $A \setminus (B \cap C) = (A \setminus B) \cup (A \setminus C)$

8. Suppose that A and B are sets, and consider the following three predicates:

(a) $P(A \cup B) = P(A) \cup P(B)$
(b) $P(A \cap B) = P(A) \cap P(B)$
(c) $P(A \times B) = P(A) \times P(B)$

One of the predicates is true for all sets A and B, one is false for all A and B, and one is true for some A and B and false for others. Decide which is which. For the predicate that may be true as well as false, find a necessary and sufficient condition for when it is true. Justify your answers.

9. Suppose that A is a set of three elements, B is a set of five elements, and C is a set of eight elements. What can you say about the number of elements in the following sets? If there are several possibilities, find all of them, but be careful to list only possibilities that may actually occur. (It might be convenient to use the notation $|X|$ for the size of the set X.)

(a) $A \cup B$, $A \cap B$, and $A \times B$
(b) $A \cup B \cup C$, $A \cap B \cap C$, $(A \cup B) \cap C$, and $A \cup (B \cap C)$
(c) $P(A \cup B)$, $P(A \cap B)$, and $P(A \times B)$
    (Hint: Use part (a).)
(d) $P(A) \cup P(B)$, $P(A) \cap P(B)$, and $P(A) \times P(B)$
    (Hint: Express the required quantities in terms of $|P(A)|$, $|P(B)|$, and $|P(A \cap B)|$ first.)

10. A special relation involving sets is that in which they have no elements in common, that is, their intersection is the empty-set $\emptyset$; in this case we say that the sets are *disjoint*. If, given a collection of sets, we find that every two of them are disjoint, then we say that the collection of sets is *pairwise disjoint*.

(a) Are the three sets $\{1, 2\}$, $\{2, 3\}$, and $\{3, 4\}$ disjoint? Are they pairwise disjoint?
(b) Are the three sets $\{\{1, 2\}, \{3, 4\}\}$, $\{\{1, 3\}, \{2, 4\}\}$, and $\{\{1, 4\}, \{2, 3\}\}$ disjoint? Are they pairwise disjoint?
(c) Find an example for three sets that are disjoint but where no two of the sets are pairwise disjoint.
(d) Find three infinite subsets of $\mathbb{N}$ (that is, subsets of $\mathbb{N}$ with infinitely many elements) that are disjoint, but their pairwise intersections are all infinite.

(e) Find infinitely many infinite subsets of $\mathbb{N}$ that are pairwise disjoint, but their union is $\mathbb{N}$.

(Hint: One possible approach is to use Problem 7 (a) in Chapter 3.)

11. For a given $n \in \mathbb{N}$, the set $\{0, 1\}^n$, sometimes referred to as the *n-dimensional hypercube*, is a set that is often used in most areas of mathematics and computer science.

(a) How many elements does $\{0, 1\}^3$ have?

(b) List the elements of $\{0, 1\}^3$ in *lexicographic order*: (a_1, a_2, a_3) will come before (b_1, b_2, b_3) if, and only if,

$$(a_1 < b_1) \vee [(a_1 = b_1) \wedge (a_2 < b_2)] \vee [(a_1 = b_1) \wedge (a_2 = b_2) \wedge (a_3 < b_3)].$$

(c) List the elements of $\{0, 1\}^3$ in *co-lexicographic order*: (a_1, a_2, a_3) will come before (b_1, b_2, b_3) if, and only if,

$$(a_3 < b_3) \vee [(a_3 = b_3) \wedge (a_2 < b_2)] \vee [(a_3 = b_3) \wedge (a_2 = b_2) \wedge (a_1 < b_1)].$$

(d) List the elements of $\{0, 1\}^3$ in a *Gray code order*: The first and the last elements, as well as any two adjacent elements, differ in exactly one position.

(Hint. There is more than one such order. A drawing of a cube, with vertices appropriately labeled, may be helpful.)

(e) List the elements of $\{0, 1\}^4$ in a Gray code order.

(Hint: We may think of the 4-dimensional hypercube as the union of two 3-dimensional hypercubes.)

12. The general position of one, two, or three sets was illustrated in Figure 6.1.

(a) Use the Gray code order of $\{0, 1\}^3$ to draw a Venn diagram that shows the general position of four sets. (Note that the diagram will have to contain curves other than circles; cf. Problem 9 in Chapter 10.)

(Hints: Start by the Venn diagram of three sets in general position, and then label each arising region by a different element of $\{0, 1\}^3$. Explain how a Gray code order of these elements generates a simple closed curve through the eight regions, dividing each region into two parts.)

(b) Use the Gray code order of $\{0, 1\}^4$ to draw a Venn diagram that shows the general position of five sets.

Chapter 7
Quantifier Mechanics

We introduced abstract mathematics in Chapter 1 with *Hackenbush* games; in particular, we analyzed one such game, the *Arion*. Here we review and further analyze this game as this will enable us to discuss quantifiers in a simple and natural way. Consider Figure 7.1.

The figure consists of six segments, labeled a_1, a_2, a_3, b_1, b_2, and b_3, arranged to form a "horse." Two players, A and B, take turns removing one segment each time. Player A can only remove segments marked a_1, a_2, and a_3; player B can only remove segments b_1, b_2, and b_3. A further restriction is that when, by the removal of some segment(s) a part of the diagram gets disconnected from the "ground," that part of the diagram becomes unavailable. For example, when the "neck" of the horse (segment a_3) is removed, then its "head" (segment b_3) gets cut off. The winner of the game is the last player who has an available option. We want to know which player has a winning strategy, that is, which player will win the game if both players play optimally? Note that we did not specify above which player starts the game, so, indeed, we have two questions: Which player wins if A starts, and which player wins if B starts?

As we have already seen in Chapter 1, *Arion* is a *win for A* no matter who starts the game, and what we mean by that is that A has a winning strategy that guarantees her a win regardless of what B does. (Unless we explicitly state otherwise, we always assume that each player plays optimally.) Therefore, we can make the following claims.

Claim 1: There is *some* first move for A that results in a win for A. (In other words: A can win if she starts the game.)

Claim 2: *Every* first move for B will result in a win for A. (In other words: B will be unable to win no matter how he starts the game.)

Below we will provide a careful analysis and a proof for each of these claims. In effect, what we will show will be slight restatements:

© Springer Nature Switzerland AG 2020
B. Bajnok, *An Invitation to Abstract Mathematics*, Undergraduate Texts in
Mathematics, https://doi.org/10.1007/978-3-030-56174-1_7

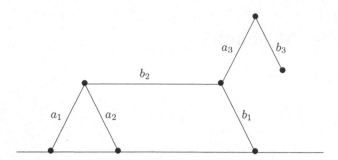

Fig. 7.1 The game *Arion*

Claim 1': There is *some* first move for *A* for which *every* response by *B* will result in a win for *A*. (In other words: *A* is able to start the game so that no response from *B* will prevent her from winning.)

Claim 2': For *every* first move by *B*, *A* has *some* response that will result in a win for *A*. (In other words: *A* can win no matter how *B* starts the game.)

It is easy to see that Claims 1 and 1' are equivalent and that Claims 2 and 2' are equivalent. Furthermore, we will prove the following strengthening of Claim 2':

Claim 2*: For *every* first move by *B* and for *every* response by *A*, the game will result in a win for *A*. (In other words: *A* can respond "blindly" and still win when *B* starts the game.)

Clearly, Claim 2* implies Claim 2' (and thus Claim 2).

To provide a more thorough analysis of our game—and to develop a better understanding of quantifiers—we will carry out further investigations. In particular, we will also establish the following:

Claim 3: There is *some* first move by *A* that results in a win for *B*. (In other words: if *A* wants to win, she must select her first move carefully.)

Or, equivalently:

Claim 3': There is *some* first move by *A* for which *B* has *some* response that will result in a win for *B*. (In other words: If *A* starts the game by a certain move, then *B* will be able to win if he responds by the right move.)

We can then ask for the exact number of possible first moves by *A* that result in a win for her and prove the following:

Claim 3*: There is a *unique* first move for *A* that results in a win for *A*. (In other words: If *A* wants to win, she only has one choice for how to start the game.)

The words "some," "every," and "unique," as well as the similar words "all," "none," and "any," are *quantifiers* that are to be discussed in this chapter.

To analyze this game, in Chapter 1 we introduced so-called *decision trees*. Figure 7.2 shows the complete decision tree of this game in the case when A starts.

The figure is read from left to right and shows all possible moves at each round. (A similar decision tree can be constructed for the case in which B makes the first move.) For example, the topmost branch of the decision tree corresponds to the following play of the game:

- A starts the game by removing a_1;
- B removes b_1;
- A wins by removing a_2.

Since this particular play ends in a win for A, we may label the rightmost endpoint of this branch of the tree with an A. We can similarly label all other endpoints.

Next we discuss how to label intermediate points in our decision tree in a way that indicates who would win if a particular play reached that position and if both players played optimally afterwards.

Obviously, if the tree does not fork beyond the intermediate point in question, then we can label the point the same way as the end of the branch was labeled. For example, if the play starts with A removing a_1, and is followed by B removing b_1 and then A removing a_3, then at that point we already know that the game will end by A winning.

Furthermore, even if the tree does fork, the label is clear if *every* branch beyond the point is labeled the same. For example, if the play starts with A removing a_3, then A will win in every case. This proves Claim 1 above.

How do we label points where two (or, perhaps, more) branches originate and where these branches are labeled differently? To answer that, we need to know whose turn it is to make a move at that point. Naturally, each player will move in the direction where the branch is favoring him or her. For example, if a particular play of our game starts with A removing a_1 and then B removing b_2, then the play, assuming both players play optimally afterwards, will end in a win for A since A will move next by removing a_3. Therefore, we can label that point in the diagram by A. Similarly, we can label the point after a_1, b_3, and a_3 have been played by B because, at that point in the play, it is B's decision, and he has the option of moving toward a branch ending in a win for him by removing b_2.

After completing the labeling process as just described, we see that the starting point is labeled A; thus Claim 3 holds. Furthermore, since only one of the three "neighbors" of the starting point is labeled by A (when A starts the game by removing a_3), we also see that Claim 3* holds. The analysis of the game when B starts is considerably easier: every point will be labeled A. This proves Claims 2 and 2*.

Let us now discuss quantifiers more formally.

Suppose that S is a given set and that $P(x)$ is a predicate that becomes a statement for every $x \in S$ (we write $P(x)$ to express the fact that P depends on x). We then may define the *truth set of $P(x)$ on S* to be the set

Fig. 7.2 The complete decision tree of *Arion* when *A* starts

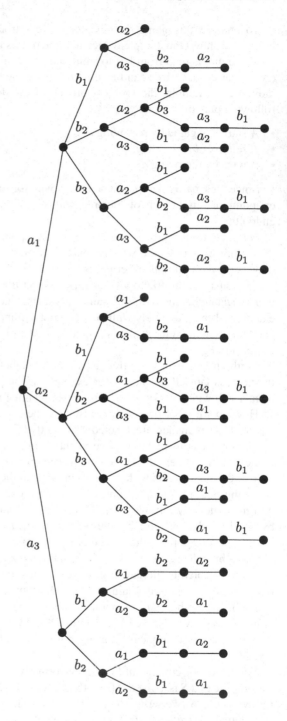

$$S_P = \{x \in S \mid P(x) \text{ is true}\};$$

this is the subset of S for which $P(x)$ holds. According to the Axiom of Separation, S_P is a well-defined set.

Many questions in mathematics essentially ask for the truth set of a certain predicate on a given set. For example, if $S = \mathbb{N}$ and $P(n)$ is the predicate that $2^n - 1$ is a prime number, then, as we have seen in Chapter 3, $7 \in S_P$, but $11 \notin S_P$. Furthermore, according to Theorem 4.7, every element of S_P is a prime number. (The Mersenne Prime Conjecture says that S_P has infinitely many elements; cf. page 50.) Another, more familiar example is when we need to solve an equation $E(x)$ in a given set, such as $x^2 - 5x + 6 = 0$ in $\mathbb{R}$. The truth set of this equation in $\mathbb{R}$ is the set $\mathbb{R}_E = \{2, 3\}$.

Often we are only interested in knowing whether the truth set contains all elements of the underlying set or if it contains at least one element. Our terminology for these cases is as follows:

Definition 7.1 *We say that $P(x)$ holds* for every $x \in S$, *and write*

$$\forall x \in S, P(x)$$

if, and only if, $S_P = S$, i.e., the predicate $P(x)$ is a true statement for every $x \in S$. The symbol $\forall$ is called the universal quantifier.

Definition 7.2 *We say that $P(x)$ holds* for some $x \in S$, *and write*

$$\exists x \in S, P(x)$$

if, and only if, $S_P \neq \emptyset$, i.e., the predicate $P(x)$ is a true statement for at least one $x \in S$. The symbol $\exists$ is called the existential quantifier.

The universal and existential quantifiers appear frequently in mathematics. For example,

$$\forall x \in \mathbb{R}, x^2 \geq 0$$

stands for the fact that the square of every real number is nonnegative; and

$$\exists x \in \mathbb{R}, x^2 = 2$$

says that there is at least one real number whose square equals 2. (Both statements are true, although the proof of the second statement is considerably harder than the first and will have to wait until Chapter 21.)

Often, we need to use the negation of a quantified statement. The negation of

$$\forall x \in S, P(x)$$

means that $P(x)$ is false for at least one $x \in S$, that is

$$\exists x \in S, \neg P(x).$$

The negation of

$$\exists x \in S, P(x)$$

occurs even more often. It is denoted by

$$\not\exists x \in S, P(x),$$

and it means that $P(x)$ is true for *none* of the elements of S or, equivalently, that it is false for every $x \in S$, that is

$$\forall x \in S, \neg P(x).$$

Thus, in short, we can write

$$\neg[\forall x \in S, P(x)] \Leftrightarrow [\exists x \in S, \neg P(x)]$$

and

$$\neg[\exists x \in S, P(x)] \Leftrightarrow [\forall x \in S, \neg P(x)].$$

Note that the existential quantifier does not specify how many elements the set S has for which $P(x)$ is true, except to claim that there is at least one such element. Sometimes we are interested in the case in which there is exactly one x for which $P(x)$ holds. The notation for this is

$$\exists! x \in S, P(x),$$

and we say that there is a *unique* $x \in S$ for which $P(x)$ holds.

Let us now return to our *Arion* game. To make our discussion more precise, we introduce some notation. Let $\mathcal{A}$ and $\mathcal{B}$ denote the sets of options that are available for A and B, respectively, so we have $\mathcal{A} = \{a_1, a_2, a_3\}$ and $\mathcal{B} = \{b_1, b_2, b_3\}$. Let $WA(x)$ (resp. $WB(x)$) denote the predicate that the game is a win for A (resp. B) after the initial move $x \in \mathcal{A} \cup \mathcal{B}$. (As always, we assume that both players play optimally in the game that results after the initial move x is made.) Our analysis of the game above revealed that $WA(a_3)$, $WA(b_1)$, $WA(b_2)$, and $WA(b_3)$ are all true statements, establishing

Claim 1: $\exists a \in \mathcal{A}, WA(a)$

and

Claim 2: $\forall b \in \mathcal{B}, WA(b)$.

We also pointed out, however, that $WA(a_1)$ and $WA(a_2)$ are false; hence, their negations $WB(a_1)$ and $WB(a_2)$ are true, and we have

Claim 3: $\exists a \in \mathcal{A}, WB(a)$.

In fact, since there are two different elements of $\mathcal{A}$ for which $WB(a)$ is true and only one for which $WA(a)$ holds, we have

Claim 3*: $\exists! a \in \mathcal{A}, WA(a)$.

When determining whether $WA(x)$ or $WB(x)$ holds for a given initial move $x \in \mathcal{A} \cup \mathcal{B}$, we had to look at the possible moves by the second player. For example, to see that $WA(a_3)$ is true, we verified that the game can be won by A regardless of how B responds to the initial move a_3. Similarly, to see that $WA(b)$ is true for every $b \in \mathcal{B}$, we had to verify that every initial move of B could be followed by some move by A that resulted in a win for A.

To do such an analysis more precisely, we need to discuss statements involving two or more quantifiers. Let $WA(x, y)$ (resp. $WB(x, y)$) stand for the predicate that the game can be won by player A (resp. B) after the initial move x by the first player is followed by the response move y by the second player ($x \in \mathcal{A} \cup \mathcal{B}$ and $y \in \mathcal{A} \cup \mathcal{B}$).

So, for example, to say that a_3 is a winning initial move for A, we have to verify that $WA(a_3, b_1)$ and $WA(a_3, b_2)$ are both true. (Note that, after moving a_3, option b_3 becomes unavailable.) That is, we had to check that $WA(a_3, b)$ held true for every $b \in \mathcal{B}$. On the other hand, to claim that b_1 was not a winning initial move for player B, we had to verify that $WA(b_1, a)$ was true for some (but not necessarily every) $a \in \mathcal{A}$. In fact, we saw that B had no winning initial moves at all; that is, there is an $a \in \mathcal{A}$ for which $WA(b_2, a)$ is true, and there is an $a \in \mathcal{A}$ for which $WA(b_3, a)$ is true.

Let us now discuss these *double quantifiers* in general. Suppose that A and B are sets, and let $P(a, b)$ be a predicate that becomes a statement once we know $a \in A$ and $b \in B$. In this case, $P(a, b)$ is a predicate on two variables; it is defined on the elements of $A \times B$, and the truth set of $P(a, b)$ on $A \times B$ will be a subset of $A \times B$, which we denote by $(A \times B)_P$. For example, if $A = \mathbb{Z}$, $B = \mathbb{N}$, and $P(a, b)$ is the predicate $a \geq b$, then $(3, 2) \in (A \times B)_P$, but $(2, 3) \notin (A \times B)_P$ and $(-2, 3) \notin (A \times B)_P$.

Using the universal quantifier $\forall$ and the existential quantifier $\exists$, we can form compounded statements such as

$$\forall a \in A, \exists b \in B, P(a, b),$$

which can be read as "for every $a \in A$, there exists some $b \in B$ for which $P(a, b)$ holds." When analyzing such compounded statements, we separate the first quantifier and regard the rest of the statement as a predicate $Q(a)$ with one free variable. For example,

$$\forall a \in A, \exists b \in B, P(a, b)$$

means that

$$\forall a \in A, Q(a),$$

where $Q(a)$ is the predicate

$$\exists b \in B, P(a, b).$$

In the example of the previous paragraph we see, for example, that $Q(5)$ is true, since we can choose $b = 3$ (for example) for which $5 \geq b$. On the other hand, $Q(-4)$ is false, since we cannot choose a $b \in \mathbb{N}$ for which $-4 \geq b$. Therefore, the statement

$$\forall a \in A, \exists b \in B, P(a, b)$$

is false in this example.

We can now rewrite our remaining claims using double quantifiers, as follows:

Claim 1': $\exists a \in \mathcal{A}, \forall b \in \mathcal{B}, WA(a, b).$
Claim 2': $\forall b \in \mathcal{B}, \exists a \in \mathcal{A}, WA(b, a).$
Claim 3': $\exists a \in \mathcal{A}, \exists b \in \mathcal{B}, WB(a, b).$
Claim 2*: $\forall b \in \mathcal{B}, \forall a \in \mathcal{A}, WA(b, a).$

We can, in fact, introduce *triple quantifiers* analogously. The triple-quantified versions of Claims 1' and 2', for example, would be (with self-explanatory notation)

Claim 1'': $\exists a \in \mathcal{A}, \forall b \in \mathcal{B}, \exists a' \in \mathcal{A}, WA(a, b, a')$

and

Claim 2'': $\forall b \in \mathcal{B}, \exists a \in \mathcal{A}, \forall b' \in \mathcal{B}, WA(b, a, b').$

Here the statement

$$\exists a \in \mathcal{A}, \forall b \in \mathcal{B}, \exists a' \in \mathcal{A}, WA(a, b, a')$$

stands for the claim that there is an initial move for player A such that no matter what the response move of player B is, player A has a winning follow-up move (this statement is true, as we have seen above). Similarly, Claim 2'' says that no matter how B starts to play, A will have a move in response so that no matter how B moves next, A will win.

Note that the quantifiers involved in a statement do not commute! Keeping our earlier example above, consider, for example, the two statements

$$\forall b \in \mathbb{N}, \exists a \in \mathbb{Z}, a \geq b,$$

and

$$\exists a \in \mathbb{Z}, \forall b \in \mathbb{N}, a \geq b.$$

The first statement claims that for every positive integer b, one can find an integer a that is greater than or equal to b; this statement is true (choose, for example, $a = b + 3$). The second statement says that there is an integer a that is greater than or equal to every positive integer b; this statement is false, because there is no largest integer.

The fact that quantifiers don't commute confuses many who are not trained in logic, and is also the source of some amusing riddles. The American comedian Sam Levenson, for example, gives us the following warning: "Somewhere on this globe, every ten seconds, there is a woman giving birth to a child. She must be found and stopped." The joke comes from the fact that the statements

$$\forall \text{ time, } \exists \text{ place, } \exists \text{ woman giving birth}$$

and

$$\exists \text{ place, } \exists \text{ woman, } \forall \text{ time giving birth}$$

are not equivalent (the first is apparently true, while the second is ridiculous).

Triple quantifiers play an extremely important role in advanced calculus (often referred to as real analysis). Namely, they provide the definition of *limits*, which in turn are the foundations of topics such as differentiation, integration, and infinite series. We will discuss limits in Chapter 18.

Problems

1. Let G be a *Hackenbush* game in which the set of available options for players A and B are $\mathcal{A}$ and $\mathcal{B}$, respectively. For each $x \in \mathcal{A} \cup \mathcal{B}$, let $WA(x)$ and $WB(x)$ denote the predicates that G is a win for A, respectively B, if the first player's move is x (and both players play optimally thereafter). Similarly, let $WA(x, y)$ and $WB(x, y)$ be the predicates that player A, respectively B, has a winning strategy after option x is removed by the first player and option y is removed by the second player. Define $WA(x, y, z)$ and $WB(x, y, z)$ analogously.

    We say that a *Hackenbush* game is *fair* if the second player wins when both players play optimally. In other words, the game is a win for A (and thus a loss for B) if B starts, and a loss for A (and thus a win for B) if A starts.

    It is easy to verify that the three games in Figure 7.3 are all fair.

    We also say that a game is *advantageous* for a particular player if that player can win the game regardless of who starts. For example, we have seen that the game *Arion* is advantageous for A.

    (a)   i. Write the definition for G to be a fair game using quantifiers and the predicates $WA(x)$ and $WB(x)$. Do the same for G being advantageous for A, and G being advantageous for B.

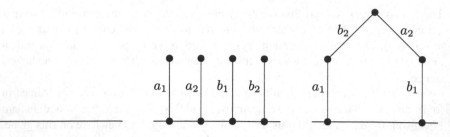

Fig. 7.3 Three fair *Hackenbush* games

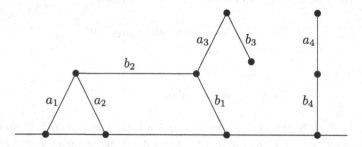

Fig. 7.4 *Arion* with a pole

      ii. Repeat part i using the predicates $WA(x, y)$ and $WB(x, y)$.

      iii. Repeat part i using the predicates $WA(x, y, z)$ and $WB(x, y, z)$.

  (b) Prove that the game in Figure 7.4 is fair by drawing as few branches of the decision tree as you deem necessary.

Remarks In Chapter 22 we will learn how to assign numerical values to all *Hackenbush* games in a way that measures just how advantageous the game is to a particular player. Fair games will have zero value, games that are advantageous to A will have a positive value (the more advantageous the game, the higher the value), and games that are advantageous to B will have a negative value. We will also see that some (in fact, most) games can only be assigned "surreal" values as no real number will do. *Surreal numbers* (invented by John H. Conway in the 1970s) form an ordered field extension of the field of real numbers. (In fact, one can show that they form the largest such extension.)

2. Let $A = \{1, 3, 5, 7\}$ and $B = \{2, 4, 6\}$.

  (a) Let $P(a, b)$ be the predicate $a \leq b$. List the elements of the truth set $(A \times B)_P$.

  (b) Let $Q(a, b)$ be a predicate that becomes a statement for every $a \in A$ and $b \in B$, and suppose that the truth set $(A \times B)_Q$ is the set

$$\{(1, 2), (1, 4), (3, 2), (3, 4), (3, 6), (5, 4), (7, 6)\}.$$

Which of the following statements are true and which are false?

    i. $\exists a \in A, \exists b \in B, Q(a, b)$
    ii. $\exists a \in A, \exists! b \in B, Q(a, b)$
    iii. $\exists! a \in A, \exists b \in B, Q(a, b)$
    iv. $\exists! a \in A, \exists! b \in B, Q(a, b)$
    v. $\exists a \in A, \forall b \in B, Q(a, b)$
    vi. $\exists! a \in A, \forall b \in B, Q(a, b)$
    vii. $\forall a \in A, \exists b \in B, Q(a, b)$
    viii. $\forall a \in A, \exists! b \in B, Q(a, b)$
    ix. $\forall a \in A, \forall b \in B, Q(a, b)$
    x. $\exists b \in B, \exists a \in A, Q(a, b)$
    xi. $\exists b \in B, \exists! a \in A, Q(a, b)$
    xii. $\exists! b \in B, \exists a \in A, Q(a, b)$
    xiii. $\exists! b \in B, \exists! a \in A, Q(a, b)$
    xiv. $\exists b \in B, \forall a \in A, Q(a, b)$
    xv. $\exists! b \in B, \forall a \in A, Q(a, b)$
    xvi. $\forall b \in B, \exists a \in A, Q(a, b)$
    xvii. $\forall b \in B, \exists! a \in A, Q(a, b)$
    xviii. $\forall b \in B, \forall a \in A, Q(a, b)$

(Hints: Recall that, by fixing the first variable, we can reduce each of these statements to an expression involving only one quantifier. For example, to analyze statement vii, we can fix $a = 1$ and ask whether the statement $\exists b \in B, Q(1, b)$ is true or not; that is, if we have an element $b \in B$ for which the ordered pair $(1, b)$ is in the truth set. Then statement vii is true if we get an affirmative answer for all choices of a.)

(c) Suppose that $R(a, b)$ is a predicate that becomes a statement for every $a \in A$ and $b \in B$, and consider the following three statements

    i. $\exists! a \in A, \exists b \in B, R(a, b)$
    ii. $\exists a \in A, \exists! b \in B, R(a, b)$
    iii. $\exists! a \in A, \exists! b \in B, R(a, b)$

For each of the three statements, find an example, if possible, for a predicate $R(a, b)$ for which that statement is true but the other two statements are false. You can give the predicate in terms of its truth set, as in part (b).

3. Let $S = \{0, 1, 2, 3, 4, 5\}$, and suppose that $P(x)$, $Q(x, y)$, and $R(x, y, z)$ are the predicates that x is divisible by 3, that x is divisible by y, and that x is divisible by $y + z$, respectively. (It is good to remember that 0 is divisible by every integer, including itself, but no nonzero integer is divisible by 0; see Definition 2.2.) Read each of the following statements in plain English. (Try to express yourself as clearly and concisely as possible.) Which of the following statements are true and which are false?

(a)   i. $\exists x \in S, P(x)$
     ii. $\forall x \in S, P(x)$

     iii. $\exists! x \in S,\, P(x)$

(b)   i. $\exists x \in S,\, \exists y \in S,\, Q(x, y)$
    ii. $\exists x \in S,\, \forall y \in S,\, Q(x, y)$
   iii. $\forall x \in S,\, \exists y \in S,\, Q(x, y)$
   iv. $\forall x \in S,\, \forall y \in S,\, Q(x, y)$

(c)   i. $\exists! x \in S,\, \exists y \in S,\, Q(x, y)$
    ii. $\exists x \in S,\, \exists! y \in S,\, Q(x, y)$
   iii. $\exists! x \in S,\, \forall y \in S,\, Q(x, y)$
   iv. $\forall x \in S,\, \exists! y \in S,\, Q(x, y)$
    v. $\exists! x \in S,\, \exists! y \in S,\, Q(x, y)$

(d)   i. $\exists! y \in S,\, \exists x \in S,\, Q(x, y)$
    ii. $\exists y \in S,\, \exists! x \in S,\, Q(x, y)$
   iii. $\exists! y \in S,\, \forall x \in S,\, Q(x, y)$
   iv. $\forall y \in S,\, \exists! x \in S,\, Q(x, y)$
    v. $\exists! y \in S,\, \exists! x \in S,\, Q(x, y)$

(e)   i. $\exists x \in S,\, \exists y \in S,\, \exists z \in S,\, R(x, y, z)$
    ii. $\exists x \in S,\, \exists y \in S,\, \forall z \in S,\, R(x, y, z)$
   iii. $\exists x \in S,\, \forall y \in S,\, \exists z \in S,\, R(x, y, z)$
   iv. $\exists x \in S,\, \forall y \in S,\, \forall z \in S,\, R(x, y, z)$
    v. $\forall x \in S,\, \exists y \in S,\, \exists z \in S,\, R(x, y, z)$
   vi. $\forall x \in S,\, \exists y \in S,\, \forall z \in S,\, R(x, y, z)$
  vii. $\forall x \in S,\, \forall y \in S,\, \exists z \in S,\, R(x, y, z)$
 viii. $\forall x \in S,\, \forall y \in S,\, \forall z \in S,\, R(x, y, z)$

(f)   i. $\exists! x \in S,\, \exists y \in S,\, \exists z \in S,\, R(x, y, z)$
    ii. $\exists x \in S,\, \exists! y \in S,\, \exists z \in S,\, R(x, y, z)$
   iii. $\exists x \in S,\, \exists y \in S,\, \exists! z \in S,\, R(x, y, z)$
   iv. $\exists! x \in S,\, \exists! y \in S,\, \exists z \in S,\, R(x, y, z)$
    v. $\exists! x \in S,\, \exists y \in S,\, \exists! z \in S,\, R(x, y, z)$
   vi. $\exists x \in S,\, \exists! y \in S,\, \exists! z \in S,\, R(x, y, z)$
  vii. $\exists! x \in S,\, \exists! y \in S,\, \exists! z \in S,\, R(x, y, z)$

4. In a certain multiple-choice test, one of the questions was illegible but the choice of answers, given below, was clearly printed. What is the right answer? Can more than one answer be correct?

(A) All of the below.
(B) None of the below.
(C) All of the above.
(D) One of the above.
(E) None of the above.
(F) None of the above.

5. Write the following statements using mathematical notations (sets and quantifiers). Denote the set of prime numbers by P, but do not use any other non-standard notation.

(a) No square number is a prime.
(b) There is a prime number between 100 and 110.
(c) There are no primes between 200 and 210.
(d) There is a unique prime between 90 and 100.
(e) There is a unique even positive prime number.
(f) 2 is the only even positive prime.
(g) 199 and 211 are consecutive primes.
(h) There are twin primes between 100 and 110.
(i) There is a perfect square between any positive integer and its double (inclusive).

   Remark We will prove this statement in Chapter 12.

(j) There is a prime between any positive integer and its double (inclusive).

   Remark This is known as Chebyshev's Theorem after the Russian mathematician Pafnuty Chebyshev (1821–1894) who provided a (remarkably complicated) proof for this statement in 1852.

(k) There is a prime between any two consecutive positive perfect squares.

   Remark This is known as Legendre's Conjecture, named after the French mathematician Adrien-Marie Legendre (1752–1833). Although it has been pursued vigorously by many mathematicians, no proof is known for the statement at this time.

6. Suppose that Γ is an irrational number given by its infinite (non-periodic) decimal representation.

   (a) Consider the following statements.

      i. Every decimal digit of Γ is 3, 5, or 7.
      ii. Every decimal digit of Γ is an odd prime.
      iii. Every decimal digit of Γ that is odd must be prime.
      iv. No decimal digit of Γ is an odd prime.
      v. No decimal digit of Γ that is odd can be prime.
      vi. If a decimal digit of Γ is odd, then it is a prime.
      vii. For every decimal digit of Γ, the digit is odd iff it is a prime.
      viii. No decimal digit of Γ that is odd can be 1 or 9.
      ix. No decimal digit of Γ is 1 or 9.

      Partition this set of statements into equivalence classes; that is, two statements should be in the same equivalence class if, and only if, they are equivalent.

   (b) State the negation of each statement (it suffices to only do this for one member of each equivalence class). Do not use the word "not."

7. Suppose that A is a given set of real numbers. Use only the mathematical symbols

$$\exists, \ \exists!, \ \forall, \ \in, \ =, \ \neq, \ >, \ <, \ \lor, \ \land, \ x, \ y, \ z, \ A, \ 0, \ (, \ \text{and} \ )$$

to write each of the following statements as well as their negations.

(a) A has a unique negative element.
(b) A has exactly two negative elements.

8. Write a logical form for each of the sentences below. Then write the negation of the sentences, first in their logical form and then in plain English. Your logical forms should use only the variables and predicates given and the quantifiers $\forall$ and $\exists$; do not use the negation operation.

(a) "Every family has its secrets."
Let $P(f, s)$ be the predicate that family f has the secret s, and let $Q(f, s)$ be the predicate that family f does not have the secret s.
(b) "There is a moment in everyone's life when nothing seems to go well."
Let $P(p, x, t)$ be the predicate that task x seems to go well for person p at time t, and let $Q(p, x, t)$ be the predicate that task x does not seem to go well for person p at time t.
(c) "Everybody loves somebody sometime." (from a Dean Martin song)
Let $P(p, q, t)$ be the predicate that person p loves person q at time t, and let $Q(p, q, t)$ be the predicate that person p does not love person q at time t.
(d) "You can fool all of the people some of the time, and you can fool some of the people all of the time, but you cannot fool all of the people all of the time." (Abraham Lincoln)
Let $P(p, t)$ be the predicate that you can fool person p at time t, and let $Q(p, t)$ be the predicate that you cannot fool person p at time t.

9. In this problem we further analyze the game *Cutcake* of Problem 5 of Chapter 1.

Let $C(m, n)$ denote the game played on an m-by-n cake (that is, a rectangular cake that is divided into little pieces by $m - 1$ horizontal lines and $n - 1$ vertical lines; $m, n \in \mathbb{N}$). Let Γ denote here the set of all such games; that is, let

$$\Gamma = \{C(m, n) \mid m, n \in \mathbb{N}\}.$$

We will introduce some additional notations for the possible outcomes of these games as follows:

- $\Gamma_{H \to H}$: the collection of games that Horizontal can win if Horizontal starts;
- $\Gamma_{H \to V}$: the collection of games that Vertical can win if Horizontal starts;
- $\Gamma_{V \to H}$: the collection of games that Horizontal can win if Vertical starts; and
- $\Gamma_{V \to V}$: the collection of games that Vertical can win if Vertical starts.

Furthermore, we set

$$\Gamma_H = \Gamma_{H \to H} \cap \Gamma_{V \to H},$$

$$\Gamma_V = \Gamma_{H \to V} \cap \Gamma_{V \to V},$$

$$\Gamma_I = \Gamma_{H \to H} \cap \Gamma_{V \to V},$$

and

$$\Gamma_{II} = \Gamma_{H \to V} \cap \Gamma_{V \to H}.$$

In other words, Γ_H and Γ_V are the collections of games that Horizontal and Vertical can win, respectively, regardless of who starts the game; Γ_I and Γ_{II} are the collections of games that are won by the first player and second player, respectively. For example, according to Problem 5 of Chapter 1, we have $C(3, 5) \in \Gamma_V$ and $C(4, 7) \in \Gamma_{II}$.

(a) Prove the following two statements.

- If

$$\exists k \in \{1, 2, \ldots, m - 1\}, \{C(k, n), C(m - k, n)\} \subset \Gamma_H \cup \Gamma_{II},$$

then $C(m, n) \in \Gamma_{H \to H}$.
- If

$$\forall k \in \{1, 2, \ldots, m - 1\}, \{C(k, n), C(m - k, n)\} \subset \Gamma_V \cup \Gamma_{II}$$

but

$$\{C(k, n), C(m - k, n)\} \not\subset \Gamma_{II},$$

then $C(m, n) \in \Gamma_{H \to V}$.

(b) Make and prove two analogous statements regarding $C(m, n) \in \Gamma_{V \to H}$ and $C(m, n) \in \Gamma_{V \to V}$.
(c) Use the statements above to decide the outcome (Γ_H, Γ_V, Γ_I, or Γ_{II}) of $C(m, n)$ for each $1 \le m \le 8$ and $1 \le n \le 8$.
(Hints: Start with the obvious claims—which also follow from parts (a) and (b) above!—that $C(1, 1) \in \Gamma_{II}$, $C(1, n) \in \Gamma_V$ for all $n \ge 2$, and $C(m, 1) \in \Gamma_H$ for all $m \ge 2$. Next, proceed to decide the outcome of $C(2, 2)$, $C(2, 3)$, etc.)
(d) Make a conjecture for the set of all values of n for which $C(1000, n) \in \Gamma_H$. (You don't need to provide a proof for your conjecture.)
(e) Conjecture a set of criteria, in terms of m and n, that determines the outcome of $C(m, n)$. (You don't need to provide a proof for your conjecture.)

Remarks As our computations demonstrate, for each $m, n \subset \mathbb{N}$,

$$C(m, n) \in \Gamma_H \cup \Gamma_V \cup \Gamma_{II}.$$

We will return to a more thorough evaluation of this game in Chapter 22.

Chapter 8
Let's Be Functional!

In this chapter we discuss another fundamental object in mathematics: functions. A *function* is described by two sets and an assignment from one of them to the other; somewhat more precisely, a function f is given by a set A, called the *domain* of f, a set B, called the *codomain* of f, and a relation that maps each element of A to an element of B. This is denoted by $f : A \to B$; the fact that an element $a \in A$ is mapped to some $b \in B$, called the *image* of a, is denoted by $a \mapsto b$ or $f(a) = b$. When $B = A$, we say that f is a *function on* A. Our description here is somewhat heuristic—we don't really define what a function is—a more precise treatment will be given in Chapter 16.

We should emphasize that, given a function $f : A \to B$, there may be elements $b \in B$ for which no element $a \in A$ exists with $f(a) = b$. We define the *image* or *range* of the function to be the subset $\mathrm{Im}(f)$ of the codomain whose elements are the images of some elements of the domain. Thus, for each element $b \in \mathrm{Im}(f)$, we must have at least one element $a \in A$ with $f(a) = b$, but we may have more than one such element. So, to summarize in terms of quantifiers, given a function $f : A \to B$, the statements

$$\forall a \in A, \exists! b \in B, f(a) = b$$

and

$$\forall b \in \mathrm{Im}(f), \exists a \in A, f(a) = b$$

must be true, but

$$\forall b \in B, \exists a \in A, f(a) = b$$

© Springer Nature Switzerland AG 2020
B. Bajnok, *An Invitation to Abstract Mathematics*, Undergraduate Texts in Mathematics, https://doi.org/10.1007/978-3-030-56174-1_8

Fig. 8.1 A diagram
representation of a function

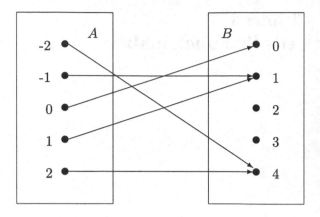

and

$$\forall b \in \operatorname{Im}(f), \exists! a \in A, f(a) = b$$

may be false.

Let us see now the various ways functions can be presented. Specifying the domain and codomain is simple enough: we only need to identify these sets by one of the methods discussed in Chapter 6. The assignments of the function can then be given in a variety of ways. An efficient method—particularly when the function has a finite domain—is to provide a *table* of two rows: the first row lists the elements of the domain and the second row lists the corresponding function values; for example:

a	-2	-1	0	1	2
$f(a)$	4	1	0	1	4

An alternative to the table description is a *diagram* in which elements of the domain and codomain are listed in two rows or two columns, with an arrow from each element of the domain to the corresponding element in the codomain; for example, the function above can be presented as in Figure 8.1.

Furthermore, some of the most useful assignments are conveniently defined by *formulas*. For instance, the function above can be given as

$$f : \{-2, -1, 0, 1, 2\} \rightarrow \{0, 1, 2, 3, 4\}$$
$$a \mapsto a^2$$

In addition, functions, especially when their domains and codomains are subsets of the real numbers, may also be represented graphically using the Descartes coordinate plane: we let the "horizontal" axis represent the domain and the "vertical" axis represent the codomain, and mark the point (a, b) if the function maps a to b.

We need to emphasize that specifying the assignments of a function is not sufficient for defining it: we must specify the domain and the codomain of the function as well. For example, the table above does not tell us what the codomain of the function is; the assignment $f(a) = a^2$ alone would not even establish its domain.

It may be useful to state explicitly when two functions are equal.

Definition 8.1 *We say that the functions f_1 and f_2 are equal if they have the same domain, the same codomain, and for each element a in the domain, we have $f_1(a) = f_2(a)$.*

For example, the functions

$$f_1 : N \to \quad \mathbb{Z}$$
$$x \mapsto x - 2$$

and

$$f_2 : N \to \quad \mathbb{Z}$$
$$x \mapsto \frac{x^2 - 4}{x+2}$$

are equal. Note that for two functions to equal each other, we assume that their codomains are equal; so, for example, the function

$$f_3 : N \to \quad \mathbb{R}$$
$$x \mapsto x - 2$$

is considered unequal to the function f_1 even though their domains and all their assigned values agree.

There are, of course, many different kinds of functions—some with well-known names, some without. Perhaps the simplest function of all is the *identity function* on a set A. It is denoted by id_A and is defined by $id_A(a) = a$ for every $a \in A$.

We now turn to three fundamentally important classes of functions.

Definition 8.2 *A function $f : A \to B$ with image $\mathrm{Im}(f)$ is called*

- injective *(or one-to-one) if no two elements of A are mapped to the same element of $\mathrm{Im}(f)$:*

$$\forall b \in \mathrm{Im}(f), \exists! a \in A, \ f(a) = b;$$

- surjective *(or onto) if $\mathrm{Im}(f) = B$:*

$$\forall b \in B, \exists a \in A, \ f(a) = b;$$

and
- bijective *(or a one-to-one correspondence) if it is both injective and surjective.*

In theory, it is easy to tell if a given function is injective, surjective, or bijective. For example, the graph of a function shows that it is

- injective if, and only if, it crosses every horizontal line at most once (this is called the *horizontal line test*);
- surjective if, and only if, it crosses every horizontal line at least once;
- bijective if, and only if, it crosses every horizontal line exactly once.

We can easily see that the function described on page 128 is not injective: it maps both 1 and -1 (and also 2 and -2) to the same element of the codomain. Neither is it surjective: the elements 2 and 3 of the codomain are not in the image of f. On the other hand, the identity function is obviously a bijection on any set. Injections, surjections, and bijections are discussed often in mathematics, and we will see a variety of examples later.

If two functions are given in such a way that the codomain of the first is the same as the domain of the second, then we are able to combine them into a new function, as follows:

Definition 8.3 *If $f : A \to B$ and $g : B \to C$, then the* composition *of g with f is the function $h : A \to C$ for which*

$$h(a) = g(f(a))$$

for all $a \in A$. In this case we write $h = g \circ f$.

Note that compositions don't commute: $g \circ f$ and $f \circ g$ are not only different, but usually they are not even both meaningful—only when the codomain of each is the same as the domain of the other can they be both defined! However, compositions satisfy the associative property: for all functions $f : A \to B$, $g : B \to C$, and $h : C \to D$, we have

$$(h \circ g) \circ f = h \circ (g \circ f),$$

since both sides map an element $a \in A$ to $h(g(f(a)))$.

Compositions can be used to define left and right inverses.

Definition 8.4 *Let $f : A \to B$. A function $g : B \to A$ is called*

- *a* left inverse *of f if $g \circ f = id_A$.*
- *a* right inverse *of f if $f \circ h = id_B$.*

As an example, let

$$f : [0, \infty) \to \mathbb{R}$$
$$x \mapsto \sqrt{x};$$

and

$$g : \mathbb{R} \to [0, \infty)$$
$$x \mapsto x^2.$$

Then f is a right inverse of g, since for all $x \in [0, \infty)$, we have

$$(g \circ f)(x) = g(f(x)) = g(\sqrt{x}) = x;$$

however, f is not a left inverse of g: for example, we have $(f \circ g)(-1) = 1$, so $f \circ g \neq id_{\mathbb{R}}$! In fact, g has no left inverse at all: such a function would need to map 1 to both -1 and 1, which is impossible. We can similarly see that g is a left inverse, but not a right inverse, of f.

We should also point out that, in general, a function may have more than one left inverse or more than one right inverse. In our last example, for instance, since it made no difference how g was defined for negative x values when calculating $(g \circ f)(x)$, f would have infinitely many left inverses. On the other hand, we have the following proposition.

Proposition 8.5 *Let $f : A \to B$. If f has both a left inverse $g : B \to A$ and a right inverse $h : B \to A$, then*

- *g is the unique left inverse of f;*
- *h is the unique right inverse of f; and*
- *$g = h$.*

Proof We will prove that for every $b \in B$, we have $g(b) = h(b)$; all three claims then follow.

Let b be an arbitrary element of B, and consider $g(f(h(b)))$. On one hand, it equals $g((f \circ h)(b))$, which, since h is a right inverse of f, can be simplified to $g(id_B(b)) = g(b)$. On the other hand, $g(f(h(b)))$ also equals $(g \circ f)(h(b))$, which, since g is a left inverse of f, becomes $id_A(h(b)) = h(b)$. But then $g(b) = h(b)$, as claimed. $\square$

By Proposition 8.5, we can make the following definition.

Definition 8.6 *Let $f : A \to B$, and suppose that f has both a left inverse and a right inverse. Then we say that f is* invertible; *the (unique) function $f^{-1} : B \to A$ for which $f^{-1} \circ f = id_A$ and $f \circ f^{-1} = id_B$ is called the* inverse *of f.*

According to Proposition 8.5, if f has both a left inverse and a right inverse, then f^{-1} exists and is unique. (Special care needs to be taken with the notation: f^{-1} here stands for the inverse of f and not for its reciprocal!)

Proposition 8.7 *Let $f : A \to B$, and suppose that f is invertible. Then f^{-1} is invertible as well, and*

$$(f^{-1})^{-1} = f.$$

Proof We need to show that f is both a left inverse and a right inverse of f^{-1}. Now for every $b \in B$, we have

$$(f \circ f^{-1})(b) = b$$

since f^{-1} is a right inverse of f; therefore, f is a left inverse of f^{-1}. The proof of the other claim is similar.   □

According to Proposition 8.7, the phrase that f and f^{-1} are "inverses of each other" is legitimate. As an example, we can easily verify that the functions

$$f : \mathbb{R} \to \mathbb{R}$$
$$x \mapsto \sqrt[3]{x}$$

and

$$g : \mathbb{R} \to \mathbb{R}$$
$$x \mapsto x^3$$

are inverses of each other: indeed, for all real numbers x we have $\left(\sqrt[3]{x}\right)^3 = x$ and $\sqrt[3]{x^3} = x$.

Functions—as the first three letters in the word indicate—are the source of much fascination in mathematics; they appear in almost all branches in fun(damental) ways.

Problems

1. Let $f : A \to B$ be a function. For a subset X of A, we define the *image of X under f* as the set $f(X) = \{f(x) \mid x \in X\}$. Let C and D be subsets of A.

   (a) Prove that $f(C \cup D) = f(C) \cup f(D)$.
   (b) Prove that $f(C \cap D) \subseteq f(C) \cap f(D)$.
   (c) Find an explicit example in which equality fails in part (b).

2. For each *piecewise-defined function* $f : \mathbb{Z} \to \mathbb{Z}$ below, decide if the function is injective and/or surjective.

   (a)

   $$f(a) = \begin{cases} a + 2 & \text{when } a \text{ is odd,} \\ 2a + 1 & \text{when } a \text{ is even;} \end{cases}$$

   (b)

   $$f(a) = \begin{cases} a + 3 & \text{when } a \text{ is odd,} \\ 2a + 1 & \text{when } a \text{ is even;} \end{cases}$$

(c)

$$f(a) = \begin{cases} a/2 & \text{when } a \text{ is even,} \\ 3a + 1 & \text{when } a \text{ is odd;} \end{cases}$$

Remark This last example is the source of what many mathematicians consider to be a particularly challenging conjecture (cf. Collatz's Conjecture on page 55). Namely, when composing this function with itself repeatedly, the image of any positive integer seems to become 1 eventually. For example, we have

$$13 \mapsto 40 \mapsto 20 \mapsto 10 \mapsto 5 \mapsto 16 \mapsto 8 \mapsto 4 \mapsto 2 \mapsto 1.$$

Paul Erdős, one of the greatest mathematicians of the twentieth century, commented that "mathematics is just not ready to solve problems like this."

3. For sets X and Y, we let:

- $I_{X \to Y}$ be the statement that there is an injection from X to Y,
- $S_{X \to Y}$ be the statement that there is an surjection from X to Y,
- $B_{X \to Y}$ be the statement that there is a bijection from X to Y.

   Suppose that A and B are nonempty finite sets. For each statement below, decide whether the statement is equivalent to $|A| \leq |B|$, to $|A| \geq |B|$, to $|A| = |B|$, or to none of them.

(a) $I_{A \to B}$
(b) $I_{B \to A}$
(c) $S_{A \to B}$
(d) $S_{B \to A}$
(e) $B_{A \to B}$
(f) $B_{B \to A}$
(g) $I_{A \to B}$ and $I_{B \to A}$
(h) $I_{A \to B}$ and $S_{A \to B}$
(i) $I_{A \to B}$ and $S_{B \to A}$
(j) $I_{B \to A}$ and $S_{A \to B}$
(k) $I_{B \to A}$ and $S_{B \to A}$
(l) $S_{A \to B}$ and $S_{B \to A}$

4. (a) Prove the following proposition.

   Proposition 8.8 *Suppose that* $f : A \to B$, $g : B \to C$, *and* $h = g \circ f$.

   i. If f and g are injective, then so is h.
   ii. If f and g are surjective, then so is h.
   iii. If f and g are bijective, then so is h.
   iv. If f and g are invertible, then so is h, and $h^{-1} = f^{-1} \circ g^{-1}$.

   (b) Suppose again that $f : A \to B$, $g : B \to C$, and $h = g \circ f$. Prove or disprove each of the following statements.

   i. If h is injective, then f is injective.
   ii. If h is injective, then g is injective.
   iii. If h is surjective, then f is surjective.
   iv. If h is surjective, then g is surjective.

5. Prove the following proposition.

   Proposition 8.9 *1. A function has a left inverse if, and only if, it is injective.*
   2. A function has a right inverse if, and only if, it is surjective.
   3. A function is invertible if, and only if, it is bijective.

6. Suppose that A and B are sets with sizes given as follows. Find the number of functions, injections, surjections, and bijections from A to B. (You do not need to give a formal proof for your answers.)

   (a) $|A| = 2$ and $|B| = 5$
   (b) $|A| = 5$ and $|B| = 2$
   (c) $|A| = 5$ and $|B| = 5$
   (d) $|A| = 3$ and $|B| = 5$
   (e) $|A| = 5$ and $|B| = 3$

7. (a) Find a function that has exactly four left inverses.
   (b) Find a function that has exactly four right inverses.
   (Hints: It is best to think of such functions in terms of their diagram descriptions. In both problems, it is helpful to determine the possible sizes of the function's domain and codomain.)

8. Prove that the function $f : \mathbb{N} \times \mathbb{N} \to \mathbb{N}$ defined by

$$f(m, n) = 2^{m-1}(2n - 1)$$

   is a bijection.
   (See Problem 7 of Chapter 3 and Problem 4 of Chapter 6.)

9. (a) Prove that every linear polynomial on $\mathbb{R}$ is a bijection.
   (b) Prove that no quadratic polynomial on $\mathbb{R}$ is a bijection.
   (c) (Note: This problem requires a basic understanding of limits and derivatives.) Find a simple necessary and sufficient condition (in terms of its coefficients) for a given cubic polynomial to be a bijection.

10. Find an explicit bijection from the set of real numbers $\mathbb{R}$ to each of the following sets.

   (a) $(0, 1)$
        (Hints: There are both simple trigonometric and algebraic invertible functions with domain $\mathbb{R}$ and range $(0, 1)$.)
   (b) $[0, 1)$
        (Hints: By part (a) and Problem 4 (a) iii above, it suffices to find a bijection from $(0, 1)$ to $[0, 1)$. Start by mapping $\frac{1}{2}$ to 0.)
   (c) $[0, 1]$

11. For a positive integer n, S_n denotes the set of all bijections on the set
$\{1, 2, \ldots, n\}$. Elements of S_n can be conveniently written in *cycle notation*; for
example, the bijection

i	1	2	3	4	5	6	7	8
$f(i)$	5	8	7	4	2	6	3	1

of S_8 can be given as

$$f = (1 \mapsto 5 \mapsto 2 \mapsto 8 \mapsto 1)(3 \mapsto 7 \mapsto 3)(4 \mapsto 4)(6 \mapsto 6)$$

or, as it is customary, $f = (1528)(37)(4)(6)$.

Here we examine S_5, the set of bijections on $A = \{1, 2, 3, 4, 5\}$.

(a) Let $f = (143)(25) \in S_5$. Find $f \circ f$ and f^{-1}.

(b) How many elements does S_5 have?

(c) For how many elements $f \in S_5$ does $f = f^{-1}$ hold?
(Hint: It may be helpful to observe that $f = f^{-1}$ is equivalent to $f \circ f = id_A$.)

(d) We define the *order* of some $f \in S_5$ to be the smallest positive integer n for
which

$$\underbrace{f \circ f \circ \cdots \circ f}_{n} = id_A$$

(for $n = 1$ the left-hand side is understood to be f).
Find the order of the element f given in part (a).

(e) How many elements of S_5 have order 6?

(f) For each positive integer m, find the number of elements of S_5 whose order
is m.

(g) What is the smallest positive integer n for which

$$\underbrace{f \circ f \circ \cdots \circ f}_{n} = id_A$$

holds for all $f \in S_5$?

(h) Given a function $f \in S_5$, we say that an element $a \in A$ is a *fixed point* of f
if $f(a) = a$. For each positive integer k, find the number of elements of S_5
that have exactly k fixed points.

Remark Elements of S_n with no fixed points are called *derangements*. We
will study derangements in Chapter 19.
(Hints: Clearly, an element of S_5 can only have 0, 1, 2, 3, or 5 fixed points.
Count the elements with 5, 3, 2, and 1 fixed points, and then use part (b) to
find the number of derangements.)

Appendix C
The Foundations of Set Theory

Set theory plays a crucial role in the development of mathematics because all ordinary mathematics can be reduced to set theory. In Chapter 21 we will learn how to construct the usual number systems—$\mathbb{N}$, $\mathbb{Z}$, $\mathbb{Q}$, $\mathbb{R}$, and $\mathbb{C}$—using only set theory. We will *define* the number 0, the natural numbers $1, 2, 3$, etc.; using the natural numbers we will define the integers, and so on. We will even define the basic operations of addition and multiplication—doing all this just relying on the basics of set theory. It is, therefore, important to make sure that the theory of sets is well founded.

Recall that we have listed the notions of a *set* and being an *element* of a set as primitives. We have also seen that not every collection of objects is a set. Thus, a key role of the axiomatic foundation of set theory is to distinguish sets from other collections.

The Zermelo–Fraenkel axioms, denoted by ZF, provide a generally accepted foundation of set theory. This system of axioms, named after Ernst Zermelo (1871–1953) and Abraham Fraenkel (1891–1965), was developed by several mathematicians, including Zermelo, Fraenkel, Thoralf Skolem (1887–1963), and John von Neumann (1903–1957).

The ZF axiom system was first formalized during the beginning of the twentieth century, but even today we do not quite have a universally accepted version. The list of seven axioms below, stated rather informally, is a variation that we find useful.

Axiom C.1 (The Axiom of Extensionality) *If two sets have the same elements, then they are equal.*

Axiom C.2 (The Axiom of Pairing) *Given two sets, there exists a set that contains both of these sets as elements.*

Axiom C.3 (The Axiom of the Union) *The union of a set of sets is a set.*

Axiom C.4 (The Axiom of the Power Set) *The power set of a set is a set.*

© Springer Nature Switzerland AG 2020
B. Bajnok, *An Invitation to Abstract Mathematics*, Undergraduate Texts in
Mathematics, https://doi.org/10.1007/978-3-030-56174-1

Axiom C.5 (The Axiom of Replacement) *If f is a definable mapping and X is a set, then $f(X) = \{f(x) \mid x \in X\}$, called the image of X under f, is a set.*

Axiom C.6 (The Axiom of Regularity) *If $\mathcal{A}$ is a nonempty set of sets, then it contains an element A that is disjoint from $\mathcal{A}$.*

Axiom C.7 (The Axiom of Infinity) *There exists an infinite set.*

Several comments are in order. First, note that the Axiom of Extensionality was stated as a definition in Chapter 6. The meaning of this axiom is that sets do not possess any additional properties beyond their elements. The next four axioms all serve a similar purpose: they list specific rules for how one can build new sets from old ones. We should note that the Axiom of Pairing is not independent from the rest of the axioms (see Problem 2 below), but we chose to state it here as it proves to be one of the most useful properties. Note also that the Axiom of Separation, stated in Chapter 6, is missing; here it is replaced by a more general version, called the Axiom of Replacement. (The Axiom of Separation was on Zermelo's original list of 1908, but, in 1922, by recommendation of Fraenkel, it was replaced by the Axiom of Replacement—no pun intended.) Next, the Axiom of Regularity works in the opposite direction; it forbids us from calling strange collections sets. For example, as a consequence (see Problem 3 (c) below), it implies that no set can contain itself as an element. Finally, the Axiom of Infinity is important; note that it is actually the only axiom on our list that explicitly states that there exists a set.

In addition to the Zermelo–Fraenkel axioms, most mathematicians accept the following axiom regarding so-called *set partitions* (the meaning of this term should be clear, but a precise definition is given in Chapter 16).

Axiom C.8 (The Axiom of Choice) *If Π is a partition of a set X, then X contains a subset A that intersects every member of Π in a single element.*

The name of the axiom refers to the fact that, according to this axiom, one can choose an element from each member of the partition. A particularly enlightening and witty example for the Axiom of Choice was given by Bertrand Russell; we paraphrase Russell's words, as follows. Suppose that a warehouse holds infinitely many distinct pairs of shoes. If we wish to separate all these shoes into two collections so that each of them contains one shoe from each pair, then we can do so easily. However, if the warehouse holds infinitely many distinct pairs of socks, then we cannot perform the separation without the Axiom of Choice! This is because, while with shoes we can just put all left shoes in one collection and all right shoes in the other, socks cannot be distinguished this way, and, therefore, an infinite number of arbitrary choices need to be made. (For finitely many pairs of socks, the Axiom of Choice is not needed.)

It has been shown that the Axiom of Choice is independent from ZF; that is, there is a model of ZF in which the Axiom of Choice is true and there is another model of ZF in which the Axiom of Choice is false. While virtually all mathematicians accept the Axiom of Choice and use it freely, some object to its nonconstructive nature and choose to reject it. One reason for this is the following particularly dramatic consequence of the Axiom of Choice.

Theorem C.9 (The Banach–Tarski Paradox) *It is possible to partition a solid ball of volume 1 cubic foot into five disjoint parts in such a way that these parts can be assembled to form two solid balls, each of volume 1 cubic foot. Here assembling means that the parts are translated and rotated until they fit together without gaps or overlaps.*

Theorem C.9, which is called a paradox only because it seems to defy our intuition, is named after Polish mathematicians Stefan Banach and Alfred Tarski who published this result in 1924. (Their paper relies on earlier work by Giuseppe Vitali and Felix Hausdorff.) Of course, the parts in the theorem are not solid pieces; in particular, they will not have well-defined volumes. The Banach–Tarski Paradox played an influential role in creating the concept of *measurable sets* (see Problem 9 below). In contrast to the Banach–Tarski Paradox, a similar statement in two dimensions cannot be made; as Banach himself showed, a rearrangement of any figure in the plane must have the same area as the original figure. We mention that it is also known that the number of pieces in Theorem C.9 cannot be less than five.

The Banach–Tarski Paradox cannot hold in two-dimensional space; however, when two regions have the same area, strange phenomena may still occur. Consider the following striking result of the Hungarian mathematician Miklós Laczkovich, published in 1989.

Theorem C.10 ("Squaring the Circle") *It is possible to partition a solid disk of area 1 square foot into a finite number of disjoint parts in such a way that these parts can be assembled to form a square of area 1 square foot. Here assembling means that the parts are translated (no rotation is necessary) so that they fit together without gaps or overlaps.*

With paradoxes such as the Banach–Tarski Paradox explained, most of mathematics today is built on the axiom system of the Zermelo–Fraenkel axioms combined with the Axiom of Choice (denoted by ZFC).

Problems and Assignments

1. Use the Axiom of Replacement to prove the Axiom of Separation.
2. Use the Axiom of Replacement and the Axiom of the Power Set to prove the Axiom of Pairing.
3. Use the Axiom of Separation and the other ZF axioms to prove the following statements.

   (a) The empty-set is a set.
   (b) If A is a set, then $\{A\}$ is also a set.
   (c) If A is a set, then $A \notin A$.

    (d) The intersection of a set of sets is a set.

    (e) The Cartesian product of two sets is a set.

    (f) The Cartesian product of an arbitrary number of sets is a set.

4. Suppose that A and B are sets.

    (a) Prove that $A \in B$ and $B \in A$ cannot happen simultaneously. (Thus, the relation $\in$ is asymmetrical.)

    (b) Prove that $A \cup \{A\} = B \cup \{B\}$ can only happen if $A = B$.

5. There are many statements that are equivalent to the Axiom of Choice; review some of these. (We discuss several of them in this book, including the Well-Ordering Theorem, cf. Theorem 17.10; the fact that the direct product of nonempty sets is nonempty, cf. Theorem 16.10; and the trichotomy of cardinals, cf. Theorem 20.23.)

6. Review some of the literature on the Banach–Tarski Paradox. A good place to start is a book of the same title by Stan Wagon (Cambridge University Press, Cambridge, UK, 1993).

7. Explain why the Banach–Tarski Paradox cannot hold in two dimensions.

8. What can one say about subsets A and B of the plane for which it is possible to partition A into countably many parts so that these parts can then be reassembled (without gaps or overlaps) to form B?

9. Explain the basics of *measure theory*, and find some explicit examples for non-measurable sets.

10. Another pillar that the foundations of mathematics rests on is *mathematical logic*; in fact, set theory is often viewed as part of mathematical logic. Review some of the concepts and results of mathematical logic.

Part III
How to Prove It

Chapter 9
Universal Proofs

We have already discussed the role of proofs in mathematics and have seen a few examples of proofs. Having learned about some of the fundamental topics of mathematics, such as logic, sets, quantifiers, and functions, we are now able to study proofs more formally and thus deepen our understanding of them.

Most mathematical statements involve quantifiers: the universal quantifier, by which we claim that something holds for all elements of a certain set, or the existential quantifier, which asserts that a statement holds for some elements. (Of course, statements may involve a combination of quantifiers as well.) In this chapter, as the title indicates, we study the proofs of universally quantified statements. Our title also has a less literal meaning: we will discuss some very general proof techniques—many of these even have names—that appear frequently in proofs. But, needless to say, we cannot possibly provide a single "universal proof" for everything; in fact, the creative process of finding proofs for statements is exactly what keeps mathematicians challenged.

Consider a statement of the form

$$\forall a \in U, P(a),$$

where U is some set and P is a predicate defined on the elements of U. To prove such a statement, we need to show that the predicate $P(a)$ becomes a true statement *for every* element a under consideration. Similarly, we may be asked to prove statements involving two or more variables, such as

$$\forall a, b \in U, P(a, b).$$

While a single counterexample would disprove such statements, constructing a proof usually involves a comprehensive argument. As we explained in Chapter 4, a proof can only contain statements that are listed in our hypotheses, are axioms, have been proven already, or follow immediately from the previous statements. Usually

© Springer Nature Switzerland AG 2020

B. Bajnok, *An Invitation to Abstract Mathematics*, Undergraduate Texts in
Mathematics, https://doi.org/10.1007/978-3-030-56174-1_9

we do not intend to trace all our theorems back to the axioms, this would be too tedious and time consuming. (We will demonstrate this method later in the book.) Nor are we attempting to make our proofs formal; in particular, we will not always be precise about how one statement in a proof follows from the one(s) before it. Our goal here, instead, is to discuss the most important proof structures and to learn some of the commonly used techniques.

We have already seen two simple logical structures in Chapter 5, the Laws of Modus Ponens and Modus Tollens:

- *Law of Modus Ponens*: For arbitrary statements P and Q, the statement

$$[P \wedge (P \Rightarrow Q)] \Rightarrow Q$$

is a true statement;
- *Law of Modus Tollens*: For arbitrary statements P and Q, the statement

$$[\neg Q \wedge (P \Rightarrow Q)] \Rightarrow \neg P$$

is a true statement.

These two laws are indeed quite obvious and are used without much thought. We will soon learn some more complicated techniques, but first let us see some examples.

Proposition 9.1 *For every integer n, the number $n^5 - n^3$ is divisible by 8.*

Proof We will separate two cases: when n is even and when n is odd.

Case 1. If n is even, then there is a $k \in \mathbb{Z}$ for which $n = 2k$; so $n^5 - n^3$ becomes

$$n^5 - n^3 = n^3(n^2 - 1) = 8k^3(4k^2 - 1) = 8 \cdot (4k^5 - k^3).$$

Since $4k^5 - k^3 \in \mathbb{Z}$, $n^5 - n^3$ is divisible by 8.

Case 2. If n is odd, then there is a $k \in \mathbb{Z}$ for which $n = 2k + 1$; in this case $n^5 - n^3$ becomes

$$n^5 - n^3 = n^3(n^2 - 1) = (2k + 1)^3(4k^2 + 4k) = 4k(k + 1)(2k + 1)^3.$$

We have two subcases: k is even and k is odd.

Case (i). If k is even, say $k = 2m$ for some $m \in \mathbb{Z}$, then $4k = 8m$ is divisible by 8, and, therefore, $4k \cdot (k + 1)(2k + 1)^3$ is divisible by 8.

Case (ii). If k is odd, say $k = 2m + 1$ for some $m \in \mathbb{Z}$, then $4(k + 1) = 8(m + 1)$ is divisible by 8, and, therefore, $4(k + 1) \cdot k(2k + 1)^3$ is divisible by 8.

Thus $n^5 - n^3$ is divisible by 8 for every integer n.   □

The example we present next could be established via truth tables or Venn diagrams, but here we present a more natural argument for it.

Proposition 9.2 *Let A, B, and C be sets inside a universal set U. Then*

$$(A \cup B) \cap \overline{C} \subseteq (A \cap \overline{B \cup C}) \cup (B \cap \overline{A \cap C}).$$

Proof We need to prove that if

$$x \in (A \cup B) \cap \overline{C},$$

then

$$x \in (A \cap \overline{B \cup C}) \cup (B \cap \overline{A \cap C}).$$

Suppose that

$$x \in (A \cup B) \cap \overline{C}.$$

Then $x \in A \cup B$ and $x \in \overline{C}$. Let us consider two cases: when $x \in B$ and when $x \notin B$.

Case 1. Assume first that $x \in B$. Since $x \in \overline{C}$, we have $x \notin C$, so we must also have $x \notin A \cap C$ and, therefore, $x \in \overline{A \cap C}$. But $x \in B$ as well, so $x \in B \cap \overline{A \cap C}$, and, therefore,

$$x \in (A \cap \overline{B \cup C}) \cup (B \cap \overline{A \cap C}).$$

Case 2. Assume now that $x \notin B$. Since $x \in A \cup B$, we have $x \in A$ or $x \in B$. But now $x \notin B$, so we must have $x \in A$. We also know that $x \in \overline{C}$ and thus $x \notin C$. Since $x \notin B$ and $x \notin C$, we get $x \notin B \cup C$, and so $x \in \overline{B \cup C}$. But $x \in A$ and $x \in \overline{B \cup C}$ imply that $x \in A \cap \overline{B \cup C}$, so

$$x \in (A \cap \overline{B \cup C}) \cup (B \cap \overline{A \cap C})$$

holds again. □

Observe that both of our proofs above consisted of two cases, and we reached our conclusions in these cases separately. This proof technique can be abstracted as follows:

- *Law of Case Separation*: For arbitrary statements P, Q, and R, the statement

$$[(P \vee Q) \Rightarrow R] \Leftrightarrow [(P \Rightarrow R) \wedge (Q \Rightarrow R)]$$

is true.

The Law of Case Separation is a common technique as it allows us to reach our conclusion (namely, that R holds) in separate cases (when P holds and when Q holds). It is worth noting that, even though our two cases were not overlapping in either proof above, this is not a necessary requirement.

A similar situation occurs when the disjunction appears in the conclusion rather than in the hypothesis. For that, we can employ the following:

- *Law of Case Exclusion*: For arbitrary statements P, Q, and R, the statement

$$[P \Rightarrow (Q \vee R)] \Leftrightarrow [(P \wedge \neg Q) \Rightarrow R]$$

is true.

The Law of Case Exclusion enables us to prove the disjunction of Q and R by assuming that one of them, say Q, is false and concluding that the other, in this case R, must then be true.

We mention two other frequently used techniques:

- *Law of Equivalence*: For arbitrary statements P and Q, the statement

$$(P \Leftrightarrow Q) \Leftrightarrow [(P \Rightarrow Q) \wedge (Q \Rightarrow P)]$$

is true;
- *Law of Transitivity*: For arbitrary statements P, Q, and R, the statement

$$[(P \Rightarrow Q) \wedge (Q \Rightarrow R)] \Rightarrow (P \Rightarrow R)$$

is true.

By the Law of Equivalence, the fact that two statements are equivalent may be established by proving that each one implies the other; the Law of Transitivity simply allows us to carry our conclusions farther. We can verify each of these laws by considering appropriate truth tables—or, more abstractly, by using the axioms governing statements; cf. Chapter 13—but most people with a bit of experience in logic find these laws quite obvious.

In our next example we revisit Problem 8 (a) of Chapter 6 and provide a proof in which several of these laws come into play.

Proposition 9.3 *Suppose that A and B are subsets of a given set U. Then*

$$P(A \cup B) = P(A) \cup P(B)$$

holds if, and only if, $A \subseteq B$ or $B \subseteq A$.

Proof First, the Law of Equivalence tells us how to reduce our proposition to two implications, namely:

(1) if $A \subseteq B$ or $B \subseteq A$, then $P(A \cup B) = P(A) \cup P(B)$, and
(2) if $P(A \cup B) = P(A) \cup P(B)$, then $A \subseteq B$ or $B \subseteq A$.

To prove our proposition, we need to prove both of these implications. According to the Law of Case Separation, the first implication can be proved in two parts:

(1.i) if $A \subseteq B$, then $P(A \cup B) = P(A) \cup P(B)$, and
(1.ii) if $B \subseteq A$, then $P(A \cup B) = P(A) \cup P(B)$.

To prove (1.i), note that $A \subseteq B$ implies that $A \cup B = B$ and $P(A) \subseteq P(B)$. Therefore, $P(A \cup B) = P(A) \cup P(B)$ as both sides equal $P(B)$. The claim (1.ii) can be proved similarly.

Turning to (2), we apply the Law of Case Exclusion and restate our claim as

(2') if $P(A \cup B) = P(A) \cup P(B)$ and $A \not\subseteq B$, then $B \subseteq A$.

To prove (2'), note that, since $A \not\subseteq B$, we have an element $a \in A$ so that $a \notin B$. Consider the set $\{a\} \cup B$. Since $\{a\} \cup B$ is a subset of $A \cup B$, it must be an element of $P(A \cup B)$. But our hypothesis is that

$$P(A \cup B) = P(A) \cup P(B),$$

so $\{a\} \cup B$ is an element of $P(A) \cup P(B)$ and thus an element of either $P(A)$ or $P(B)$. Now $\{a\} \cup B$ cannot be an element of $P(B)$, since that would mean that $\{a\} \cup B$ is a subset of B, which is not true since $a \notin B$. Therefore, $\{a\} \cup B$ is an element of $P(A)$, which means that $\{a\} \cup B$ is a subset of A. But then we have

$$B \subseteq \{a\} \cup B \subseteq A,$$

and so $B \subseteq A$, as claimed. Our proof is now complete.   □

We should note that, with a method that we describe below, we will be able to find a substantially simpler proof for Proposition 9.3 in Problem 7. This is also a good time to point out that, if the two cases we are using in the Law of Case Separation are completely symmetrical, as (1.i) and (1.ii) were above (switching A and B turns one statement into the other), then it is enough to prove one of the cases—but, of course, this needs to be accomplished without relying on the other. In such situations it is customary to assume that one of the cases holds *without loss of generality* (or *wlog*). For example, in our proof we could have written "without loss of generality, assume that $A \subseteq B$." But care needs to be taken to assure that the two cases are indeed similar; if this is not the case, we need to treat both cases separately. ("Wlog" does *not* stand for "without lots (!) of generality.")

Notice that several of our arguments involved the negations of statements. Proof techniques that employ negations in a fundamental way are called *indirect* proofs. We used this method as early as in Theorem 4.7 when proving that if $2^n - 1$ is a prime number for some positive integer n, then n must be prime as well: Recall that we established this claim by proving that n could be neither 1 nor a composite number.

The two most common logical structures for indirect proofs are as follows:

• *Law of Contraposition*: For arbitrary statements P and Q, the statement

$$(P \Rightarrow Q) \Leftrightarrow (\neg Q \Rightarrow \neg P)$$

is true;

- *Law of Contradiction*: For arbitrary statements P and Q and for an arbitrary false statement F, the statement

$$(P \Rightarrow Q) \Leftrightarrow [(P \wedge \neg Q) \Rightarrow F]$$

is true.

The Law of Contraposition says that, instead of proving that "P implies Q," we may prove the equivalent *contrapositive*: "If Q is false then P has to be false as well." The contrapositive of a statement is not to be confused with the converse of the statement: The *converse* of "If P then Q" is the statement "If Q then P." While the contrapositive of a statement is equivalent to the statement, its converse is not (cf. Problem 1 of Chapter 5).

The Law of Contradiction lets us prove that "P implies Q" by proving instead that if P is true and Q is false, then we can find a false statement F. It allows us a bit more freedom than the Law of Contraposition; we may choose any false statement F (in our system)!

The following proposition allows us to compare three proof methods: direct proof, proof by contraposition, and proof by contradiction.

Proposition 9.4 *The sum of a positive number and its reciprocal is at least 2.*

I: Direct Proof Let x be an arbitrary positive real number. Then we have

$$x + \frac{1}{x} = \frac{x^2 + 1}{x} = \frac{(x-1)^2 + 2x}{x} = \frac{1}{x} \cdot (x-1)^2 + 2.$$

Since the first term is at least zero, we have

$$x + \frac{1}{x} \geq 2,$$

as claimed.   □

II: Proof by Contraposition Let us assume that x is an arbitrary real number for which

$$x + \frac{1}{x} < 2.$$

Note that, since its reciprocal is defined, x is not zero, hence x^2 is positive. Therefore, multiplying the inequality by x^2 will yield

$$x^3 + x < 2x^2,$$

or, equivalently,

$$x^3 - 2x^2 + x < 0,$$

or

$$x(x-1)^2 < 0.$$

This last inequality implies that $x \neq 1$, hence $1/(x-1)$ is a real number; since it's nonzero, its square, $1/(x-1)^2$ must be positive. Therefore, multiplying

$$x(x-1)^2 < 0$$

by $1/(x-1)^2$, we get $x < 0$, the negation of our hypothesis. □

III: Proof by Contradiction Let us assume that x is an arbitrary positive real number. Assume indirectly that

$$x + \frac{1}{x} < 2.$$

Since $x > 0$, multiplying by x we get

$$x^2 + 1 < 2x,$$

which can be rewritten as

$$(x-1)^2 < 0,$$

which is a contradiction. □

Note that, between the two indirect proofs, the proof by contradiction is simpler. This is perhaps no surprise; the form $(P \wedge \neg Q) \Rightarrow F$ lets us assume two statements, P and $\neg Q$, and allows us to derive any false statement, while the form $\neg Q \Rightarrow \neg P$ used in a proof by contraposition has only one hypothesis and a more restrictive conclusion. Thus, in a certain sense, proofs by contradiction allow us more to work with and more freedom than proofs by contraposition.

Our last proof of this chapter illustrates the astonishing power of indirect proofs and establishes one of the most unforeseen results in mathematics. During the fifth century BCE came the shocking discovery that not every number can be written as a fraction of two integers, as was believed by the Greeks of the time. In particular, the diagonal of a square is *incommensurable* with its sides: there is no unit length (no matter how small) such that both the side and the diagonal of the square have lengths that are integer multiples of this unit length. Applied to the square with side length 1 and using today's terminology, we can say that $\sqrt{2}$ is not a rational number. This was quite a setback in ancient Greece, where irrational numbers were not accepted. The proper theory of real numbers, including both rationals and irrationals, was not fully developed until the nineteenth century. We will discuss this theory in Chapter 21; in particular, we will prove that $\sqrt{2}$—defined as the unique positive real number whose square equals 2 (cf. Problem 1 of Chapter 2)—indeed exists.

Taking the existence of $\sqrt{2}$ now for granted, we may state our claim as follows:

Theorem 9.5 *The number $\sqrt{2}$ is irrational.*

Proof Suppose, indirectly, that $\sqrt{2}$ is a rational number, so there are integers a and b for which $\sqrt{2} = \frac{a}{b}$. Let d be the greatest common divisor of a and b. Then there are integers a' and b' for which $a = da'$, $b = db'$, and a' and b' are relatively prime. Furthermore, we have $\sqrt{2} = \frac{a'}{b'}$.

Now squaring both sides and multiplying by $(b')^2$ yields $2(b')^2 = (a')^2$, and so $(a')^2 = a' \cdot a'$ is divisible by 2. By Euclid's Principle (cf. page 14), since 2 is a prime number, we get that a' is divisible by 2 as well. Therefore, there exists an integer $\hat{a}$ for which $a' = 2\hat{a}$. Substituting this into $2(b')^2 = (a')^2$ and dividing by 2 yields $(b')^2 = 2(\hat{a})^2$. This means that $(b')^2$ is even from which, as above, we can conclude that b' is even; thus there exists an integer $\hat{b}$ for which $b' = 2\hat{b}$. Therefore, we now get that a' and b' have a common factor of 2, contradicting that they are relatively prime. Hence $\sqrt{2}$ cannot be a rational number.   □

We have now seen a variety of proof techniques: direct proofs, indirect proofs; the Laws of Case Separation, Case Exclusion, and Equivalence; and several variations of these—in subsequent chapters, we will see quite a few others. Naturally, one cannot give a complete list of all proof methods and techniques; in fact, the same statement may often be proved via entirely different methods, can be presented giving different emphases, or may be explained providing various levels of detail. As we have already discussed in Chapter 4, there are many different ways to write a correct proof, depending, for example, on our audience and our emphasis.

Nevertheless, some proofs are generally accepted as more beautiful than others. What exactly makes a proof beautiful is hard to say—yet most mathematicians would recognize it when they see it. There is a wonderful book entitled *Proofs from THE BOOK* (by Martin Aigner, Günter Ziegler, and K. H. Hofmann. Springer, Berlin, 6th ed. 2018) that attempts to collect the most beautiful proofs; the title refers to the notion of the great late Hungarian mathematician Paul Erdős who maintained that God kept a book containing the single best proof of every mathematical statement and that the task of mathematicians was to discover these proofs.

While our repertoire of proof techniques is now quite extensive, it does *not*, however, include the proof technique of convenience ("it would be very nice if it were true"), imagination ("let's pretend it's true"), plausibility ("it sounds good, so it must be true"), or profanity (example omitted). These and other humorous "proof techniques" are listed in the October 1998 issue of the *Mathematics Teacher* magazine.

Problems

1. The following examples come from the English writer and mathematician Lewis Carroll (1832–1898), author of *Alice's Adventures in Wonderland*. Justify that the arguments are correct.

   (a) • Babies are illogical.
       • Nobody is despised who can manage a crocodile.
       • Illogical persons are despised.
       Therefore, babies cannot manage crocodiles.
   (b) • No ducks waltz.
       • No officers ever decline to waltz.
       • All my poultry are ducks.
       Therefore, my poultry are not officers.

2. (a) Suppose that n denotes an arbitrary integer and $P(n)$ and $Q(n)$ are predicates that become statements for each value of n. Put the following claims in *equivalence classes*: Two claims should be in the same class if, and only if, they are equivalent.
   Claim 1. $\forall n \in \mathbb{Z}, P(n) \Rightarrow Q(n)$.
   Claim 2. $\forall n \in \mathbb{Z}, \neg P(n) \Rightarrow \neg Q(n)$.
   Claim 3. $\forall n \in \mathbb{Z}, Q(n) \Rightarrow P(n)$.
   Claim 4. $\forall n \in \mathbb{Z}, \neg Q(n) \Rightarrow \neg P(n)$.
   Claim 5. $\forall n \in \mathbb{Z}, P(n) \Leftrightarrow Q(n)$.
   Claim 6. $\forall n \in \mathbb{Z}, \neg P(n) \Leftrightarrow \neg Q(n)$.

   (b) Suppose that $P(n)$ is the predicate that "$3|n$" and $Q(n)$ is the predicate that "$3|n^2$." For each of the following arguments, decide if it provides a proof for any of the claims in the previous part.

      i. *Argument A.* Suppose that n is divisible by 3. Then n is of the form $n = 3k$ for some integer k and, therefore, $n^2 = 9k^2 = 3 \cdot (3k^2)$, a number that is clearly divisible by 3.
      ii. *Argument B.* Suppose that n is of the form $n = 3k + 1$ for some integer k. Then $n^2 = 9k^2 + 6k + 1 = 3 \cdot (3k^2 + 2k) + 1$, a number that is clearly not divisible by 3.
      iii. *Argument C.* Suppose that n is of the form $n = 3k - 1$ for some integer k. Then $n^2 = 9k^2 - 6k + 1 = 3 \cdot (3k^2 - 2k) + 1$, a number that is clearly not divisible by 3.
      iv. *Argument D.* Suppose that n is not divisible by 3. Then n is either of the form $n = 3k + 1$ or of the form $n = 3k - 1$ for some integer k. In the first case, follow Argument B; in the second case, follow Argument C.
      v. *Argument E.* If n is divisible by 3, follow Argument A; otherwise, follow Argument D.
      vi. *Argument F.* If $n = 3$, then $n^2 = 9$, which is divisible by 3.

    vii. *Argument G.* If $n^2 = 9$, then $n = 3$ or $n = -3$, both of which are
        divisible by 3.

3. (a) Prove that the product of any two consecutive integers is even.
  (b) Prove that the product of any three consecutive integers is divisible by 6.
    (Hints: First prove that if an integer is divisible by both 2 and 3, then it is
    divisible by 6. Then use part (a).)
  (c) Prove that the product of any four consecutive integers is divisible by 24.
    (Hints: Note that a method similar to the one suggested for part (b) does
    not work here, since an integer divisible by both 6 and 4 is not necessarily
    divisible by 24. Instead, follow the approach that builds on the fact—and
    prove this fact—that divisibility by 3 and 8 implies divisibility by 24.)
  (d) Prove that the product of any five consecutive integers is divisible by 120.

Remark We will prove a generalization of these statements in Chapter 11.

4. Prove that for every integer n at least one of the following statements holds:

- n is even,
- n is divisible by 3,
- $n^2 - 1$ is divisible by 24.

(Hints: First verify and then use the fact that for all statements P, Q, and R,

$$(P \vee Q \vee R) \Leftrightarrow [(\neg P \wedge \neg Q) \Rightarrow R]$$

holds. What can an integer be congruent to mod 6 if it is odd and not divisible
by 3?)
5. Suppose that A, B, and C are arbitrary sets. Prove that

$$(A \cup \overline{B} \cup C) \cap \overline{(A \cup C)} = \overline{A} \cap \overline{B} \cap \overline{C}.$$

(Hint: Follow the approach used in the proof of Proposition 9.2, but note that
here we claim equality.)
6. Provide a proof like the one given for Proposition 9.2 for the two identities of
Problem 7 of Chapter 6.

Remark We will provide yet another proof for these identities in Problem 8 of
Chapter 14.

7. In the proof of Proposition 9.3, we used the Law of Case Exclusion to verify
statement (2). Use the Law of Contradiction to simplify the proof.
(Hints: Assume that $A \not\subseteq B$ and $B \not\subseteq A$, then consider the set $\{a, b\}$ with
$a \in A \setminus B$ and $b \in B \setminus A$.)
8. Prove that the following numbers are irrational. You may use, without proof,
Euclid's Principle stated on page 14.

(a) $\sqrt{3}$

(b) $\sqrt{6}$

(c) $\sqrt{12}$

(d) $\sqrt{2} + \sqrt{3}$

(e) $\sqrt[3]{2}$

(Hint: Be careful when using Euclid's Principle as it applies only to the product of *two* integers.)

(f) $\log_2(3)$

(Hint: Recall that $\log_2(3) = a$ means that $2^a = 3$.)

9. Consider the equation $2x^3 + 6x + 1 = 0$.

(a) Prove that the equation has no integer solutions.
(Hint: While this statement is, of course, an immediate corollary to part (b), it also has a simple one-sentence proof.)

(b) Prove that the equation has no rational number solutions.
(Hint: Assume, indirectly, that the equation has a rational number solution, and get a contradiction as in Problem 8 above.)

(c) We know that the equation has a unique real number solution (as well as two non-real solutions). According to Cardano's Formula, this solution is

$$x = \sqrt[3]{\frac{\sqrt{17}-1}{4}} - \sqrt[3]{\frac{\sqrt{17}+1}{4}}.$$

Verify that this number indeed satisfies the given equation.
(Hint: The computation can be greatly simplified by first proving and then using the fact that the two terms in the expression are reciprocals of one another.)

10. Prove or disprove each of the following statements. You may use the axioms that the set of rational numbers are closed for addition and multiplication; that is, if $x \in \mathbb{Q}$ and $y \in \mathbb{Q}$, then $x + y \in \mathbb{Q}$ and $x \cdot y \in \mathbb{Q}$.

(a) If $x \in \mathbb{Q}$ and $y \in \mathbb{R} \setminus \mathbb{Q}$, then $x + y \in \mathbb{R} \setminus \mathbb{Q}$.

(b) If $x \in \mathbb{R} \setminus \mathbb{Q}$ and $y \in \mathbb{R} \setminus \mathbb{Q}$, then $x + y \in \mathbb{R} \setminus \mathbb{Q}$.

(c) If $x \in \mathbb{R} \setminus \mathbb{Q}$ and $y \in \mathbb{R}$, then $x + y \in \mathbb{R} \setminus \mathbb{Q}$ or $x - y \in \mathbb{R} \setminus \mathbb{Q}$.

(d) If $x \in \mathbb{Q}$ and $y \in \mathbb{R} \setminus \mathbb{Q}$, then $x \cdot y \in \mathbb{R} \setminus \mathbb{Q}$.

(e) If $x \in \mathbb{R} \setminus \mathbb{Q}$ and $y \in \mathbb{R} \setminus \mathbb{Q}$, then $x \cdot y \in \mathbb{R} \setminus \mathbb{Q}$.

(f) If $x \in \mathbb{R} \setminus \mathbb{Q}$ and $y \in \mathbb{R} \setminus \mathbb{Q}$, then $x^y \in \mathbb{R} \setminus \mathbb{Q}$.

(Hint: Use Theorem 9.5 and Problem 8 (f) above.)

Remarks By a highly nontrivial proof—certainly beyond our scope here—one can establish that the number $\sqrt{2}^{\sqrt{2}}$ is irrational. However, it is interesting to note that, even without relying on this result, one can easily see that either $\sqrt{2}^{\sqrt{2}}$ or $\left(\sqrt{2}^{\sqrt{2}}\right)^{\sqrt{2}}$ provides a counterexample for (f) above.

Indeed, if $\sqrt{2}^{\sqrt{2}} \in \mathbb{Q}$, then we can set $x = y = \sqrt{2}$, and if $\sqrt{2}^{\sqrt{2}} \notin \mathbb{Q}$, then $x = \sqrt{2}^{\sqrt{2}}$ and $y = \sqrt{2}$ provide a counterexample. Following the hint above, we can construct a counterexample that avoids this ambiguity.

11. Prove the following:

Theorem 9.6 (The Triangle Inequality) *For arbitrary real numbers x and y we have*

$$|x + y| \leq |x| + |y|.$$

(Hint: Recall the definition of $|x|$ from Problem 1 (m) of Chapter 2.)

Remark The theorem is named after a generalization: Given a triangle ABC in $\mathbb{R}^2$ (or higher dimensions), if $\mathbf{x}$ and $\mathbf{y}$ denote the vectors pointing from A to B and from B to C, respectively, then the vector pointing from A to C is given by $\mathbf{x} + \mathbf{y}$. The inequality above expresses the fact that no side of a triangle can be longer than the sum of the lengths of the other two sides.

12. We started this problem set with two of Lewis Carroll's puzzles; we end with a third one. Do the nine hypotheses imply the conclusion?

- All, who neither dance on tight-ropes nor eat penny-buns, are old.
- Pigs that are liable to giddiness are treated with respect.
- A wise balloonist takes an umbrella with him.
- No one ought to lunch in public who looks ridiculous and eats penny-buns.
- Young creatures who go up in balloons are liable to giddiness.
- Fat creatures who look ridiculous may lunch in public if they do not dance on tight-ropes.
- No wise creatures dance on tight-ropes if they are liable to giddiness.
- A pig looks ridiculous carrying an umbrella.
- All, who do not dance on tight-ropes and who are treated with respect, are fat.

Therefore, a wise young pig will not become a balloonist.
(Hint: Argue indirectly. Make the assumption—that, apparently, Carroll did too—that an old creature is not young.)

Chapter 10
The Domino Effect

In Chapter 9 we studied universal statements of the form

$$\forall a \in U, \; P(a)$$

for given sets U and predicates P. Here we continue this discussion by examining the case in which U is the set of natural numbers.

Recall from Chapter 5 the Law of Modus Ponens, one of the most commonly used steps in writing proofs: This law says that if we know that the statements P and $P \Rightarrow Q$ are true, then we can conclude that statement Q is also true.

Suppose now that we need to prove not one but infinitely many statements, and that these statements can be arranged in a sequence: $P(1)$, $P(2)$, $P(3)$, etc. If we know that $P(1)$ and $P(1) \Rightarrow P(2)$ are true, then Modus Ponens guarantees that $P(2)$ is true. If we suppose further that we also know that $P(2) \Rightarrow P(3)$ is true, then again by Modus Ponens we conclude that $P(3)$ is true. Suppose now that we can carry on this argument indefinitely, that is, we know that $P(k) \Rightarrow P(k+1)$ is true for every positive integer k. Does this imply that all the statements on our list are true?

This method of reasoning is called mathematical induction, and is one of the most often-used proof techniques in mathematics. It can be stated formally as follows:

Theorem 10.1 (The Principle of Mathematical Induction) *Suppose that $P(n)$ is a predicate that becomes a statement for all $n \in \mathbb{N}$. If*

- *$P(1)$ and*
- *$\forall k \in \mathbb{N}, \; P(k) \Rightarrow P(k+1)$*

both hold, then $P(n)$ is true for every positive integer n.

So, when using mathematical induction to prove the statement $P(n)$ for every $n \in \mathbb{N}$, we first verify that $P(1)$ is true; this is called the *base step*. We then prove that, for every positive integer k, $P(k)$ implies $P(k+1)$; this is referred to as the *inductive*

© Springer Nature Switzerland AG 2020

B. Bajnok, *An Invitation to Abstract Mathematics*, Undergraduate Texts in Mathematics, https://doi.org/10.1007/978-3-030-56174-1_10

step, consisting of the *inductive assumption* $P(k)$ and the *inductive conclusion* $P(k + 1)$. The method of mathematical induction can be imaginatively described: If our infinitely many dominoes are lined up in a queue, then toppling the first one will topple all of them!

The Principle of Induction seems quite self-evident; indeed, it follows immediately from the following axiom. We call a subset of the positive integers *inductive* if it has the property that, whenever it contains an element k, it also contains $k + 1$.

Axiom 10.2 (The Induction Axiom) *If S is an inductive subset of $\mathbb{N}$ and $1 \in S$, then $S = \mathbb{N}$.*

We accept this statement without proof. We will revisit the Induction Axiom— and the other axioms of the natural number system—in Chapter 21.

Let us now see some well-known examples for proofs by induction. Recall that we discussed recursively defined sequences in Chapter 2. For example, in Problem 8 (a) (ii), we saw that the sequence

$$41, 43, 47, 53, 61, 71, 83, 97, 113, 131, 151, 173, 197, 223, 251, 281, 313, \ldots$$

can be defined recursively by $a_1 = 41$ and $a_{k+1} = a_k + 2k$ for $k \geq 1$; the Principle of Induction guarantees that the recursive formula with the initial value uniquely determines all values in the sequence. Later, in Problem 2 of Chapter 3, we stated an explicit formula for the sequence; we now use induction to prove this formula. (This sequence, discovered by Euler, yields a prime number value for every positive integer up to $k = 40$; see discussion in Chapter 3.)

Proposition 10.3 *Let us define the sequence $\mathbf{a} = (a_1, a_2, \ldots)$ recursively by $a_1 = 41$ and $a_{k+1} = a_k + 2k$ for $k \geq 1$. Then $a_n = n^2 - n + 41$ for all $n \in \mathbb{N}$.*

Proof We use induction. First we need to check that the formula holds for $n = 1$: Indeed, $a_1 = 41 = 1^2 - 1 + 41$.

Next we need to show that, for any given value of $k \in \mathbb{N}$, if $a_k = k^2 - k + 41$, then $a_{k+1} = (k+1)^2 - (k+1) + 41$. To see that this is so, we start with $a_k = k^2 - k + 41$ and add $2k$ to both sides. The left-hand side is then $a_k + 2k = a_{k+1}$ by the definition of the sequence, and the right-hand side becomes

$$(k^2 - k + 41) + 2k = (k + 1)^2 - (k + 1) + 41,$$

which proves our claim.   □

One of the most well-known examples for an inductive proof is the following:

Proposition 10.4 *For all positive integers n we have*

$$\sum_{i=1}^{n} i = \frac{n(n + 1)}{2}.$$

Thus, the sum of the first n positive integers equals $n(n + 1)/2$. (Recall from Problem 3 (a) of Chapter 9 that the quantity $n(n + 1)/2$ is an integer; cf. also Theorem 11.13.)

Proof We use induction on n. For $n = 1$ both sides of our equation equal 1, hence the claim holds.

Suppose now that

$$1 + 2 + \cdots + k = \frac{k(k + 1)}{2}$$

for some positive integer k. We need to prove that

$$1 + 2 + \cdots + k + (k + 1) = \frac{(k + 1)(k + 2)}{2}.$$

We can use our inductive assumption to rewrite the sum of the first k terms; this yields

$$1 + 2 + \cdots + k + (k + 1) = \frac{k(k + 1)}{2} + (k + 1).$$

The latter expression clearly equals

$$\frac{(k + 1)(k + 2)}{2}$$

as claimed. □

Induction is a very powerful tool to prove that a formula holds for all values of a variable. A drawback to this proof technique is that it only works if one already knows (or at least conjectures) the formula to be verified. But how does one come up with the right formula in the first place?

The answer to this question is not at all easy in general, but we can shed some light on it at least in the case of Proposition 10.4. We will demonstrate this—without *lots* of generality—for $n = 5$. Consider Figure 10.1.

The diagram contains 5 rows and 6 columns and, therefore, a total of $5 \cdot 6$ symbols: some open circles and some full circles. Clearly, exactly half of the symbols are open circles and half are full circles; in particular, the number of open circles is $(5 \cdot 6)/2$. Now looking at only the open circles in the diagram and counting them in each row, we see that we have $1 + 2 + 3 + 4 + 5$ open circles. This yields the identity

$$1 + 2 + 3 + 4 + 5 = (5 \cdot 6)/2;$$

the statement of Proposition 10.4 can then be easily arrived at as a simple generalization.

Fig. 10.1
$1 + 2 + 3 + 4 + 5 = (5 \cdot 6)/2$

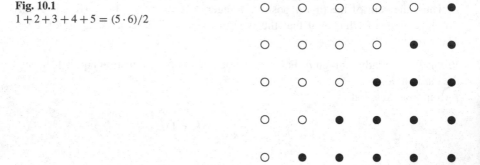

According to a charming anecdote, the German mathematician Carl Friedrich Gauss (1777–1855) discovered this argument as a young child. At school one day, to keep the class quietly occupied, his teacher asked the pupils to add up the whole numbers from 1 to 100. The young Gauss was done almost immediately, as he discovered that the sum equaled $(100 \cdot 101)/2$, or $5,050$. (After this episode, his teacher was, rightfully, so impressed with the young prodigy that he called Gauss to the attention of the Duke of Brunswick, whose continued financial assistance helped Gauss become one of the greatest mathematicians of all time.)

With our next example we fulfill a promise made in Chapter 4 and prove Lemma 4.2. We repeat the statement here, this time as a proposition.

Proposition 10.5 *If a and b are arbitrary real numbers and n is a positive integer, then*

$$(a - b) \cdot \left(a^{n-1}b^0 + a^{n-2}b^1 + a^{n-3}b^2 + \cdots + a^1 b^{n-2} + a^0 b^{n-1} \right) = a^n - b^n.$$

More concisely,

$$(a - b) \cdot \sum_{i=0}^{n-1} a^{n-1-i} b^i = a^n - b^n.$$

Proof We use induction on n. Our claim obviously holds for $n = 1$ as the sum in the equation has only one term and that term equals 1.

Suppose now that the statement holds for $n = k$:

$$(a - b) \cdot \sum_{i=0}^{k-1} a^{k-1-i} b^i = a^k - b^k,$$

and consider

$$(a - b) \cdot \sum_{i=0}^{k} a^{k-i} b^i.$$

Since

$$\sum_{i=0}^{k} a^{k-i} b^i = \sum_{i=0}^{k-1} a^{k-i} b^i + b^k = a \cdot \sum_{i=0}^{k-1} a^{k-1-i} b^i + b^k,$$

we have

$$(a - b) \cdot \sum_{i=0}^{k} a^{k-i} b^i = (a - b) \cdot \left(a \cdot \sum_{i=0}^{k-1} a^{k-1-i} b^i + b^k \right)$$

$$= a \cdot (a - b) \cdot \sum_{i=0}^{k-1} a^{k-1-i} b^i + (a - b) \cdot b^k$$

$$= a \cdot (a^k - b^k) + (a - b) \cdot b^k$$

$$= a^{k+1} - b^{k+1}.$$

Therefore, the statement is true for $n = k + 1$.   □

Next we prove an important property of the natural numbers.

Theorem 10.6 *Every nonempty set of natural numbers contains a minimum (least) element.*

Formally, Theorem 10.6 says that if $A \subseteq \mathbb{N}$ and $A \neq \emptyset$, then

$$\exists m \in A, \forall a \in A, m \leq a.$$

For example, there is a smallest positive prime (namely, 2), and there is a smallest positive prime n for which the n-th Mersenne number, $2^n - 1$, is not prime (namely, 11, see Chapter 3). Furthermore, by Theorem 10.6 we can say that if there is an integer $n \geq 5$ for which the n-th Fermat number, $2^{2^n} + 1$, is prime, then there is a smallest such integer as well. (According to the Fermat Prime Conjecture, there is no such integer; cf. Problem 6 (b) in Chapter 2.) While Theorem 10.6 is intuitively obvious, note that this property of $\mathbb{N}$ is not shared by $\mathbb{Z}$, $\mathbb{Q}$, or $\mathbb{R}$; even the set of positive real (or rational) numbers does not contain a minimum element!

Proof Let A be a subset of $\mathbb{N}$ and suppose that $A \neq \emptyset$. We need to prove that A has a minimum element. We use an indirect argument and assume that A has no minimum element; our goal is then to prove that A cannot contain any elements.

Let $P(n)$ be the statement that, for a given positive integer n,

$$A \cap \{1, 2, \ldots, n\} = \emptyset;$$

that is, A contains no elements less than or equal to n. We use induction to prove that for every $n \in \mathbb{N}$, $P(n)$ is true.

First observe that 1 is not in A, otherwise it would clearly be its minimum element. Therefore, $A \cap \{1\} = \emptyset$; thus $P(1)$ holds.

Assume now that for some fixed $k \in \mathbb{N}$, $P(k)$ is true; that is,

$$A \cap \{1, 2, \ldots, k\} = \emptyset.$$

This then implies that $k+1 \notin A$, since otherwise $k+1$ would clearly be the minimum element of A. Thus,

$$A \cap \{1, 2, \ldots, k, k+1\} = \emptyset,$$

which means that $P(k + 1)$ is true. Therefore, by the Principle of Induction, $P(n)$ is true for every $n \in \mathbb{N}$.

But, if

$$A \cap \{1, 2, \ldots, n\} = \emptyset,$$

then n itself cannot be in A. Since we just proved that this holds for every $n \in \mathbb{N}$, A cannot contain any $n \in \mathbb{N}$. Thus $A = \emptyset$, which is a contradiction. $\square$

We proved Theorem 10.6 using the Principle of Mathematical Induction (Theorem 10.1 above), but it is easy to see that the two theorems are, actually, equivalent. Indeed, we can easily establish Theorem 10.1 from Theorem 10.6 by arguing that if $P(n)$ were to be false for some $n \in \mathbb{N}$, then there would have to be a smallest such value of n, say m. However, m could not be 1 as $P(1)$ is assumed to be true, and if $m \geq 2$, then $P(m-1) \Rightarrow P(m)$ could only hold if $P(m-1)$ were false, contradicting our choice of m. We repeat this argument below in the proof of Theorem 10.7.

Sometimes we use the following generalized version of the Principle of Induction.

Theorem 10.7 (The Generalized Induction Principle) *Let $P(n)$ be a predicate that becomes a statement for all $n \in \mathbb{N}$. Suppose further that n_0 is a positive integer for which the statements*

- $P(n_0)$ *and*
- $\forall k \in \mathbb{N} \cap [n_0, \infty)$, $P(k) \Rightarrow P(k + 1)$

both hold. Then $P(n)$ is true for every integer $n \geq n_0$.

Proof We prove our claim by using Theorem 10.6. Consider the set A of positive integers n for which $n \geq n_0$ and for which $P(n)$ is false. Establishing our claim is equivalent to proving that $A = \emptyset$.

Assume indirectly that $A \neq \emptyset$. By Theorem 10.6, A has a minimum element, say m. By assumption, $P(n_0)$ is true, thus we must have $m \geq n_0 + 1$ and, therefore, $m - 1 \geq n_0$. Since m is the minimum element of A, $m - 1 \notin A$; therefore $P(m - 1)$

holds. Then by the inductive hypothesis, $P(m)$ must hold as well, which means that $m \notin A$, a contradiction.   □

Our final pair of examples in this chapter gives us a good opportunity to compare and contrast the Induction Principle and the Generalized Induction Principle.

Proposition 10.8 *1. For all positive integers n, we have*

$$1.1^n \geq 1 + \frac{n}{10}.$$

2. For all integers n $\geq$ 100, we have

$$1.1^n \geq 1 + \frac{n^2}{1000}.$$

Proof We use induction to prove 1. The inequality obviously holds for $n = 1$. Assume now that k is an arbitrary positive integer for which

$$1.1^k \geq 1 + \frac{k}{10};$$

we need to prove that

$$1.1^{k+1} \geq 1 + \frac{k+1}{10}.$$

To accomplish this, we rewrite 1.1^{k+1} as $1.1 \cdot 1.1^k$ and apply the inductive hypothesis. This yields

$$1.1^{k+1} = \left(1 + \frac{1}{10}\right) \cdot 1.1^k \geq \left(1 + \frac{1}{10}\right) \cdot \left(1 + \frac{k}{10}\right) = 1 + \frac{k+1}{10} + \frac{k}{100},$$

from which our claim clearly follows since $\frac{k}{100} > 0$.

We prove 2 using the Generalized Induction Principle. To verify the inequality for the base case, we need to show that

$$1.1^{100} \geq 1 + \frac{100^2}{1000} = 11.$$

But applying inequality 1 for $n = 100$, we get

$$1.1^{100} \geq 1 + \frac{100}{10} = 11;$$

thus inequality 2 holds for $n = 100$.

Suppose now that we know

$$1.1^k \geq 1 + \frac{k^2}{1000}$$

for some positive integer $k \geq 100$; we need to prove that

$$1.1^{k+1} \geq 1 + \frac{(k+1)^2}{1000} = 1 + \frac{k^2 + 2k + 1}{1000}.$$

Proceeding as above, we have

$$1.1^{k+1} = \left(1 + \frac{1}{10}\right) \cdot 1.1^k \geq \left(1 + \frac{1}{10}\right) \cdot \left(1 + \frac{k^2}{1000}\right) = 1 + \frac{k^2}{1000} + \frac{100 + k^2/10}{1000}.$$

It remains to be shown that the last term is not smaller than $(2k + 1)/1000$ or, equivalently, that

$$100 + \frac{k^2}{10} - (2k + 1) \geq 0.$$

With a bit of elementary algebra, we see that this inequality is equivalent to

$$k^2 - 20k + 100 \geq -890,$$

which clearly holds since the left-hand side is the square of $k - 10$ and thus is never negative.   □

We should mention that the first inequality in Proposition 10.8 is a special case of the well-known *Bernoulli Inequality*, which says that for every positive real number a and positive integer n, we have

$$(1 + a)^n \geq 1 + na.$$

The second inequality does not have a name, but it is, in fact, a special case of the far-reaching fact that an exponential sequence with base greater than 1 is "eventually" always bigger than any polynomial.

Induction is a very powerful proof technique that is used frequently in mathematics. We will see variations of the Induction Principle in Chapter 11.

Problems

1. Suppose that $P(n)$ is a predicate that becomes a statement for every positive integer n. What can you say about the truth set of $P(n)$ in the following cases?

Determine all values of n that are guaranteed to be in the truth set and all values of n that are certainly not in the truth set.

(a) • $\forall k \in \mathbb{N}, P(k) \Rightarrow P(k+1)$.
(b) • $P(1)$ is true and
    • $\forall k \geq 2, P(k) \Rightarrow P(k+1)$.
(c) • $P(7)$ is true and
    • $\forall k \in \mathbb{N}, P(k) \Leftrightarrow P(k+1)$.
(d) • $P(7)$ is true and
    • $\forall k \geq 4, P(k) \Rightarrow P(k+1)$.
(e) • $P(7)$ is true and
    • $\forall k \geq 10, P(k) \Leftrightarrow P(k+1)$.
(f) • $P(1)$ is true and
    • $\forall k \in \mathbb{N}, P(k) \Rightarrow P(2k)$.
(g) • $P(1)$ is true and
    • $\forall k \in \mathbb{N}, P(k) \Rightarrow P(2k+1)$.
(h) • $P(1)$ is true and
    • $\forall k \in \mathbb{N}, P(k) \Rightarrow P(2k) \wedge P(2k+1)$.
(i) • $P(10)$ is false and
    • $\forall k \in \mathbb{N}, P(k) \Rightarrow P(k+1)$.
(j) • $P(10)$ is false and
    • $\forall k \in \mathbb{N}, P(k+1) \Rightarrow P(k)$.
(k) • $P(10)$ is true and
    • $\forall k \in \mathbb{N}, \neg P(k+1) \Rightarrow \neg P(k)$.
(l) • $P(10)$ is true and
    • $\forall k \in \mathbb{N}, \neg P(k) \Rightarrow \neg P(k+1)$.
(m) • $P(10)$ is true and
    • $\forall k \in \mathbb{N}, P(k) \Rightarrow \neg P(k+1)$.

2. Let n_0 be any (positive, negative, or zero) integer, and suppose that $P(n)$ is a predicate that becomes a statement for every integer $n \geq n_0$. State and prove an extension of the Generalized Induction Principle that could be used to prove a theorem of the form

$$\forall n \geq n_0, P(n).$$

3. For a set $S \subseteq \mathbb{R}$, let $P(S)$ denote the predicate that every nonempty subset of S contains a least element. According to Theorem 10.6, $P(\mathbb{N})$ holds. For each of the following sets S, decide if $P(S)$ is true or false. Justify your answer.

(a) $S = (5, 6)$
(b) $S = [5, 6)$
(c) $S = \left\{ \frac{a}{5} \mid a \in \mathbb{N} \right\}$
(d) $S = \left\{ \frac{a}{b} \mid a \in \mathbb{N}, b \in \{5, 6\} \right\}$

(Hints: Observe that $S = S_1 \cup S_2$ where $S_1 = \{\frac{a}{5} \mid a \in \mathbb{N}\}$ and $S_2 = \{\frac{a}{6} \mid a \in \mathbb{N}\}$, and consider a nonempty subset T of S in the form $T = T_1 \cup T_2$ accordingly.)

(e) $S = \{\frac{a}{b} \mid a \in \mathbb{N}, b \in \mathbb{N}\}$

(Hint: Show by way of a counterexample that $P(S)$ is false.)

4. The claims below are all false. Create a counterexample for each and then find the mistakes in their arguments. Be as specific as possible.

(a) **Claim.** Every positive integer is a perfect square.

Argument. Let $P(n)$ be the predicate that the positive integer n is a perfect square; we need to prove that $P(n)$ is true for all $n \in \mathbb{N}$. Since $1 = 1^2$ is a perfect square, $P(1)$ is true.

To prove the inductive step, let $k \in \mathbb{N}$ be an arbitrary positive integer for which $P(k)$ is true. Since k was arbitrary, we could let $k = n$, and, therefore, $P(n)$ is true as claimed.

(b) **Claim.** For every $n \in \mathbb{N}$ we have

$$1 + 2 + \cdots + n = \frac{n^2 + n + 2}{2}.$$

Argument. We will use induction on n. The claim holds for $n = 1$. Suppose that

$$1 + 2 + \cdots + k = \frac{k^2 + k + 2}{2}$$

for some $k \in \mathbb{N}$. We then have

$$1 + 2 + \cdots + k + (k + 1) = \frac{k^2 + k + 2}{2} + (k + 1);$$

a simple calculation verifies that this expression equals

$$\frac{(k + 1)^2 + (k + 1) + 2}{2},$$

which completes the proof.

(c) **Claim.** If we have a finite number of lines in space arranged so that each pair intersects in a point, then they all go through the same point.

Argument. We use induction on the number of lines, n. The claim is clearly true for $n = 2$ (the claim is meaningless for $n = 1$).

Now let $k \in \mathbb{N}$ and suppose that the claim is true for k, that is, if any k lines in space are such that each pair intersects in a point, then they go through the same point. To prove our claim for $k + 1$, take an arbitrary collection $\{l_1, l_2, \ldots, l_{k+1}\}$ of $k + 1$ lines so that each pair intersects in a point. Then both sets $\{l_1, l_2, \ldots, l_k\}$ and $\{l_2, l_3, \ldots, l_{k+1}\}$ are collections of

k lines in space such that each pair intersects in a point. Therefore, by our inductive hypothesis, there is a point P such that the lines $l_1, l_2, \ldots, l_k$ as well as the lines $l_2, l_3, \ldots, l_{k+1}$ go through P. Therefore, every line in the set $\{l_1, l_2, \ldots, l_{k+1}\}$ goes through P as claimed.

(d) **Claim.** If S is a finite set of real numbers then all the elements in S have the same sign (positive, zero, or negative).

Argument. We will use induction on the number of elements in the set (called the *size* of the set). Clearly in every set containing exactly 1 element, all the elements have the same sign.

Now let $k \in \mathbb{N}$, and suppose that in every set of k elements all the elements have the same sign. Consider a set S of $k + 1$ elements, $S = \{a_1, a_2, \ldots, a_{k+1}\}$. Since $S \setminus \{a_1\}$ has k elements, these elements have the same sign by our inductive hypothesis. Similarly, all numbers in $S \setminus \{a_2\}$ have the same sign. Choose an element $a \in S \setminus \{a_1, a_2\}$. Since $a \in S \setminus \{a_1\}$, every element in $S \setminus \{a_1\}$ has the same sign as a (since all elements must share their signs); similarly, every element of $S \setminus \{a_2\}$ must also have the same sign as a.

Therefore, all numbers in

$$(S \setminus \{a_1\}) \cup (S \setminus \{a_2\})$$

must have the same sign, but, clearly,

$$(S \setminus \{a_1\}) \cup (S \setminus \{a_2\}) = S,$$

so all numbers in S have the same sign.

(e) Recall from Problem 10 of Chapter 4 that in the Plutonian alphabet there are only four letters—A, B, C, and D—and that every finite string containing these letters is a Plutonian word. We say that a Plutonian word is a "D-lite" if at most half of its letters are Ds.

Claim. If a Plutonian word contains the letter D at all, then it cannot be a "D-lite."

Argument. We will use induction on the length n of the words. The claim is clearly true for $n = 1$: The four 1-letter words are A, B, C, and D; of these, only the word D contains a letter D and, since it contains no other letters, the claim holds.

Now let $k \in \mathbb{N}$ and suppose that our claim is true for all Plutonian words of length k. Consider a word of length $k + 1$. If this word contains no Ds, we are done. Assume, therefore, that it contains exactly m Ds for some $1 \leq m \leq k + 1$; it will then contain $k + 1 - m$ other letters. We will need to prove that $m > k + 1 - m$.

Deleting one of the Ds from our word will then result in a word of length k that has exactly $m - 1$ Ds and $k + 1 - m$ other letters. By our inductive hypothesis, $m - 1 > k + 1 - m$. But this inequality implies that $m > k + 1 - m$, which was our claim.

5. Prove that the following identities hold for every natural number n.

   (a) $\displaystyle\sum_{i=1}^{n} i^2 = \frac{n(n+1)(2n+1)}{6}$

   (b) $\displaystyle\sum_{i=1}^{n} i^3 = \left(\frac{n(n+1)}{2}\right)^2$

6. Suppose that $n \in \mathbb{N}$. Find and prove a closed formula for each of the following expressions.

   (a) $1 + 2 + 4 + 8 + \cdots + 2^n$
   (b) $1 \cdot 1! + 2 \cdot 2! + 3 \cdot 3! + \cdots + n \cdot n!$
   (c) $\frac{1}{2} + \frac{1}{6} + \frac{1}{12} + \cdots + \frac{1}{n(n+1)}$
   (d) $\frac{1}{2!} + \frac{2}{3!} + \frac{3}{4!} + \cdots + \frac{n}{(n+1)!}$
   (e) $\left(1 - \frac{1}{4}\right) \cdot \left(1 - \frac{1}{9}\right) \cdots \cdots \left(1 - \frac{1}{(n+1)^2}\right)$

7. For each of the following, determine (with proof) the set of *all* natural numbers n for which the statement holds.

   (a) $2^n < n!$
   (b) $2^n > n^2$
   (c) $3^n + 5$ is divisible by 8.
      (Hints: Let $P(n)$ be the predicate that $3^n + 5$ is divisible by 8 for $n \in \mathbb{N}$. Use the identity

      $$3^{k+2} + 5 = (3^k + 5) + 8 \cdot 3^k$$

      to show that $P(k) \Leftrightarrow P(k+2)$ holds for every $k \in \mathbb{N}$.)
   (d) $3^n + 4^n$ is divisible by 13.
      (Hints: Let $P(n)$ be the predicate that $3^n + 4^n$ is divisible by 13 for $n \in \mathbb{N}$. Show that $P(k) \Leftrightarrow P(k+6)$ holds for every $k \in \mathbb{N}$.)

8. Given an 8-by-8 board with one square missing (any square), use induction to prove that the remaining 63 squares can be covered with 21 L-shaped trominoes (tiles covering three squares).
   (Hint: Prove the more general statement for a 2^n-by-2^n board.)

9. (a) Define the infinite sequence $(a_1, a_2, a_3, \ldots)$ recursively by $a_1 = 2, a_{n+1} = a_n + 2n$ for $n \in \mathbb{N}$. Find and prove an explicit formula for a_n (i.e., a formula in terms of n only).

   (b) Find a formula for the number of regions created by n pairwise intersecting circles in the plane in general position (i.e., every pair of circles intersects in two points and no three of the circles intersect at the same point).
      (Hint: Use part (a).)

   (c) Prove that it is possible to draw the Venn diagram of n sets in general position using only circles if, and only if, $n \leq 3$.
      (Hint: Use Problem 7 (b).)

10. (a) Define the infinite sequence $(a_1, a_2, a_3, \dots)$ recursively by $a_1 = 2, a_{n+1} = a_n + n + 1$ for $n \in \mathbb{N}$. Find and prove an explicit formula for a_n (i.e., a formula in terms of n only).

(b) Find a formula for the number of regions created by n lines in the plane in general position (i.e., no two of the lines are parallel and no three pass through the same point).

(Hint: Use part (a).)

(c) Consider the planar map created by n lines in the plane in general position. (Note that some of the regions are unbounded.) Find (with proof) the chromatic number of this map. (Recall that the chromatic number of a map is the minimum number of colors needed to color all regions so that regions sharing a boundary edge receive different colors.)

(Hint: Use induction to prove that the chromatic number equals 2.)

11. Given positive integers $a_1, a_2, \dots, a_n$ (here $n \in \mathbb{N}$), let $\mathcal{Z}$ be the set of positive integers that can be written as a linear combination of $a_1, a_2, \dots, a_n$ over the set of integers; that is,

$$\mathcal{Z} = \mathbb{N} \cap \{a_1 x_1 + a_2 x_2 + \dots + a_n x_n \mid x_1, x_2, \dots, x_n \in \mathbb{Z}\}.$$

Let $d(a_1, a_2, \dots, a_n)$ be the minimum element of $\mathcal{Z}$.

(a) Explain how Theorem 10.6 guarantees that $d(a_1, a_2, \dots, a_n)$ exists for every $a_1, a_2, \dots, a_n$.

(b) Find, with proof, the values of $d(3, 10)$, $d(6, 10)$, and $d(6, 10, 15)$.

(c) Formulate a conjecture for $d(a_1, a_2, \dots, a_n)$ in general. (You do not need to prove your conjecture.)

(d) Find a way to describe the elements of the set $\mathcal{Z}$. (You do not need to prove your assertion.)

Chapter 11
More Domino Games

In this chapter we discuss three variations of induction: strong induction, split induction, and double induction. This arsenal of induction techniques will allow us to prove a variety of fundamental and powerful mathematical statements.

Let us start by recalling that, according to Theorem 10.6, every nonempty set of natural numbers contains a minimum element. (As we explained in Chapter 10, this result is actually equivalent to the Principle of Mathematical Induction, Theorem 10.1.) A frequent way of employing Theorem 10.6 in order to prove that a predicate $P(n)$ is true for all $n \in \mathbb{N}$ is to assume indirectly that it is not; by Theorem 10.6 there is then a smallest value of $m \in \mathbb{N}$ for which $P(m)$ is false. If we know that $P(1)$ is true (thus $m \geq 2$) and that

$$(P(1) \wedge P(2) \wedge \cdots \wedge P(m-1)) \Rightarrow P(m),$$

then this contradicts the fact that m is the least natural number for which the predicate is false, thus $P(n)$ is true for all $n \in \mathbb{N}$. This particular proof technique is referred to as the Principle of Strong Induction and is stated formally as follows:

Theorem 11.1 (The Principle of Strong Induction) *Suppose that $P(n)$ is a predicate that becomes a statement for all $n \in \mathbb{N}$. If*

* $P(1)$ *and*
* $\forall k \in \mathbb{N}, (P(1) \wedge P(2) \wedge \cdots \wedge P(k)) \Rightarrow P(k+1)$

both hold, then $P(n)$ is true for every positive integer n.

To prove the statement $P(n)$ for every $n \in \mathbb{N}$ using the Principle of Strong Induction, we first verify that $P(1)$ is true (the *base step*). We then prove that for every positive integer k, if $P(1), P(2), \ldots, P(k)$ are all true, then so is $P(k+1)$ (the *inductive step*). Note that the Principle of Strong Induction is indeed a strengthening of the Principle of Induction in that when proving $P(k+1)$ in the inductive step, we can use any or all of the inductive assumptions $P(1), P(2), \ldots, P(k)$.

© Springer Nature Switzerland AG 2020

B. Bajnok, *An Invitation to Abstract Mathematics*, Undergraduate Texts in
Mathematics, https://doi.org/10.1007/978-3-030-56174-1_11

The generalized version of the Principle of Strong Induction is the following:

Theorem 11.2 (Generalized Strong Induction) *Suppose that $P(n)$ is a predicate that becomes a statement for all $n \in \mathbb{N}$ and that n_0 is a positive integer. If*

- $P(n_0)$ *and*
- $\forall k \in \mathbb{N} \cap [n_0, \infty), (P(n_0) \wedge P(n_0 + 1) \wedge \cdots \wedge P(k)) \Rightarrow P(k + 1)$

both hold, then $P(n)$ is true for every integer $n \geq n_0$.

The following proposition nicely illustrates the similarities and differences between the Principle of Induction and the Principle of Strong Induction. We prove that the following three definitions for Mersenne numbers (cf. Problem 6 (a) of Chapter 2) are equivalent.

Proposition 11.3 *Let us define three sequences as follows:*

1. *For all $n \in \mathbb{N} \cup \{0\}$, let $M_n = 2^n - 1$.*
2. *Let $M_0' = 0$ and for all $n \in \mathbb{N}$, let $M_n' = 2M_{n-1}' + 1$.*
3. *Let $M_0'' = 0$ and for all $n \in \mathbb{N}$, let $M_n'' = n + \sum_{i=0}^{n-1} M_i''$.*

Then the three sequences are identical; that is, for every $n \in \mathbb{N} \cup \{0\}$, we have $M_n = M_n' = M_n''$.

Proof First we use induction to prove that for all $n \in \mathbb{N} \cup \{0\}$, we have $M_n = M_n'$. This is obviously true for $n = 0$. Let k be an arbitrary nonnegative integer and assume that $M_k' = 2^k - 1$. (Technically, we are using the generalized version of induction with $n_0 = 0$; we could easily just verify that $M_1 = M_1'$ and stick with $n_0 = 1$.) We need to show that $M_{k+1}' = 2^{k+1} - 1$.

But, according to the recursive formula, $M_{k+1}' = 2M_k' + 1$; since $M_k' = 2^k - 1$, we immediately get $M_{k+1}' = 2^{k+1} - 1$, as claimed.

Next we use strong induction to prove that for all $n \in \mathbb{N} \cup \{0\}$, we have $M_n = M_n''$. Again, this is obviously true for $n = 0$. Let k be an arbitrary nonnegative integer and assume that the equality also holds for all $i = 0, 1, 2, \ldots, k$. Therefore, we assume that $M_i'' = 2^i - 1$ holds for $i = 0, 1, 2, \ldots, k$, and we need to show that $M_{k+1}'' = 2^{k+1} - 1$.

We use the definition of the sequence and our hypotheses to write

$$M_{k+1}'' = k+1+\sum_{i=0}^{k} M_i'' = k+1+\sum_{i=0}^{k}(2^i - 1) = k+1+\sum_{i=0}^{k} 2^i - \sum_{i=0}^{k} 1 = \sum_{i=0}^{k} 2^i;$$

from which we arrive at the desired $M_{k+1}'' = 2^{k+1} - 1$ using Lemma 4.2.

Thus we proved that $M_n = M_n' = M_n''$ holds for all $n \in \mathbb{N} \cup \{0\}$, and thus the three definitions yield the same sequence.   □

We can also prove that the three different definitions for Fermat numbers given in Problem 6 (b) of Chapter 2 are equivalent; cf. Problem 6.

We now proceed to establish some famous results whose proofs we have been postponing since Chapter 2.

Theorem 11.4 (The Factorization Theorem) *Every integer n with $n \geq 2$ can be expressed as a product of (positive) primes.*

Note that we have defined the product of one number as the number itself (cf. Problem 4 of Chapter 2), so our claim holds trivially if n itself is prime.

Proof We use generalized strong induction. The statement clearly holds for $n = 2$ since 2 is a prime.

Let k be an integer with $k \geq 2$, and suppose that every integer between 2 and k (inclusive) can be expressed as a product of primes. Consider the integer $k + 1$. If it is prime, we are done. If it is not prime, then it must have a divisor a such that $1 < a < k + 1$. Therefore, $k + 1 = a \cdot b$ for some positive integer b; furthermore, both a and b are between 2 and k (inclusive). Applying our inductive hypothesis, both a and b factor into a product of primes, and the product of these yields a prime factorization of $k + 1$.   □

The theorem we just proved has an important sibling that states that the prime factorization is essentially unique. First, however, we prove Euclid's Principle, that Definitions 2.1 and 2.1b are equivalent. The fact that every number satisfying Definition 2.1b also satisfies Definition 2.1 is the easy part of this claim and was done in Problem 3 of Chapter 2. We now prove the other direction, known as Euclid's First Theorem, which we have been postponing since page 14. There are many proofs known for this famous result; here we present a particularly concise argument.

Theorem 11.5 (Euclid's First Theorem) *If a prime number divides the product of two integers then it must divide one of them.*

Proof It clearly suffices to prove the result for positive integers. Suppose indirectly that there are some positive primes for which the claim fails. By Theorem 10.6, there is then a smallest one; let this prime be p. Then we have positive integers a and b so that ab is divisible by p but neither a nor b is; using Theorem 10.6 again, we can assume that we made our selection in such a way that ab is the minimum number with these properties. Our assumptions imply that $a \neq 1$; by Theorem 11.4, we can then let q be any positive prime divisor of a. We then set $c = a/q$ and $n = (ab)/p$. Note that a not being divisible by p implies that neither c nor $a - p$ is divisible by p—we will use both of these facts momentarily.

Observe that $p(n - b) = (a - p)b$ is divisible by p but neither $a - p$ nor b is. By the minimum property of ab, this can only happen if $a - p$ is negative, so $a < p$. Therefore, $q < p$ as well, so q cannot divide a product of two integers without dividing one of them. Since q divides $qcb = np$ and q does not divide p, q must divide n; let $m = n/q$.

Now p divides $pm = cb$ and cb is less than ab, so p must divide c or b, which is a contradiction.   □

Using induction, one can easily prove the following generalized version.

Theorem 11.6 *If a prime number divides the product of any finite number of integers then it must divide one of them.*

We are now ready to prove the following famous result.

Theorem 11.7 (The Unique Factorization Theorem) *No positive integer can be factored into a product of primes in two* essentially different *ways; that is, if* $p_1, p_2, \ldots, p_k$ *and* $q_1, q_2, \ldots q_l$ *are two lists of positive primes in increasing order (i.e.,* $p_1 \leq p_2 \leq \cdots \leq p_k$ *and* $q_1 \leq q_2 \leq \cdots \leq q_l$*) for which*

$$p_1 \cdot p_2 \cdots \cdots p_k = q_1 \cdot q_2 \cdots \cdots q_l,$$

then $k = l$ *and* $p_1 = q_1, p_2 = q_2, \ldots, p_k = q_k$.

Proof Assume indirectly that at least one positive integer violates unique factorization, as stated above. By Theorem 10.6, there is then a smallest positive integer N that has two essentially different prime factorizations, say

$$N = p_1 \cdot p_2 \cdots \cdots p_k$$

with $2 \leq p_1 \leq p_2 \leq \cdots \leq p_k$ and

$$N = q_1 \cdot q_2 \cdots \cdots q_l$$

with $2 \leq q_1 \leq q_2 \leq \cdots \leq q_l$. Note that the sets $P = \{p_1, p_2, \cdots, p_k\}$ and $Q = \{q_1, q_2, \cdots, q_l\}$ are disjoint since, if they had a common element p, then N/p would also have two essentially different prime factorizations, contradicting that N is the smallest such positive integer.

Next observe that p_1 divides $N = q_1 \cdot q_2 \cdots \cdots q_l$, so by Theorem 11.6, p_1 must also divide q_i for some $1 \leq i \leq l$. But q_i is prime, so it only has two positive divisors, 1 and itself. Since p_1 cannot be 1, we have $p_1 = q_i$, a contradiction with $P \cap Q = \emptyset$.   □

Theorems 11.4 and 11.7 together are referred to as the *Fundamental Theorem of Arithmetic*.

Theorem 11.8 (The Fundamental Theorem of Arithmetic) *Every positive integer n with $n \geq 2$ can be factored into a product of primes in an essentially unique way (as described in the Unique Factorization Theorem).*

As a corollary, we immediately get statements such as those in Problem 7 of Chapter 3; in fact, for any positive prime p, every positive integer can be written as the product of a (nonnegative) power of p and an integer that is not divisible by (and, therefore, relatively prime to) p. We also get that the expression is unique. While the Fundamental Theorem of Arithmetic among positive integers seems to state the obvious, it is far from it. In fact, one can construct and study instances of other sets of numbers where the factorization or the uniqueness part fails.

The next variations of inductive proofs that we will discuss are the Principle of Split Induction and its generalized version.

Theorem 11.9 (The Principle of Split Induction) *Let $P(n)$ be a predicate that becomes a statement for all $n \in \mathbb{N}$, and suppose that there exists a positive integer d such that the statements*

- $P(1), P(2), \ldots, P(d)$*, and*
- $\forall k \in \mathbb{N}, P(k) \Rightarrow P(k + d)$

all hold. Then $P(n)$ is true for every positive integer n.

Theorem 11.10 (Generalized Split Induction) *Let $P(n)$ be a predicate that becomes a statement for all $n \in \mathbb{N}$, and suppose that n_0 and d are positive integers such that the statements*

- $P(n_0), P(n_0 + 1), \ldots, P(n_0 + d - 1)$*, and*
- $\forall k \geq n_0, P(k) \Rightarrow P(k + d)$

all hold. Then $P(n)$ is true for every integer $n \geq n_0$.

The proofs of Theorems 11.9 and 11.10 can be easily established using, for example, Theorem 10.6. The next proposition uses the Principle of Split Induction.

Proposition 11.11 *Let S be the set of all positive integer values of n for which it is not possible to tile a 1-by-n board using a combination of 1-by-4 tetrominoes and 1-by-5 pentominoes. Then $S = \{1, 2, 3, 6, 7, 11\}$.*

Proof Instead, let us find the set T of all positive integer values of n for which the tiling is possible. Algebraically, T is the set of all $n \in \mathbb{N}$ for which the equation $4x + 5y = n$ has nonnegative integer solutions.

Experimenting with small nonnegative integer values of x and y, we see that $n = 4, 5, 8, 9, 10, 12, 13, 14,$ and 15 are all members of T, but other positive integers under 15 are not. We now use the generalized version of split induction with $d = 4$ and $n_0 = 12$ to prove that all values of n with $n \geq 12$ are in T.

We already know that the equations $4x + 5y = 12, 4x + 5y = 13, 4x + 5y = 14$, and $4x + 5y = 15$ all have nonnegative integer solutions, namely $(x, y) = (3, 0)$, $(x, y) = (2, 1), (x, y) = (1, 2),$ and $(x, y) = (0, 3)$, respectively.

Let k be an integer with $k \geq 12$, and suppose that the equation $4x + 5y = k$ has a solution $(x, y) = (a, b)$ where a and b are nonnegative integers. Consider the equation $4x + 5y = k + 4$. Then it is easy to see that $(x, y) = (a + 1, b)$ provides a solution to $4x + 5y = k + 4$ and that $a + 1$ and b are nonnegative. This completes our proof.

So, in summary, $T = \{4, 5, 8, 9, 10\} \cup \{12, 13, 14, 15, \ldots\}$ and thus $S = \{1, 2, 3, 6, 7, 11\}$, as claimed.   □

The third version of induction we discuss here is double induction: it applies to predicates with two variables. There are several alternate ways of stating the principle; we find the following version most helpful.

Theorem 11.12 (Double Induction) *Let $P(m, n)$ be a predicate that becomes a statement for all $m \in \mathbb{N}$ and $n \in \mathbb{N}$. Suppose that the statements*

- $\forall m \in \mathbb{N}, P(m, 1)$,
- $\forall n \in \mathbb{N}, P(1, n)$, *and*
- $\forall k \in \mathbb{N}, \forall l \in \mathbb{N}, (P(k, l+1) \wedge P(k+1, l)) \Rightarrow P(k+1, l+1)$

all hold. Then $P(m, n)$ is true for every pair of positive integers (m, n).

We use double induction to prove the following important theorem (a generalization of Problem 3 of Chapter 9).

Theorem 11.13 *For any $m \in \mathbb{N}$, the product of an arbitrary m consecutive positive integers is divisible by $m!$. In particular, if n and m are positive integers, then*

$$\frac{n(n+1)\cdots(n+m-1)}{m!}$$

is an integer.

Proof Let us set

$$f(m, n) = \frac{n(n+1)\cdots(n+m-1)}{m!}.$$

We will use double induction, as specified by Theorem 11.12, to prove that $f(m, n)$ is an integer for all $m, n \in \mathbb{N}$.

Since $f(m, 1) = 1$ and $f(1, n) = n$ are integers, the base cases hold.

Suppose now that k and l are positive integers for which $f(k, l+1)$ and $f(k+1, l)$ are integers. We will need to show that these two assumptions imply that $f(k+1, l+1)$ is also an integer. We will accomplish this by simply verifying that

$$f(k+1, l+1) = f(k, l+1) + f(k+1, l).$$

To see that this is true, we write

$$
\begin{aligned}
f(k+1, l+1) &= \frac{(l+1)(l+2)\cdots(l+k)(l+k+1)}{(k+1)!} \\
&= \frac{(l+1)(l+2)\cdots(l+k)l}{(k+1)!} + \frac{(l+1)(l+2)\cdots(l+k)(k+1)}{(k+1)!} \\
&= \frac{l(l+1)(l+2)\cdots(l+k)}{(k+1)!} + \frac{(l+1)(l+2)\cdots(l+k)}{k!} \\
&= f(k+1, l) + f(k, l+1).
\end{aligned}
$$

$$
\begin{array}{ccccccccccccccc}
& & & & & & & 1 & & & & & & & \\
& & & & & & 1 & & 1 & & & & & & \\
& & & & & 1 & & 2 & & 1 & & & & & \\
& & & & 1 & & 3 & & 3 & & 1 & & & & \\
& & & 1 & & 4 & & 6 & & 4 & & 1 & & & \\
& & 1 & & 5 & & 10 & & 10 & & 5 & & 1 & & \\
& 1 & & 6 & & 15 & & 20 & & 15 & & 6 & & 1 & \\
1 & & 7 & & 21 & & 35 & & 35 & & 21 & & 7 & & 1
\end{array}
$$

Fig. 11.1 Pascal's Triangle

By our inductive hypothesis, both terms are integers, so we have proved that $f(k + 1, l + 1)$ is also an integer. □

We should note that Theorem 11.13 is valid for all integers n—however, we will only need it and prove it here for the case in which n is positive.

We now introduce two notations that are often used in combinatorics and many other areas of mathematics. For nonnegative integers n and m, we let

$$
\binom{n}{m} = f(m, n - m + 1) = \frac{n(n-1)\cdots(n-m+1)}{m!}
$$

and

$$
\left\lfloor \begin{array}{c} n \\ m \end{array} \right\rfloor = f(m, n) = \frac{n(n+1)\cdots(n+m-1)}{m!}.
$$

According to the theorem we just proved, these quantities denote integers.

The symbols $\binom{n}{m}$ and $\left\lfloor \begin{array}{c} n \\ m \end{array} \right\rfloor$ are pronounced "n choose m" and "n multichoose m," respectively—the reason for these names will become clear in Chapter 19. Here we just mention that *Pascal's Triangle*, whose first few rows are in Figure 11.1, tabulates these values. (Entries with $m > n$, which yield $\binom{n}{m} = 0$, are skipped.)

Namely, if we label the rows, the left diagonals, and the right diagonals 0, 1, 2, etc. (we start with 0), then $\binom{n}{m}$ appears as the entry where row n and right diagonal m intersect; $\left\lfloor \begin{array}{c} n \\ m \end{array} \right\rfloor$ is the entry where left diagonal $n - 1$ and right diagonal m intersect. For example, we see that $\binom{5}{2} = 10$ and $\left\lfloor \begin{array}{c} 5 \\ 2 \end{array} \right\rfloor = 15$. We will study the entries in Pascal's Triangle—the so-called *binomial coefficients*—in more detail in Chapter 19.

Problems

1. Suppose that $P(n)$ is a predicate that becomes a statement for every positive integer n. What can you say about the truth set of $P(n)$ in the following cases?

Determine all values of n that are guaranteed to be in the truth set and all values of n that are certainly not in the truth set.

(a) • $P(3)$ is true and
  • $\forall k \in \mathbb{N}, (P(k) \wedge P(3)) \Rightarrow P(k+1)$.

(b) • $P(3)$ is true and
  • $\forall k \in \mathbb{N}, (P(k) \wedge P(3)) \Rightarrow \neg P(k+1)$.

(c) • $P(6)$ is true and
  • $\forall k \in \mathbb{N}, (P(k-2) \wedge P(k-1)) \Rightarrow P(k)$.

(d) • $P(6)$ and $P(7)$ are true and
  • $\forall k \in \mathbb{N}, (P(k-2) \wedge P(k-1)) \Rightarrow P(k)$.

(e) • $P(5)$ and $P(7)$ are true and
  • $\forall k \in \mathbb{N}, (P(k-2) \wedge P(k-1)) \Rightarrow P(k)$.

(f) • $P(1)$ is true and
  • $\forall k \geq 2, [P(1) \wedge P(2) \wedge \cdots \wedge P(k)] \Rightarrow P(k+1)$.

(g) • $P(2)$ is true and
  • $\forall k \geq 3, [P(1) \wedge P(2) \wedge \cdots \wedge P(k)] \Rightarrow P(k+1)$.

(h) • $P(1)$ and $P(2)$ are true and
  • $\forall k \in \mathbb{N}, P(k) \Rightarrow P(k+3)$.

(i) • $P(1)$ is true,
  • $\forall k \in \mathbb{N}, P(k) \Rightarrow P(2k)$, and
  • $\forall k \in \mathbb{N}, P(k+1) \Rightarrow P(k)$.

2. State the generalized version (in the sense of Theorems 11.2 and 11.10) of the Principle of Double Induction.

3. Suppose that $P(m, n)$ is a well-defined mathematical statement for every $m \in \mathbb{N}$ and $n \in \mathbb{N}$. For each of the following cases, determine all values of (m, n) that are guaranteed to be in the truth set of $P(m, n)$.

(a) • $P(1, 1)$ and
  • $P(k, l) \Rightarrow P(k+1, l+1)$ for every $k, l \in \mathbb{N}$.

(b) • $P(1, 1)$ and
  • $P(k, l) \Rightarrow P(k+1, l) \wedge P(k, l+1)$ for every $k, l \in \mathbb{N}$.

(c) • $P(1, 1)$,
  • $P(k, 1) \Rightarrow P(k, 2)$ for every $k \in \mathbb{N}$, and
  • $P(k, l) \Rightarrow P(k+1, l)$ for every $k, l \in \mathbb{N}$.

(d) • $P(1, 1)$ and
  • $[P(1, k) \wedge P(2, k-1) \wedge \cdots \wedge P(k, 1)] \Rightarrow [P(1, k+1) \wedge P(2, k) \wedge \cdots \wedge P(k+1, 1)]$ for every $k \in \mathbb{N}$.

(e) • $P(m, 1)$ for every $m \in \mathbb{N}$,
  • $P(1, n)$ for every $n \in \mathbb{N}$, and
  • $[P(k, l+1) \wedge P(k+1, l) \wedge P(l, k+1) \wedge P(l+1, k)] \Rightarrow P(k+1, l+1)$ for every $k, l \in \mathbb{N}$.

(f) • $P(m, 1)$ for every $m \in \mathbb{N}$,
  • $P(1, n)$ for every $n \in \mathbb{N}$,
  • $(P(k, l+1) \wedge P(k+3, l)) \Rightarrow P(k+3, l+1)$ for every $k, l \in \mathbb{N}$, and

- $(P(k+1, l) \wedge P(k, l+4)) \Rightarrow P(k+1, l+4)$ for every $k, l \in \mathbb{N}$.
(g) • $P(1, 1) \wedge P(1, 2) \wedge P(1, 3) \wedge P(2, 1) \wedge P(2, 2) \wedge P(2, 3)$ and
- $P(k, l) \Rightarrow P(k+2, l+3)$ for every $k, l \in \mathbb{N}$.
(h) • $P(1, 1)$,
- $P(k, l) \Rightarrow P(2k, 3l)$ for every $k, l \in \mathbb{N}$, and
- $P(k+1, l+1) \Rightarrow P(k+1, l) \wedge P(k, l+1)$ for every $k, l \in \mathbb{N}$.

4. For each of the claims below, decide if the claim is true or false. In either case, analyze the arguments given; if the argument is incorrect, find the mistakes. Be as specific as possible.

   (a) **Claim.** The inequality $2^n < m!$ holds for every pair of positive integers (m, n) with $m \geq 5$ and $n \geq 5$. (Cf. Problem 7 (a) of Chapter 10.)
   Argument. The inequality clearly holds for $n = 5$ and $m = 5$ since $32 < 120$.
   Assume now that $2^k < l!$ holds for some $k \geq 5$ and $l \geq 5$; we will then prove that $2^{k+1} < (l+1)!$. Note that

   $$\frac{2^{k+1}}{(l+1)!} = \frac{2^k}{(l)!} \cdot \frac{2}{l+1}.$$

   The first factor is less than 1 by our inductive hypothesis; the second factor is less than 1 for all $l \geq 2$ (and thus for $l \geq 5$). Therefore,

   $$\frac{2^{k+1}}{(l+1)!} < 1,$$

   from which our claim follows.

   (b) **Claim.** Let $n \in \mathbb{N}$. Then $a^n = a$ for every positive real number a.
   Argument. We will use strong induction on n. The claim holds for $n = 1$, since $a^1 = a$ for every positive real number a. Let $k \in \mathbb{N}$ and assume that $a^t = a$ for all $t \in \mathbb{N}$, $t \leq k$. We need to show that $a^{k+1} = a$ for every positive real number a. But we have

   $$a^{k+1} = \frac{a^k a^k}{a^{k-1}}.$$

   Since we have assumed that our statement holds for $t = k$ and $t = k - 1$, we have $a^k = a$ and $a^{k-1} = a$, so

   $$a^{k+1} = \frac{a \cdot a}{a} = a,$$

   which completes the proof.

   (c) **Claim.** No positive integer can be factored into a product of primes in two *essentially different* ways. Namely, if some positive integer n has a factorization into a product of positive primes (or is itself a prime) and this factorization contains the prime p exactly m times, then every other factorization of n into a product of positive primes will contain p exactly m times.

Argument. We use strong induction on n. The claim is trivially true for $n = 1$ since 1 has no prime factorizations at all.

Suppose now that the claim holds for all positive integers up to k and consider $k + 1$. If $k + 1$ itself is a prime, then it clearly has no other prime factorizations, hence our claim holds.

If $k + 1$ is composite, then there are integers a and b with $2 \leq a \leq k$ and $2 \leq b \leq k$ so that $k + 1 = a \cdot b$. Let us apply our inductive hypothesis for a and b: If a prime p is contained in a (positive) prime factorization of a exactly m_1 times, then it will appear in every other prime factorization of a exactly m_1 times. Similarly, the number of times p appears in any prime factorization of b is a fixed m_2. But then the prime p has to appear exactly $m_1 + m_2$ times in every prime factorization of $a \cdot b$, which is what we needed to prove.

5. Recall that the Fibonacci sequence is the sequence defined recursively by $F_1 = 1$, $F_2 = 2$, and $F_{n+2} = F_n + F_{n+1}$ for $n \geq 1$.

   (a) Prove that

$$F_n = \frac{1}{\sqrt{5}} \left[\left(\frac{1 + \sqrt{5}}{2} \right)^{n+1} - \left(\frac{1 - \sqrt{5}}{2} \right)^{n+1} \right]$$

   for every natural number n.

   Remark Note that, while the formula involves irrational numbers, it gives an integer value (namely, F_n) for every n.

   (b) Prove that F_n is the closest integer to

$$\frac{1}{\sqrt{5}} \left(\frac{1 + \sqrt{5}}{2} \right)^{n+1}.$$

   Remark This fact enables us to compute F_n even faster than the previous part.

6. Prove that the three definitions given for Fermat numbers in Problem 6 (b) of Chapter 2 are equivalent.

7. For each of the following, determine (with proof) the set of *all* ordered pairs of natural numbers (n, m) for which the statement holds. (Cf. Problem 7 of Chapter 10.)

   (a) $3^n + 5^m$ is divisible by 8.

(Hints: Let $P(m, n)$ be the predicate that $3^n + 5^m$ is divisible by 8 ($m, n \in \mathbb{N}$). Prove that $P(k, l) \Leftrightarrow P(k + 2, l)$ and $P(k, l) \Leftrightarrow P(k, l + 2)$ hold for all positive integers k and l.)

(b) $3^n + 4^m$ is divisible by 13.

8. Use Problem 1 (h) above to prove Lemma 4.10.
   (Hints: To prove that for every positive integer k, $P(k) \Rightarrow P(k + 3)$, use the identity

   $$a^{2k+6} + a^{k+3}b^{k+3} + b^{2k+6} = a^3 b^3 (a^{2k} + a^k b^k + b^{2k}) + (a^{2k+3} - b^{2k+3})(a^3 - b^3),$$

   and note that $a^3 - b^3 = (a - b)(a^2 + ab + b^2)$.)

9. Let $m \geq 2$ and $n \geq 2$ be integers. Find the maximum number of rooks that one can place on an m-by-n board so that no rook is under attack both horizontally and vertically.
   (Hints: Note that rooks may move any number of spaces in a row or column of the board, and thus this problem is asking for the maximum number of squares that can be selected so that no three of them form a right triangle. Make a conjecture, then use double induction to prove it.)

   Remark This problem was created by undergraduate student Peter Francis as a lemma for Problem 11 (c) of Chapter 20.

10. (a) Prove that if an integer has divisors a and b with $\gcd(a, b) = 1$, then it is divisible by $a \cdot b$.
    (Hint: Use the Fundamental Theorem of Arithmetic.)
    (b) Generalize part (a) to the case of more than two divisors.
    (Hint: Note that 30 is divisible by 6, 10, and 15 and $\gcd(6, 10, 15) = 1$, but 30 is not divisible by $6 \cdot 10 \cdot 15$!)

11. Prove that if the product of integers a and b is divisible by c and $\gcd(a, c) = 1$, then c must divide b.
    (Hint: Use the Fundamental Theorem of Arithmetic.)

12. (a) Prove that if an integer greater than 1 has no divisors between 2 and its square root (inclusive), then it is a prime. (Cf. Problem 5 in Chapter 2.)
    (b) Find all positive integers that are divisible by every positive integer below their square roots; in particular, prove that 24 is the largest such integer.

    Remark According to our claim, we may say that 24 is the "most composite" number.

    (Hints: Use Problem 10 to claim that integers with this property between 25 and 48, inclusive, would have to be divisible by $3 \cdot 4 \cdot 5$, but there is no such integer; and integers between 49 and 99, inclusive, would have to be divisible by $5 \cdot 6 \cdot 7$, but there is no such integer either. To rule out integers n that are 100 or higher, prove that if $k = \lfloor \sqrt{n} \rfloor$ is odd, then n would have to be divisible by $k(k - 1)(k - 2)$ but that $k(k - 1)(k - 2) > n$; similarly, if k is even, then n would have to be divisible by $(k-1)(k-2)(k-3)$ but that $(k-1)(k-2)(k-3) > n$.)

13. Given positive integers $a_1, a_2, \ldots, a_n$ (here $n \in \mathbb{N}$), let $\mathcal{N}$ be the set of nonnegative integers that can be written as a linear combination of $a_1, a_2, \ldots, a_n$ over the set of nonnegative integers; that is, let

$$\mathcal{N} = \mathcal{N}(a_1, a_2, \ldots, a_n) = \{a_1 x_1 + a_2 x_2 + \cdots + a_n x_n \mid x_1, x_2, \ldots, x_n \in \mathbb{N} \cup \{0\}\}.$$

The *Frobenius number* of $a_1, a_2, \ldots, a_n$, denoted by $g(a_1, a_2, \ldots, a_n)$, is defined to be the maximum element of $\mathbb{N} \setminus \mathcal{N}$, if it exists. For example, Proposition 11.11 shows that $g(4, 5) = 11$.

(a) Explain why $g(a_1, a_2, \ldots, a_n)$ can exist only if $a_1, a_2, \ldots, a_n$ are relatively prime.

Remark The converse of this statement is true as well: If the integers are relatively prime, then their Frobenius number exists.

(b) Find, with proof, $g(3, 10)$.
(c) Find, with proof, $g(3, 5, 7)$.
(d) Find, with proof, $g(7, 19, 37)$.
(e) Assuming that a and b are relatively prime, formulate a conjecture for $g(a, b)$ (in terms of $a, b \in \mathbb{N}$). (You do not need to prove your claim.)
(f) Assuming again that a and b are relatively prime, conjecture a formula for the size of the set $\mathbb{N} \setminus \mathcal{N}(a, b)$. (You do not need to prove your claim.)

Remark There is no nice formula known for the Frobenius number of three or more numbers.

14. (a) Prove that for every positive integer $n \geq 6$, a square can be divided into n (not necessarily congruent) squares.
(Hints: First show that one can increase the number of squares by 3, 5, or 7, then use Problem 13 (c) above.)

Remarks It is true—though not entirely straightforward to prove—that the claim is false for $n = 5$.

In a variation of this problem, one asks for all values of $n > 1$ for which a square can be divided into n squares that are pairwise incongruent, as shown in Figure 11.2.
This famous problem is called *squaring the square* (or *perfect square dissection*). As a first attempt, one may try to arrange squares of side lengths 1, 2, 3, et cetera, to form one large square, but it turns out that this is impossible. Indeed, there is just one pair of integers m and n (with $n > 1$) for which

$$1^2 + 2^2 + \cdots + n^2 = m^2;$$

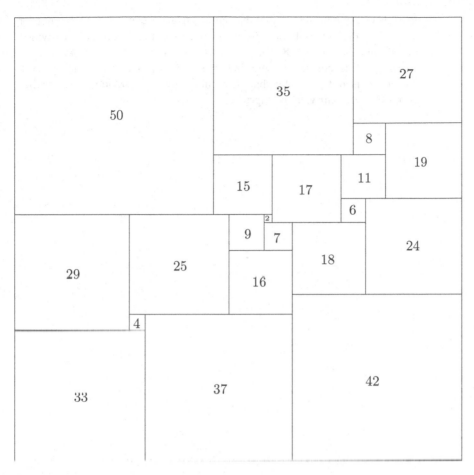

Fig. 11.2 Squaring a square

namely, $n = 24$ and $m = 70$, but, as it turns out, a square of side length 70 cannot be divided into 24 squares of side lengths $1, 2, \ldots, 24$, respectively. For a long time it was conjectured that squaring the square is not possible for any n, until four University of Cambridge mathematicians discovered in the 1930s that the answer is affirmative for $n = 69$. (Their methods used electrical network theory and Kirchoff's Laws.) Then, in 1978, A. J. W. Duijvestijn proved that the smallest value of n for which a square can be divided into n pairwise incongruent squares is $n = 21$. Figure 11.2 exhibits this amazing (and unique) example.

(b) Prove that for every positive integer $n \geq 71$, a cube can be divided into n (not necessarily congruent) cubes.

(Hint: Use Problem 13 (d) above.)

Remarks It can be shown that the claim is true for every $n \geq 48$ but not for $n = 47$; in fact, the only positive integers n for which a cube can be divided into n cubes are $n = 1, 8, 15, 20, 22, 27, 29, 34, 36, 38, 39, 41, 43, 45, 46,$ and $n \geq 48$. In contrast to squaring the square, one can prove that "cubing the cube" is not possible; that is, a cube cannot be divided into smaller cubes that are pairwise incongruent!

Chapter 12
Existential Proofs

In the last several chapters we discussed proof techniques for *universal statements* of the form

$$\forall a \in U, P(a);$$

in this chapter we focus on the existential quantifier and analyze *existential statements* of the form

$$\exists a \in U, P(a).$$

For instance, we may claim that a certain equation has a real number solution (the existence of $\sqrt{2}$, to be formally proven only in Chapter 21, is a prime example), or we may claim that a certain set has a minimum element (by Theorem 10.6, every nonempty set of natural numbers does). Quite often, we deal with variations such as

$$\forall a \in U, \exists b \in V, P(a, b);$$

for example, when in Chapter 1 we claimed that a certain game had a winning strategy for Player 2, we made an existential statement that for any sequence of moves by Player 1, there was a response by Player 2 that resulted in a win for Player 2.

Obviously, the proof of an existential statement requires establishing that a certain object exists. This can be done in two ways: we may provide a *constructive proof* in which we find the required object explicitly (and prove that it satisfies the requirements), or we might find a *nonconstructive* method to verify the existence of the object without actually specifying it. In this chapter we will see examples for both of these techniques.

As a matter of fact, we have already seen examples of constructive and nonconstructive proofs. We gave a constructive proof for the statement in Problem 6

© Springer Nature Switzerland AG 2020

B. Bajnok, *An Invitation to Abstract Mathematics*, Undergraduate Texts in Mathematics, https://doi.org/10.1007/978-3-030-56174-1_12

(b) of Chapter 3 that "Every odd positive integer can be written as the sum or the difference of two perfect squares": We showed that, if n is an odd integer, then

$$\left(\frac{n+1}{2}\right)^2 \text{ and } \left(\frac{n-1}{2}\right)^2$$

are squares of integers whose difference is n. We have also seen numerous nonconstructive proofs: For example, in Problem 10 (a) of Chapter 3 we proved that "In every group of five people, there are two people who know the same number of people in the group" by proving that the negation of the statement (that all five of them know a different number of people) could not be true.

Occasionally, it is not even clear if a particular proof for an existential statement is constructive or not. For example, in what one may call a semi-constructive proof, we can prove that the set of irrational numbers is not closed for exponentiation by exhibiting two numbers,

$$\sqrt{2}^{\sqrt{2}} \text{ and } \left(\sqrt{2}^{\sqrt{2}}\right)^{\sqrt{2}},$$

and proving that one of them (although it is not clear which one) provides an example of irrational numbers x and y for which x^y is rational. Indeed, we have two possibilities: If $\sqrt{2}^{\sqrt{2}}$ is rational, we are done; otherwise,

$$\left(\sqrt{2}^{\sqrt{2}}\right)^{\sqrt{2}}$$

(which equals 2) provides an example. (It has actually been proved that $\sqrt{2}^{\sqrt{2}}$ is irrational, but our point is that we don't need to know this very complicated proof to establish our claim. See also Problem 10 (f) of Chapter 9.) The proof of part (c) of Problem 2 at the end of this chapter follows a similar outline.

In this chapter we present some additional examples for existential proofs. First, we employ a typical constructive technique to make good on our promise made earlier and provide a proof for the well-known existential statement of Problem 5 (i) of Chapter 7.

Proposition 12.1 *There is a perfect square between any positive integer and its double (inclusive).*

Proof Let a be an arbitrary positive integer; we will construct a perfect square number b that satisfies $a \leq b \leq 2a$.

Note that by the definition of the ceiling function,

$$\sqrt{a} \leq \lceil \sqrt{a} \rceil < \sqrt{a} + 1$$

and therefore

$$\sqrt{a}^2 \le \lceil \sqrt{a} \rceil^2 < (\sqrt{a} + 1)^2$$

or

$$a \le \lceil \sqrt{a} \rceil^2 < a + 2\sqrt{a} + 1.$$

We claim that $b = \lceil \sqrt{a} \rceil^2$ satisfies all required conditions. By definition, b is a perfect square; from above, $b \ge a$. It remains to be shown that $b \le 2a$. This can be verified easily for $a = 1, 2, 3, 4$, and 5; for example, for $a = 5$ we have $b = 9 < 10 = 2a$.

Suppose now that $a \ge 6$. Since $a \ge 6$, we have $a - 1 > 0$, $a^2 - 2a = a(a-2) > 0$, and $a^2 - 6a = a(a - 6) \ge 0$. Thus, we have

$$\begin{aligned}
b &= \lceil \sqrt{a} \rceil^2 \\
&< a + 2\sqrt{a} + 1 \\
&= a + \sqrt{4a} + 1 \\
&= a + 1 + \sqrt{(a^2 - 2a) - (a^2 - 6a)} \\
&\le a + 1 + \sqrt{a^2 - 2a} \\
&< a + 1 + \sqrt{a^2 - 2a + 1} \\
&= a + 1 + a - 1 \\
&= 2a,
\end{aligned}$$

as claimed.   □

Our next example for an existential statement is a result from antiquity; its famous proof, given by Euclid around 300 BCE, is what many mathematicians consider to be the most elegant proof of all time.

Theorem 12.2 *There are infinitely many prime numbers.*

Proof We will prove that there are infinitely many positive primes by showing that no finite list of positive primes can contain all of them. Suppose that we are given finitely many, say n, positive primes, namely $p_1, p_2, \ldots,$ and p_n. Consider the quantity

$$K_n = p_1 \cdot p_2 \cdots p_n + 1.$$

Since K_n is an integer greater than 1, by the Fundamental Theorem of Arithmetic (in fact, it suffices to rely on the Factorization Theorem as presented in Theorem 11.4), it must have at least one positive prime factor (whether K_n itself is prime or not). We will prove that this prime cannot equal any of $p_1, p_2, \ldots, p_n$; therefore, there must exist at least one additional positive prime.

Indeed, if K_n were divisible by p_j (where $j = 1, 2, \ldots,$ or n), then we could write K_n as $p_j \cdot c$ for some positive integer c. Clearly, the product $p_1 \cdot p_2 \cdot \cdots \cdot p_n$ is also divisible by p_j; say it equals $p_j \cdot d$ for some positive integer d. But then we have

$$1 = K_n - p_1 \cdot p_2 \cdot \cdots \cdot p_n = p_j \cdot c - p_j \cdot d = p_j \cdot (c - d),$$

which is impossible, since 1 cannot be divisible by the prime p_j. This establishes our claim.   □

Why is this proof so elegant? Because it establishes the fact that the number of primes is indeed infinite without ever exhibiting more than finitely many primes. Furthermore, the argument establishing that no finite list of primes can possibly contain all of them does not yield explicit additional primes. It is important to point out that we do not claim that K_n itself is prime, only that it has some prime divisor that is different from those listed; as we saw in Problem 4 (b) of Chapter 3, the n-th Euclid number K_n is prime for $n \leq 5$, but composite for $n = 6$. (At this time we do not know whether there are finitely many or infinitely many Euclid numbers that are prime.)

Our next example also employs a nonconstructive technique, and establishes a well-known result that we will return to later in this book.

Proposition 12.3 *If positive integers a and b are relatively prime, then the equation*

$$ax + by = 1$$

has integer solutions (x, y).

We can observe that the equation having an integer solution at all implies that there are infinitely many: Indeed, if $ax_0 + by_0 = 1$ for some integers x_0 and y_0, then the ordered pair $(x_0 + bk, y_0 - ak)$ is a solution for any integer value of k, since

$$a(x_0 + bk) + b(y_0 - ak) = 1.$$

Note also that the assumption that a and b are relatively prime is necessary: If they had a common divisor $d > 1$, then $ax + by$ would be divisible by d for all integers x and y and thus could not equal 1.

Proof Consider the set

$$A = \{0 \cdot a, 1 \cdot a, 2 \cdot a, \ldots, (b - 1) \cdot a\}.$$

Our goal is to prove that one of the elements of A is congruent to 1 mod b (that is, has a remainder of 1 when divided by b). This will yield our claim easily, since if $k \cdot a$ is congruent to 1 mod b for some integer k, then $k \cdot a = q \cdot b + 1$ for some integer q; setting $x = k$ and $y = -q$ will then give

$$a \cdot x + b \cdot y = a \cdot k - q \cdot b = 1,$$

as desired.

In order to prove this, we first show that the elements of A are pairwise incongruent mod b. Indeed, if for some integers i and j with $0 \le i \le b - 1$ and $0 \le j \le b-1$, the elements $i \cdot a$ and $j \cdot a$ leave the same remainder when divided by b, then their difference, $(i - j) \cdot a$, is divisible by b. But, since a and b are relatively prime, by Problem 11 of Chapter 11 we find that $i - j$ is divisible by b, which for $1 \le i \le b - 1$ and $1 \le j \le b - 1$ is only possible if $i = j$.

So we have found that the b elements of A all leave different remainders when divided by b. This means that these b different remainders are (in some order) $0, 1, 2, \ldots$, and $b-1$; in particular, one of them leaves a remainder of 1, as claimed. $\square$

The key idea in our proof is worth a bit of discussion. We constructed a set A of size b, then considered the remainders of the elements of A when divided by b. If these remainders, forming a subset of $\{0, 1, 2, \ldots, b - 1\}$, wouldn't include 1, then two of them would be the same, and we proved that this cannot happen. This seemingly simple method is used quite often in mathematics; it can be stated in general as follows:

Theorem 12.4 (The Pigeonhole Principle) *Let m and n be positive integers with $m > n$. If m objects are placed into n boxes, then at least one box will contain at least two objects.*

The name of this principle comes from an historic formulation in which the objects are pigeons and the boxes are pigeonholes. Note that we cannot claim that exactly one box will have at least two objects in it or that some box will have exactly two objects in it. For example, with $m = 6$ and $n = 3$, we could leave one box empty and place three objects in both of the other boxes.

We can generalize the Pigeonhole Principle as follows:

Theorem 12.5 (The Generalized Pigeonhole Principle) *Let m, n, and k be positive integers with $m > kn$. If we place m objects into n boxes, then at least one box will contain at least $k + 1$ objects.*

The proof of these principles can be established by an easy indirect argument: If the claims were to be false, then the total number of objects in the n boxes would be less than m, a contradiction. We will discuss this in a more general and precise manner in Chapter 19.

Our next example states a well-known result—often referred to as "the first result in *Ramsey theory*." We have already assigned this proof in Chapter 1 Problem 8 (c) as well as in Chapter 3 Problem 10 (c), but it is a beautiful enough proof to be repeated here.

Proposition 12.6 *In any group of six people, there are either three people who all know each other or three people so that no two of them know each other.*

Proof Let the six people be A, B, C, D, E, and F. Consider person A. By the Generalized Pigeonhole Principle (with the five objects being B, C, D, E, and F and the two boxes being the ones whom A knows and the ones whom A does not know, respectively), we can claim that either A knows three or more of the other people or there are three or more of them whom A does not know.

Consider the first case; assume wlog that A knows B, C, and D (and perhaps also E and/or F). Again there are two cases: either B, C, and D are three people such that none of them know each other; or there is (at least) one pair of them, wlog B and C, who do know each other. In either case we are done: either we have B, C, and D forming three people who are all strangers, or we have A, B, and C who all know each other.

The case in which there are three people whom A does not know is the reverse of the first case; replacing "knowing each other" by "not knowing each other" completes the proof.   □

Note that Proposition 12.6 implies that the conclusion holds for any group of n people with $n \geq 6$ (first choose an arbitrary six of the n people, then use Proposition 12.6). The conclusion fails, however, when $n = 5$, as Chapter 1 Problem 8 (b) and Chapter 3 Problem 10 (d) demonstrate.

Proposition 12.6 motivates the following famous concept.

Definition 12.7 *The* Ramsey number $R(h, k)$ *of integers* $h \geq 2$ *and* $k \geq 2$ *is the smallest positive integer value of n for which it is true that in any group of n people there are either h people who all know each other or k people so that no two of them know each other.*

It is easy to see that $R(2, k) = k$ for every $k \geq 2$ and $R(h, 2) = h$ for every $h \geq 2$ (see Problem 7). According to Proposition 12.6, we have $R(3, 3) \leq 6$; since the conclusion of Proposition 12.6 fails for $n = 5$, we actually have $R(3, 3) = 6$. In Problem 7 we prove that $R(3, 4) = 9$. One can prove that the Ramsey number $R(h, k)$ exists for all integers $h \geq 2$ and $k \geq 2$ but, at the present time, exact values of $R(h, k)$ are only known for very small values of h and k.

We now turn to an important variation of existential statements: uniqueness statements. A typical *uniqueness statement* is of the form

$$\exists! a \in U, P(a)$$

or

$$\forall a \in U, \exists! b \in V, P(a, b).$$

Uniqueness proofs can naturally be broken up into two parts: we need to prove that the desired object exists, and then we need to prove that we cannot have two or more such objects (this latter part is frequently accomplished by an indirect argument).

Let us see an important example of a uniqueness statement. While the claim is quite evident (in fact, one might be inclined to use this result without proof), it provides a good example for uniqueness proofs.

Theorem 12.8 (The Division Theorem) *Suppose that a and b are integers and $b > 0$. Then there are unique integers q and r such that $a = bq + r$ and $0 \le r \le b - 1$.*

Here q is called the *quotient* and r is called the *remainder*; the notation for the quotient and the remainder are "a div b" and "a mod b," respectively. According to the Division Theorem, when we divide an integer by a positive integer, there is a unique quotient and a unique remainder.

Proof We first prove the existence. Let us set

$$q = \left\lfloor \frac{a}{b} \right\rfloor \text{ and } r = a - b \cdot \left\lfloor \frac{a}{b} \right\rfloor.$$

With these choices, it is clear that $a = bq + r$; we can verify that r falls into the required range, as follows: By the definition of the floor function, we have

$$\frac{a}{b} - 1 < \left\lfloor \frac{a}{b} \right\rfloor \le \frac{a}{b}.$$

Multiplying by $-b$ (note that $-b < 0$) and then adding a implies

$$0 \le a - b \cdot \left\lfloor \frac{a}{b} \right\rfloor < b,$$

as required.

To prove the uniqueness, suppose that $a = bq_1 + r_1$ with $0 \le r_1 < b$, and $a = bq_2 + r_2$, with $0 \le r_2 < b$; we need to prove that $q_1 = q_2$ and $r_1 = r_2$. Note that the two equations yield

$$r_1 - r_2 = b(q_2 - q_1),$$

and the two pairs of inequalities imply that

$$-b < r_1 - r_2 < b.$$

Since the only number strictly between $-b$ and b that is divisible by b is 0, we must have $r_1 - r_2 = 0$, from which $q_2 - q_1 = 0$ as well. Therefore, $q_1 = q_2$ and $r_1 = r_2$, as claimed. □

An immediate—and rather obvious—consequence of the Division Theorem is that every integer is either of the form $2k$ or $2k + 1$ ($k \in \mathbb{Z}$); similarly, every integer is of the form $3k$, $3k+1$, or $3k+2$, etc. A representation of the integers in these forms is frequently used (cf., for example, congruences introduced in Definition 3.2).

Another application of the Division Theorem is the following important result for the greatest common divisor of integers.

Recall from Problem 11 of Chapter 10 that, for given positive integers $a_1, a_2, \ldots, a_n$, we defined $\mathcal{Z}$ as the set of positive integers that can be written as a linear combination of $a_1, a_2, \ldots, a_n$ over the set of integers; that is,

$$\mathcal{Z} = \mathbb{N} \cap \{a_1 x_1 + a_2 x_2 + \cdots + a_n x_n \mid x_1, x_2, \ldots, x_n \in \mathbb{Z}\}.$$

According to Theorem 10.6, the set $\mathcal{Z}$ has a minimum element.

Theorem 12.9 *Given positive integers $a_1, a_2, \ldots, a_n$, let $\mathcal{Z}$ be the set of positive integers that can be written as a linear combination of $a_1, a_2, \ldots, a_n$ over the set of integers. Then the minimum element of $\mathcal{Z}$ is the greatest common divisor of $a_1, a_2, \ldots, a_n$.*

Proof Let $d = \min \mathcal{Z}$; as we explained, d exists.

First we prove that d is a common divisor of $a_1, a_2, \ldots, a_n$; wlog we show that d divides a_1.

By the Division Theorem, we have integers q and r so that $a_1 = dq + r$ and $0 \le r \le d - 1$. We need to show that $r = 0$. We proceed indirectly and assume that $1 \le r \le d - 1$.

Since $d \in \mathcal{Z}$, there are some integers $x_1, x_2, \ldots, x_n$ for which

$$d = a_1 x_1 + a_2 x_2 + \cdots + a_n x_n.$$

Therefore, we have

$$r = a_1 - dq$$
$$= a_1(1 - x_1 q) + a_2(-x_2 q) + \cdots + a_n(-x_n q).$$

Therefore, r is a linear combination of $a_1, a_2, \ldots, a_n$ over the set of integers; since we also assume that $r \in \mathbb{N}$, we have $r \in \mathcal{Z}$. Since $d = \min \mathcal{Z}$, we have $d \le r$, which is a contradiction with $1 \le r \le d - 1$. Therefore, d is a divisor of a_1; similarly, d is a divisor of $a_2, \ldots, a_n$ as well, and thus it is a common divisor of $a_1, a_2, \ldots, a_n$.

To prove that d is the greatest common divisor, we let c be any common divisor of $a_1, a_2, \ldots, a_n$. But then c is also a divisor of any linear combination of $a_1, a_2, \ldots, a_n$; in particular, c divides d. Therefore, $c \le d$, proving that d is indeed the greatest common divisor.   □

Existence statements and uniqueness statements are ubiquitous in mathematics, and we will certainly see further examples later in this book.

Problems

1. Prove that the first player has a winning strategy for the *Divisor* game (cf. Problem 2 of Chapter 1) no matter what the chosen value of n is.
   (Hints: Assume indirectly that for some integer n, the second player has a winning strategy; that is, no matter what divisor of n the first player names at the beginning, the second player can win the game. What if the first player names n in the first turn?)

   Remark Although we can provide a nonconstructive argument to prove that the first player always has a winning strategy, such a strategy (for arbitrary n) has not yet been found.

2. We say that a subset S of the real numbers is *dense* if for any two given distinct elements of S, there is an element of S that is strictly between them.

   (a) Prove that the set of real numbers $\mathbb{R}$ is dense.
   (b) Prove that the set of rational numbers $\mathbb{Q}$ is dense.
   (c) Prove that the set of irrational numbers $\mathbb{R} \setminus \mathbb{Q}$ is dense.
       (Hint: Given two distinct irrational numbers, construct two real numbers strictly between them with the property that at least one of them is irrational.)

3. (a) Suppose that a and b are integers and $b > 0$. Prove that there are unique integers q, r_1, and r_0 such that $a = b^2 q + b r_1 + r_0$, $0 \leq r_1 \leq b - 1$, and $0 < r_0 \leq b - 1$.
   (b) Prove the following theorem.

   Theorem 12.10 *Suppose that a and b are positive integers and $b > 1$. Then there exist unique nonnegative integers $m, r_m, r_{m-1}, \ldots, r_1, r_0$ so that*

   $$a = r_m b^m + r_{m-1} b^{m-1} + \cdots + r_1 b + r_0,$$

   $0 \leq r_i \leq b - 1$ *holds for each* $0 \leq i \leq m$, *and* $r_m \geq 1$.

   Remarks This is called the *base b (for $b = 2$: binary; for $b = 3$: ternary; and for $b = 10$: decimal) representation* of a. When the base b is clear, we usually write simply $a = r_m r_{m-1} \ldots r_1 r_0$.
   (Hints: For the existence, use strong induction on a. Set

   $$m = \max\{l \in \mathbb{N} \cup \{0\} \mid b^l \leq a\}$$

   and $r_m = \lfloor a/b^m \rfloor$.)

4. (a) Let n be a positive integer. Prove that there are unique nonnegative integers m and r such that $n = m^2 + r$ and $0 \leq r \leq 2m$.
   (b) For a positive integer m, the m-th *triangular number* t_m is defined as

$$t_m = \frac{m(m+1)}{2}.$$

Suppose that n is a positive integer. Prove that there are unique integers m and r such that $n = t_m + r$ and $0 \le r \le m$.

(Hint: Suppose that n is a positive integer and choose m so that t_m is the largest triangular number not larger than n. What can be said about $n - t_m$?)

Remarks The reason for the name is clear: Rather than arranging m^2 dots in an m by m array to yield squares, one can place t_m dots in a triangular pattern so that m dots are in the first row, $m - 1$ in the second row, and so on, all the way to a single dot in the last row. By Proposition 10.4, the total number of points in the arrangement is then t_m.

5. A college has 2,500 students. We do not have any information about their birthdays.

   (a) Find all positive integer values of a for which we can be sure that there are a students who all have the same birthday.
   (b) Find all positive integer values of b for which we can be sure that there are b male students who all have the same birthday or there are b female students who all have the same birthday.
   (c) Find all positive integer values of c and d for which we can be sure that there are c male students who all have the same birthday or there are d female students who all have the same birthday.

6. (a) Use the Pigeonhole Principle to prove that no matter how 7 points are placed within an 8 by 9 rectangle, there will always be a pair whose distance is at most 5. You may use, without proof, the following:

   Lemma 12.11 *If two points are placed within a rectangle with side lengths a and b, then their distance is at most $\sqrt{a^2 + b^2}$.*

   (b) Use the Pigeonhole Principle to prove that no matter how 7 points are placed within an 8 by 9 rectangle, there will always be three that form a triangle of area at most 12. You may use, without proof, the following:

   Lemma 12.12 *If three points are placed within a rectangle with side lengths a and b, then the triangle that they determine has area at most $ab/2$.*

7. In this problem we evaluate the Ramsey number $R(h, k)$ when h or k equals 2, and establish that $R(3, 4)$ equals 9.

   (a) Prove that $R(2, k) = k$ for every $k \ge 2$ and $R(h, 2) = h$ for every $h \ge 2$.
   (b) Prove that $R(3, 4) \le 10$.
   (Hint: Follow the approach of Proposition 12.6. Consider one of the people in the group, and explain first why that person must either know at least four other people or not know at least six.)
   (c) Prove that $R(3, 4) \le 9$.

(Hint: Explain first why there must be at least one person in the group who either knows at least four other people or does not know at least six.)

(d) Prove that $R(3, 4) \geq 9$.

(Hint: Provide an example for a group of eight people without three among them who all know each other or four so that no two of them know each other.)

8. Let n be an integer with $n \geq 2$ and suppose that S is a set of n positive integers so that none of them is more than $2n - 2$.

   (a) Prove that S has two elements that are relatively prime.
   (Hint: Use the Pigeonhole Principle to argue that, in fact, S will contain two consecutive numbers.

   (b) Prove that S has two elements so that one of them is divisible by the other.
   (Hint: Use Problem 7 (a) of Chapter 3.)

   (c) Does our claim in part (a) remain valid if we replace $2n - 2$ by $2n - 1$?

   (d) Does our claim in part (b) remain valid if we replace $2n - 2$ by $2n - 1$?

9. In Chapters 3 and 4 we learned that if n is not prime, then the n-th Mersenne number $M_n = 2^n - 1$ definitely cannot be prime as it must have more than two positive divisors. In fact, even when n is prime, M_n may have non-trivial divisors (e.g., M_{11} is divisible by 23 and 89 in addition to being divisible by 1 and itself). Use the Pigeonhole Principle and the Division Theorem to prove that, in fact, for every odd integer d, there is a Mersenne number M_n that is divisible by d.
   (Hint: Use the Division Theorem with $a = M_n$ and $b = d$, then use the Pigeonhole Principle for the possible remainders.)

10. A landmark theorem in number theory is Dirichlet's Theorem (proved in 1837), stated as follows:

Theorem 12.13 (Dirichlet's Theorem) *Suppose that q and r are relatively prime positive integers and $r < q$. Then there are infinitely many positive primes that leave a remainder of r when divided by q.*

For $q = 2$, Dirichlet's Theorem simply claims that there are infinitely many odd primes; of course, this clearly holds as all positive primes except for 2 are odd. For $q = 3$, Dirichlet's Theorem says that there are infinitely many positive primes of the form $3k + 1$ and also of the form $3k + 2$. (Of course, 3 is the only positive prime that is in neither of these forms.) Similarly, there are infinitely many positive primes that leave a remainder of 1 when divided by 4 and there are infinitely many positive primes that leave a remainder of 3 (with 2 being the single prime that is in neither group).
In this problem we prove two cases of Dirichlet's Theorem (other cases are considerably more difficult).

(a) Prove that there are infinitely many positive primes that leave a remainder of 3 when divided by 4.
(Hints: The first few such positive primes are 3, 7, 11, 19, 23, 31, 43, and 47. Suppose, indirectly, that there are finitely many such primes, say $p_1, p_2, \ldots,$

and p_n, then consider the quantity

$$K_n = 4p_1 \cdot p_2 \cdots p_n - 1.$$

Use Problem 5 (d) of Chapter 3.)

(b) Prove that there are infinitely many positive primes that leave a remainder of 5 when divided by 6.

11. We call an integer d a *primary number* if $d \geq 2$ and for every positive integer n that is relatively prime to d, the number $n^{d-1} - 1$ is divisible by d.

(a) Find, with proof, the first four primary numbers.
(b) Prove the following theorem.

Theorem 12.14 (Fermat's Little Theorem) *Every positive prime is a primary number; that is, for every positive prime p and every positive integer n, if n is not divisible by p, then the number $n^{p-1} - 1$ is divisible by p.*

(Hints: Use the method of Proposition 12.3. Start by showing that for each positive integer n that is not divisible by p, the $p - 1$ numbers in the set

$$A = \{1 \cdot n, 2 \cdot n, \ldots, (p-1) \cdot n\}$$

all leave different remainders mod p and that none of these remainders is 0. Use this to prove that

$$n^{p-1} \cdot (p-1)! - (p-1)!$$

is divisible by p, then conclude that $n^{p-1} - 1$ is divisible by p.)

(c) The converse of Fermat's Little Theorem is false: even though the first 102 primary numbers happen to agree with the first 102 positive primes, the 103rd primary number is 561, which is not a prime. Prove that 561 is a primary number.

(Hints: $561 = 3 \cdot 11 \cdot 17$. Let n and 561 be relatively prime. To prove that $n^{560} - 1$ is divisible by 561, by problem 10 (a) of Chapter 11, it suffices to show that it is divisible by each of 3, 11, and 17. Since 3 is a primary number, $n^2 - 1$ is divisible by 3. Since 560 is even, $n^{560} - 1$ is divisible by $n^2 - 1$. Apply this method to the other two factors.)

Remarks A composite primary number such as 561 is called a *Carmichael number*. It was a famous open problem, until proved in 1994, that there are infinitely many Carmichael numbers.

12. In a letter to Leonhard Euler, written in 1730, Christian Goldbach noted that the infinitude of primes follows from the fact that the Fermat numbers are pairwise relatively prime. Provide the details to Goldbach's proof; namely, verify that the Fermat numbers—as defined recursively in Chapter 2 Problem 6 (b)—are pairwise relatively prime, and explain why this implies that there are infinitely many primes.

Appendix D
Ten Famous Problems

In Appendices A and B we discussed our top ten lists of most famous conjectures and most famous theorems, respectively. Here we feature a similar collection of ten famous problems. We will not justify our selection; indeed, there really is no clear distinction between theorems and solved problems or, similarly, between conjectures and unsolved problems. We should just note that we selected problems whose solutions, while considered elementary—that is, not requiring any knowledge beyond what we already have at this point—provide challenges beyond the typical problems of the earlier chapters. We hope that our collection will prove to be interesting.

Problem 1: Nobody Knows Nothing!

Alvin is thinking of two distinct integers, each more than 1 but less than 100. He tells their sum to Sam, and the product to Preston, then leaves the room. Sam and Preston are trying to figure out what the two numbers are. We overhear the following conversation.

> Preston: "I don't know what the two numbers are."
> Sam: "I knew that. I don't know what the numbers are either."
> Preston: "In that case I know what the two numbers are."
> Sam: "Well, then I do too."

What were the two numbers?
(Hints: Let P and S denote the product and the sum of the two numbers, respectively. Preston's first claim implies that P cannot be the product of two primes, and it cannot have a prime divisor that is more than 50. Sam's announcement that he knew that Preston cannot immediately find out what the two numbers are tells us that S could not be the sum of two primes (thus S cannot be even or 2 larger than a prime),

© Springer Nature Switzerland AG 2020
B. Bajnok, *An Invitation to Abstract Mathematics*, Undergraduate Texts in Mathematics, https://doi.org/10.1007/978-3-030-56174-1

and neither can S be greater than or equal to 55. Therefore, S can only be 11, 17, 23, 27, 29, 35, 37, 41, 47, 51, or 53. Find a way to rule out all but one of these possibilities, then proceed to find the two numbers.)

Problem 2: Mathematical Survivor

The game begins with n people on an island. The people are numbered 1 through n. Each day, the remaining islanders vote on whether the remaining islander with the highest number can stay on the island. If half or more of them say the person with the highest number must leave, then that person leaves the island and the game continues. Otherwise, the game ends and the remaining islanders split a million dollars equally. Assume the islanders act independently, are perfectly rational, and will vote in whatever way will give them the most money at the end. How long will the game last and how many people will remain on the island at the end?

(Hints: Start by making a conjecture about the set of n values for which nobody gets voted off the island and the game ends on the first day, and prove your conjecture using strong induction.)

Problem 3: From the White House

The President and the First Lady are hosting a party at the White House for n invited couples.

(a) When the couples arrived, a certain number of handshakes took place. The President asked each of the other people present how many people they shook hands with (the members of the Secret Service and other staff in the room were not involved in handshakes). Amazingly, all answers were different. How many hands did the First Lady shake? (We don't know who shook hands with whom, only that spouses did not shake each other's hands.)

(Hints: Note that the numbers of handshakes reported to the President must be the nonnegative integers up to $2n$. Can you decide who is married to whom?)

(b) After getting acquainted, everybody took a seat around a round table. Protocol maintains that for each course, servers serve the First Lady first, then move around the table clockwise and serve the second person they come to who has not been served yet, continuing this way, until only one person remains who then gets served last. (For example, with $n = 2$ and numbering the 6 people at the table clockwise as 1, 2, 3, 4, 5, and 6, with the First Lady being 1, the order is 1, 3, 5, 2, 6, and finally 4.) Where shall the President sit in order to be served last?

(Hint: Consider first the case in which the number of people around the table is a (nonnegative integer) power of 2.)

Remark This problem is a variation of the well-known *Josephus Problem*.

Problem 4: The Towers of Hanoi

(a) The towers of Hanoi consist of n disks of all different sizes with holes in the middle, placed on three pegs in such a way that no disk is placed on a disk of smaller size. Initially, all n disks are on a single peg. Prove that it is possible to move all disks onto another peg one disk at a time, so that at no time is a disk sitting on a smaller disk. What is the minimum number of moves required?
(Hints: Note that, once you make a conjecture about the minimum number of moves, you must prove that it is possible to carry out the task with that many moves, and that it is not possible to do this with fewer moves. Use induction for both claims.)

(b) How does the answer change if initially all disks are on peg #1, they must be moved to peg #3, and only moves to "neighboring" pegs are allowed?

Remark These questions have not been answered in general when there are more than three pegs available.

Problem 5: Checkerboard Tilings and Primino Numbers

Let's call a positive integer p a *primino number* if it is impossible to tile an m-by-n board ($m, n \in \mathbb{N}$) with a collection of 1-by-p tiles unless at least one of m or n is divisible by p. By Euclid's Principle, every positive prime is a primino number. Find all other primino numbers.
(Hints: Suppose that it is possible to cover an m-by-n board with a collection of 1-by-p tiles. Color all squares of the board in columns $1, p + 1, 2p + 1, \ldots$ green, and color all squares in columns $p, 2p, 3p, \ldots$ yellow. Let a be the number of horizontal tiles, b be the number of vertical tiles in a green column, and let c be the number of vertical tiles in a yellow column. Express the number of green squares of the board as a function of a, b, c, and p; do the same for the number of yellow squares. Consider two cases: when the number of green squares equals the number of yellow squares and when it does not.)

Problem 6: Watching Your Weights

You are a grocer at the market, and you are required to measure various items using a balance scale with two pans and some certified measuring weights. (A balance scale cannot show the actual weight of items; it is only able to tell if the total weight of items placed on the left pan equals, is more than, or is less than the total weight of items placed on the right pan.)

(a) What is the minimum number of measuring weights that you will need to measure each integer weight up to n units if the weights must all be placed on the same pan?
    (Hint: Note that for $n = 3$, two measuring weights are needed: one of 1 unit and another of 2 units.)
(b) How many measuring weights are needed if they can be placed on either pan?
(c) Suppose that you know that the item you need to measure has an integer weight somewhere between 1 and n units. What is the minimum number of measuring weights that you will need to have available so that you are guaranteed to be able to identify the weight of the item? (As in part (b), the weights can be placed on either pan.)
    (Hint: Observe that for $n = 3$, a single measuring weight of 2 units suffices.)

Problem 7: Some Are Sums

(a) Let us say that a positive integer is *accommodating* if it can be written as the sum of (at least two) consecutive positive integers. For example, $13 = 6 + 7$, $14 = 2+3+4+5$, and $15 = 4+5+6$ are accommodating. Prove that a number is accommodating if, and only if, it is not a (nonnegative integer) power of 2.
(b) Let us say that a positive integer is *yielding* if it can be written as the sum of (at least two) consecutive odd positive integers or the sum of consecutive even positive integers. For example, $14 = 6+8$, $15 = 3+5+7$, and $16 = 1+3+5+7$ are yielding. Prove that a number is yielding if, and only if, it is a composite (i.e., not 1 and not prime).

Remarks We can generalize the previous parts as follows. For a positive integer d, we define the positive integer n to be a *d-sum* if it can be written as the sum of (at least two) terms of an arithmetic sequence of positive integers with difference d, that is, if

$$n = a + (a + d) + (a + 2d) + \cdots + (a + kd)$$

for some positive integers a and k. There is no pretty characterization of all d-sums when $d \geq 3$.

Problem 8: A Really "Mean" Problem

Suppose that a and b are positive real numbers. The *arithmetic mean*, the *geometric mean*, and the *harmonic mean* of a and b are defined as

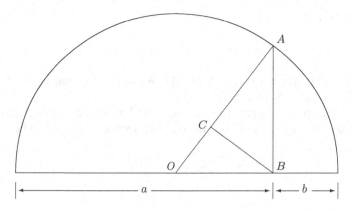

Fig. D.1 A geometric representation of the arithmetic, geometric, and harmonic means of a an b

$$A(a, b) = \frac{a + b}{2},$$

$$G(a, b) = \sqrt{ab},$$

and

$$H(a, b) = \frac{2ab}{a + b},$$

respectively.

(a) Consider Figure D.1 of a semi-circle with center O and right triangles OAB and ABC, as shown.
    Prove that the lengths of the segments OA, AB, and AC are $A(a, b)$, $G(a, b)$, and $H(a, b)$, respectively.
(b) Use part (a) to conclude that for all positive real numbers a and b, we have

$$H(a, b) \le G(a, b) \le A(a, b).$$

(c) Use algebraic methods to prove the inequalities of part (b).
(d) The arithmetic and geometric means of the positive real numbers $a_1, a_2, \ldots, a_n$ ($n \in \mathbb{N}$) are defined as

$$A(a_1, a_2, \ldots, a_n) = \frac{a_1 + a_2 + \cdots + a_n}{n}$$

and

$$G(a_1, a_2, \ldots, a_n) = \sqrt[n]{a_1 a_2 \cdots a_n},$$

respectively. Prove that

$$G(a_1, a_2, \ldots, a_n) \leq A(a_1, a_2, \ldots, a_n).$$

(Hints: Use Problem 1 (i) of Chapter 11. When proving that $P(k + 1)$ implies $P(k)$, let $a_{k+1} = A(a_1, a_2, \ldots, a_k)$.)

(e) (Note: This part requires a basic understanding of derivatives.) Let r be a nonzero real number. We define the r-th mean value of a and b to be

$$M_r(a, b) = \left(\frac{a^r + b^r}{2} \right)^{1/r}.$$

Note that

$$M_{-1}(a, b) = H(a, b)$$

and

$$M_1(a, b) = A(a, b).$$

Prove that for any pair of nonzero real numbers r_1 and r_2 with $r_1 \leq r_2$, we have

$$M_{r_1}(a, b) \leq M_{r_2}(a, b).$$

(f) Where does the geometric mean of a and b fit into the chain of inequalities of part (e)?

Problem 9: Waring's Problem with Negativity

In Problem 9 of Chapter 3, for $k \in \mathbb{N}$, we introduced the quantity $g(k)$ as the smallest positive integer m for which every $n \in \mathbb{N}$ can be written as

$$n = x_1^k + x_2^k + \cdots + x_m^k$$

with some nonnegative integers $x_1, x_2, \ldots, x_m$. Analogously, $v(k)$ is defined as the smallest positive integer m for which every $n \in \mathbb{N}$ can be written as

$$n = \pm x_1^k \pm x_2^k \pm \cdots \pm x_m^k$$

with some nonnegative integers $x_1, x_2, \ldots, x_m$. Here we investigate $v(k)$ for small values of k.

(a) Prove that $v(2) = 3$.
  (Hint: Problem 6 of Chapter 3.)

(b) Prove that $\nu(3) \le 5$.
  (Hint: Start with the identity $(k+1)^3 - 2k^3 + (k-1)^3 = 6k$.)
(c) Prove that $\nu(3) \ge 4$.
  (Hint: Look at integers of the form $9k + 4$.)
(d) Prove that $\nu(4) \le 12$.
  (Hint: Use the identity $(k+3)^4 - 3(k+2)^4 + 3(k+1)^4 - k^4 = 24k + 36$.)
(e) Prove that $\nu(4) \ge 9$.
  (Hint: Look at integers of the form $16k + 8$.)

Remarks Of course, we have $\nu(k) \le g(k)$ for all k, but the exact value of $\nu(k)$ is
only known for $k = 2$. This is in sharp contrast to $g(k)$ that is known for "almost all"
k; see remarks on page 35. As parts (b) and (c) above show, we have $\nu(3) = 4$ or 5.
With a more clever identity than in the hint for (d), we get $\nu(4) \le 10$, so the current
question is whether $\nu(4)$ is 9 or 10. At this time we only know that $5 \le \nu(5) \le 10$,
and even less is known about $\nu(k)$ for $k \ge 6$.

We should also note that the identities in the hints above will make more sense
once we discuss difference sequences in Problem 11 of Chapter 19. For example,
the identity used in part (d) above comes from the third difference sequence of the
sequence of fourth powers:

$$\Delta\Delta\Delta(k^4) = \Delta\Delta((k+1)^4 - k^4)$$
$$= \Delta((k+2)^4 - 2(k+1)^4 + k^4)$$
$$= (k+3)^4 - 3(k+2)^4 + 3(k+1)^4 - k^4.$$

Problem 10: Coloring Lines, Planes, and Space

The *chromatic number* of the n-dimensional space $\mathbb{R}^n$, denoted by $\chi(\mathbb{R}^n)$, is the
minimum number of colors needed to color all points of $\mathbb{R}^n$ in such a way that no
two points of distance 1 have the same color.

(a) Prove that the chromatic number of the line is 2; that is, $\chi(\mathbb{R}) = 2$.
(b) Prove that the chromatic number of the plane satisfies $\chi(\mathbb{R}^2) \ge 4$.
  (Hints: Our goal is to find a small number of points in the plane that
  require four colors. We provide hints for two possible approaches. The first,
  given by Canadian mathematicians Leo and William Moser, consists of two
  diamonds attached to the same point. A different solution, given by American
  mathematician Solomon Golomb, consists of the wheel graph W_6 (cf. Problem 7
  of Appendix B) together with an equilateral triangle.)
(c) Prove that the chromatic number of the plane satisfies $\chi(\mathbb{R}^2) \le 7$.
  (Hints: We need to partition the plane into seven parts so that any two points that
  are exactly distance 1 apart are in different parts. We again provide hints for two
  possible approaches. The first, given by Swiss mathematician Hugo Hadwiger,

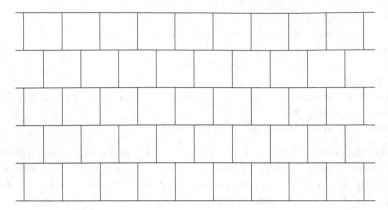

Fig. D.2 Hitting a brick wall

is based on a tiling of the plane by regular hexagons. The second partition, given by Hungarian mathematician László Székely, looks like a brick wall that is partially shown in Figure D.2.)

(d) Prove that $\chi(\mathbb{R}^3) \geq 5$.
(e) Prove that $\chi(\mathbb{R}^3) \leq 21$.

Remarks The question of finding the chromatic number of $\mathbb{R}^n$ has been open for many years. After about six decades of very little progress, in 2018 the biologist and amateur mathematician Aubrey de Grey proved that the chromatic number of the plane is at least 5, and thus $\chi(\mathbb{R}^2)$ equals 5, 6, or 7. The best bounds for three dimensions currently are

$$6 \leq \chi(\mathbb{R}^3) \leq 15,$$

which is quite a wide gap.

Part IV
Advanced Math for Beginners

Chapter 13
Mathematical Structures

In this chapter we study some of the main structures that form the building blocks of mathematics. As a way of introduction, we revisit a point we made earlier about statements and sets.

In Chapter 6 we made the somewhat heuristic claim that properties and identities of statements can easily be altered so that they also hold true for sets. As an example, we considered the claim that

$$(P \vee \neg Q \vee R) \wedge (\neg(P \vee R)) \Leftrightarrow \neg P \wedge \neg Q \wedge \neg R$$

holds for all statements P, Q, and R. Then we made certain replacements and arrived at a claim about sets, namely, that

$$(A \cup \overline{B} \cup C) \cap \overline{(A \cup C)} = \overline{A} \cap \overline{B} \cap \overline{C}$$

holds for all sets A, B, and C. We proved both of these claims by considering their truth tables. (We also illustrated the proof of the claim for sets using Venn diagrams.)

We are about to learn a more sophisticated approach, one that not only avoids having to examine (sometimes rather large) truth tables, but also carries out the two proofs simultaneously; furthermore, the method is powerful enough to be applicable not only to statements and sets but to a large variety of other objects. Our goal is to make these notions more precise. Namely, we will follow the following three-step approach.

- Step 1: We "abstract" some of the properties that both sets and statements satisfy. (There is a reason higher mathematics is often referred to as "abstract mathematics!") As we have already pointed out, there is a strong analogy between negation of statements and complementation of sets, between disjunction of statements and union of sets, and between conjunction of statements and intersection of sets. We will, in fact, see that, if done efficiently, our abstract properties will be applicable in a variety of other settings.

B. Bajnok, *An Invitation to Abstract Mathematics*, Undergraduate Texts in Mathematics, https://doi.org/10.1007/978-3-030-56174-1_13

- Step 2: We then collect some of these properties and say that any setting that satisfies these properties is a particular structure. We will, in fact, form several different structures this way. Like with primitives (cf. Chapter 2) and axioms (cf. Chapter 3), our aim is to keep the number of properties small enough so as to make them applicable in a variety of settings, yet large enough so that, cumulatively, they help us make far-reaching conclusions.
- Step 3: Our third step will be to formulate general claims about our various structures, and to prove these claims using only the properties that any setting of the structure satisfies. For example, we will provide a single unified proof for the equivalence of the logical formulas

$$(P \vee \neg Q \vee R) \wedge (\neg(P \vee R))$$

and

$$\neg P \wedge \neg Q \wedge \neg R$$

and for the equality of the expressions

$$(A \cup \overline{B} \cup C) \cap \overline{(A \cup C)}$$

and

$$\overline{A} \cap \overline{B} \cap \overline{C}$$

about sets. Once such statements are proved in a general structure, they can be applied to any particular model of the structure.

In this chapter we carry out Steps 1 and 2 in a variety of settings: We systematically study some of the familiar and less familiar operation properties, and form certain structures that possess certain ones of them. We will then see how Step 3—proving statements about these structures—can be done in Chapter 14.

We start with the operation properties. Let us assume that $\mathcal{X}$ is a collection of objects (e.g., numbers, statements, sets, polynomials, etc.). A *unary operation* on $\mathcal{X}$ (e.g., taking the negative of a number, the negation of a statement, the complement of a set, etc.) assigns to each object in $\mathcal{X}$ another object; if this other object is also in $\mathcal{X}$, we say that the operation is *closed*. Similarly, a *binary operation* (e.g., addition or multiplication of two numbers, taking the union or intersection of sets) assigns an object to each pair of objects of $\mathcal{X}$. Formally:

- *Closure property (for unary operations):* The unary operation $^{-}$ is closed if, for any element $x \in \mathcal{X}$, we have $\overline{x} \in \mathcal{X}$;
- *Closure property (for binary operations):* The binary operation $*$ is closed, if for any pair of elements $x \in \mathcal{X}$ and $y \in \mathcal{X}$, we have $x * y \in \mathcal{X}$.

For example, we see that addition and multiplication on the set of integers are closed binary operations; multiplication is also closed on the set of odd integers, but addition is not: the sum of two odd integers is not odd.

Two fundamental properties one needs to know about a binary operation are whether the order of the terms can be interchanged and if the terms can be grouped in any way when we have more than two of them. More precisely:

- *Commutative property:* The binary operation $*$ is commutative, if for any pair of elements $x \in \mathcal{X}$ and $y \in \mathcal{X}$, we have

$$x * y = y * x;$$

- *Associative property:* The binary operation $*$ is associative, if for any ordered triple of elements $x \in \mathcal{X}$, $y \in \mathcal{X}$, and $z \in \mathcal{X}$, we have

$$(x * y) * z = x * (y * z).$$

Most familiar operations—such as adding or multiplying numbers or functions, taking unions or intersections of sets, applying disjunctions or conjunctions on statements—are both commutative and associative. Composition of functions is associative but not commutative: $(f \circ g) \circ h = f \circ (g \circ h)$, but usually $f \circ g$ is different from $g \circ f$ (see page 130). Subtraction of real numbers is neither commutative nor associative: $a - b \neq b - a$ unless $a = b$, and $(a - b) - c \neq a - (b - c)$ unless $c = 0$.

Another important property is the following:

- *Identity property:* The binary operation $*$ has an identity in $\mathcal{X}$ if there is an element $e \subset \mathcal{X}$ for which $a * e = a$ and $e * a = a$ hold for every $a \in \mathcal{X}$.

For example, on the set of integers (or the set of real numbers), both addition and multiplication satisfy the identity property, with 0 being the identity for addition (since $a + 0 = 0 + a = a$ for all integers a) and 1 being the identity for multiplication (since $a \cdot 1 = 1 \cdot a = a$ for all integers a). However, multiplication on the set of even integers fails this property as there is no even integer e for which $e \cdot a = a \cdot e = a$ holds for all even integers a. We should point out that if the identity element exists then it is unique: if e_1 and e_2 were both identity elements for the operation $*$ on a set $\mathcal{X}$, then $e_1 * e_2$ would need to equal both e_1 (because e_2 is an identity) and e_2 (because e_1 is an identity), implying that $e_1 = e_2$.

Once the identity property holds, one can ask whether the operation is invertible:

- *Inverse property:* Assume that binary operation $*$ has an identity e in $\mathcal{X}$. We say that $*$ satisfies the inverse property if for each $a \in \mathcal{X}$, there is an $x \in \mathcal{X}$ for which the equations $a * x = e$ and $x * a = e$ hold.

As we just observed, if the identity property holds in $\mathcal{X}$, then the identity element is uniquely determined, hence we don't need to worry about inverses with respect to different identity elements. Clearly, addition has the inverse property on both the integers and the real numbers, with $-a$ being the additive inverse of a. Multiplication fails the inverse property on the integers since only 1 and -1 have multiplicative inverses, and it fails on the set of real numbers as well since 0 does not have a multiplicative inverse (a reciprocal).

We now turn to properties involving more than one operation. The most familiar of these is distributivity:

- *Distributive property:* The binary operation $\diamond$ is distributive with respect to the binary operation $*$ if

$$a \diamond (b * c) = (a \diamond b) * (a \diamond c)$$

  and

$$(a * b) \diamond c = (a \diamond c) * (b \diamond c)$$

  hold for every $a, b, c \in \mathcal{X}$.

As we know, multiplication on the integers (or rationals or reals) is distributive with respect to addition, since $a \cdot (b + c)$ equals $(a \cdot b) + (a \cdot c)$ for all integers a, b, and c. But addition is not distributive with respect to multiplication: $a + (b \cdot c)$ does not equal $(a + b) \cdot (a + c)$ unless $a = 0$ or $a + b + c = 1$.

As we pointed out above, the set of real numbers does not quite have the inverse property for multiplication since 0 has no reciprocal, but, indeed, all other real numbers do. Keeping this in mind, we introduce the following property.

- *Nonzero inverse property:* The binary operation $\diamond$ satisfies the nonzero inverse property excepting e_* (the identity element of the binary operation $*$ in $\mathcal{X}$) if for every $a \in \mathcal{X}$, either $a = e_*$ or there is an $x \in \mathcal{X}$ for which the equations $a \diamond x = e_\diamond$ and $x \diamond a = e_\diamond$ both hold (here $e_\diamond$ is the identity element of $\diamond$).

A fundamental property of the set of real numbers (that we will prove in Chapter 14) is that the product of two real numbers equals zero if, and only if, (at least) one of them is zero. More generally, if $*$ and $\diamond$ are two binary operations on a set $\mathcal{X}$ and e_* denotes the identity element for $*$, we may wonder which elements a and b of $\mathcal{X}$ satisfy the equation $a \diamond b = e_*$. It turns out that in a wide variety of situations (we will make this more precise in Chapter 14), if at least one of a or b equals the identity element e_*, then $a \diamond b$ equals e_* as well; however, the converse of this, that the equation $a \diamond b = e_*$ holds only if $a = e_*$ or $b = e_*$, is true less often. As an example, consider subsets of a universal set U, with $*$ denoting unions and $\diamond$ denoting intersections. The identity element for unions is the empty-set: indeed, $A \cup \emptyset = \emptyset \cup A = A$ holds for all $A \subseteq U$. We can then consider the equation $A \cap B = \emptyset$; this equation holds if $A = \emptyset$ or $B = \emptyset$, but it also holds whenever A

and B are disjoint. Therefore, it is useful to study situations in which this property holds:

- *Nonzero product property:* The binary operation $\diamond$ satisfies the nonzero product property with respect to the binary operation $*$ if for every $a \in \mathcal{X}$ and $b \in \mathcal{X}$, $a \diamond b = e_*$ can only happen if either $a = e_*$ or $b = e_*$.

Finally, we define a property that involves three operations:

- *Complementation property:* Given a unary operation $^-$, the binary operation $*$ has the complementation property with respect to the binary operation $\diamond$ if $a * \bar{a} = e_\diamond$ and $\bar{a} * a = e_\diamond$ hold for every $a \in \mathcal{X}$ (here e_* and $e_\diamond$ denote the identity elements for $*$ and $\diamond$, respectively, and $\bar{a}$ denotes the result of the given unary operation applied to a).

For example, with $^-$, $*$, and $\diamond$ denoting the complementation, union, and intersection of subsets of a universal set U, we have $A \cup \bar{A} = U$ and $\bar{A} \cup A = U$ for all $A \subseteq U$.

We are now ready to define some of the most common mathematical structures. We start with the structures that are most familiar, the ones formed by the number sets $\mathbb{Z}$, $\mathbb{Q}$, $\mathbb{R}$, and $\mathbb{C}$.

Definition 13.1 *Suppose that $\mathcal{R}$ is any collection of objects on which two binary operations are defined: $*$ and $\diamond$. We say that $\mathcal{R}$ is a* ring *for the operations $*$ and $\diamond$ if all of the following properties hold.*

- *($*1$) $*$ has the closure property;*
- *($*2$) $*$ has the commutative property;*
- *($*3$) $*$ has the associative property;*
- *($*4$) $*$ has the identity property;*
- *($*5$) $*$ has the inverse property;*
- *($\diamond1$) $\diamond$ has the closure property;*
- *($\diamond3$) $\diamond$ has the associative property; and*
- *($\diamond D*$) $\diamond$ has the distributive property with respect to $*$.*

If, in addition, in our ring

- *($\diamond2$) $\diamond$ has the commutative property,*

then we say that $\mathcal{R}$ is a commutative ring *for $*$ and $\diamond$; and if in our ring*

- *($\diamond4$) $\diamond$ has the identity property,*

then we call $\mathcal{R}$ a ring with identity *for $*$ and $\diamond$.*

Furthermore, if $\mathcal{R}$ is a commutative ring with identity that has at least two elements and for which

- *($\diamond*$) $\diamond$ has the nonzero product property with respect to $*$,*

then we say that $\mathcal{R}$ is an integral domain *for $*$ and $\diamond$.*

Finally, if $\mathcal{R}$ is a commutative ring with identity that has at least two elements and for which

- $(\diamond 5')$ $\diamond$ *has the nonzero inverse property excepting* e_*,

then we say that $\mathcal{R}$ *is a* field *for* $*$ *and* $\diamond$.

(We placed labels next to the properties above so we can refer to them later.)

We must note that the operations $*$ and $\diamond$ are not interchangeable in Definition 13.1; they satisfy different axioms. Even in the case of a field, when the two operations come closest to symmetry, the first operation, $*$, has the inverse property while the second operation, $\diamond$, only has the nonzero inverse property (and, according to Problem 5 of Chapter 14, this cannot be remedied). Also, $\diamond$ is distributive with respect to $*$, but not vice versa. Therefore, to be precise, we should say that $\mathcal{R}$ is a ring or field for the ordered pair $(*, \diamond)$—however, we will just say that $\mathcal{R}$ is a ring (or field) for $*$ and $\diamond$ and assume that the two operations play their respective roles in order. We should also call attention to the fact that integral domains and fields must have at least two elements (which is equivalent to saying that the identity elements for the two operations are distinct). For traditional reasons, while we see that a set $X = \{a\}$ of a single element is a ring with $a * a = a$ and $a \diamond a = a$, we do not consider this one-element structure an integral domain or a field.

Now let us see some standard examples. We all know that the number sets $\mathbb{Z}$, $\mathbb{Q}$, $\mathbb{R}$, and $\mathbb{C}$ form integral domains for addition and multiplication. Of these, $\mathbb{Q}$, $\mathbb{R}$, and $\mathbb{C}$ are all fields. However, $\mathbb{Z}$ is not a field as it does not satisfy property $(\diamond 5')$ above: for example, there is no integer x for which $2 \cdot x = 1$. This shows that not every integral domain is a field. On the other hand, as we will prove in Chapter 14, every field is an integral domain; that is, the field axioms imply the nonzero product property.

Besides the infinite fields $\mathbb{Q}$, $\mathbb{R}$, and $\mathbb{C}$, there exist some *finite fields*—fields with finitely many elements—as well. Consider first $\mathbb{Z}_2$, the set $\{0, 1\}$ on which the operations $*$ and $\diamond$ are defined as addition mod 2 and multiplication mod 2, respectively. The operation tables are given in Table 13.1.

It takes only a few seconds to check that $\mathbb{Z}_2$ is a field for these operations. We can also verify that $\mathbb{Z}_3$, the structure built on the set $\{0, 1, 2\}$ with the operations defined as addition and multiplication mod 3, is also a field (Table 13.2).

However, moving on to $\mathbb{Z}_4$, we find that the resulting operation tables do not give a field (Table 13.3). We see that 2 has no multiplicative inverse, so this structure is not a field!

Table 13.1 The operation tables of $\mathbb{Z}_2$

$*$	0	1
0	0	1
1	1	0

$\diamond$	0	1
0	0	0
1	0	1

Table 13.2 The operation tables of $\mathbb{Z}_3$

$*$	0	1	2
0	0	1	2
1	1	2	0
2	2	0	1

$\diamond$	0	1	2
0	0	0	0
1	0	1	2
2	0	2	1

Table 13.3 The operation
tables of $\mathbb{Z}_4$

*	0	1	2	3
0	0	1	2	3
1	1	2	3	0
2	2	3	0	1
3	3	0	1	2

◇	0	1	2	3
0	0	0	0	0
1	0	1	2	3
2	0	2	0	2
3	0	3	2	1

Table 13.4 The operation
tables for *Nim* addition and
multiplication on $\{0, 1, 2, 3\}$

⊕	0	1	2	3
0	0	1	2	3
1	1	0	3	2
2	2	3	0	1
3	3	2	1	0

⊗	0	1	2	3
0	0	0	0	0
1	0	1	2	3
2	0	2	3	1
3	0	3	1	2

In spite of our most promising candidate not cooperating, a field on four elements does exist. Recall *Nim* addition and multiplication from Problem 12 of Chapter 2. The operation tables on $\{0, 1, 2, 3\}$ are as in Table 13.4.

Quite amazingly, it turns out that all field axioms are satisfied. For example, we see that 0 is the identity element for $\oplus$, 1 is the identity element for $\otimes$, all four elements have an inverse for $\oplus$, and all elements with the exception of 0 have an inverse for $\otimes$. Some of the other properties are harder to verify: for example, to check that multiplication is distributive with respect to addition, we would need to verify that

$$x \otimes (y \oplus z) = (x \otimes y) \oplus (x \otimes z)$$

holds for all $x, y, z \in \{0, 1, 2, 3\}$. It may take a few minutes, but we can indeed verify that this identity always holds as do all other field axioms. This example generalizes to fields of size $n = 2^{2^m}$ for every $m \in \mathbb{N}$, thus, the next size for which the *Nim* operations provide a field is $n = 16$; see page 429 in Appendix F. However, using different constructions, we can find fields of certain other sizes (see also Problem 3); in fact, we have the following result (which we present here without proof):

Theorem 13.2 *A finite field of order n exists if, and only if, n is a positive integer power of a prime.*

Furthermore, any two fields of the same size are essentially the same: we can get one from the other by a relabeling of the elements. (Clearly, it should not matter how we denote the elements or in what order we list them.) Finite fields have a beautiful and elegant theory, and they play an important role in several branches of mathematics.

Let us now return to infinite fields. Perhaps the most important example of a field is the field of real numbers $\mathbb{R}$. It might be useful to explicitly list the field axioms for this set. Since we will not prove these properties here, we treat them as axioms. The properties for $\mathbb{Q}$ and $\mathbb{C}$ can be listed similarly.

Axiom 13.3 (The Field Axioms of $\mathbb{R}$) *The set $\mathbb{R}$ of real numbers and the usual addition and multiplication operations on $\mathbb{R}$ satisfy the following:*

- *($\neq 0$) (Non-triviality)*
  There are at least two real numbers.
- *(+1) (Closure)*
  For all real numbers a and b, $a + b$ is a real number.
- *(+2) (Commutativity)*
  For all real numbers a and b, we have $a + b = b + a$.
- *(+3) (Associativity)*
  For all real numbers a, b, and c, we have $(a + b) + c = a + (b + c)$.
- *(+4) (Identity)*
  For every real number a, we have $0 + a = a + 0 = a$.
- *(+5) (Inverse)*
  For every real number a, there exists a real number x, such that $a + x = x + a = 0$.
- *($\cdot 1$) (Closure)*
  For all real numbers a and b, $a \cdot b$ is a real number.
- *($\cdot 2$) (Commutativity)*
  For all real numbers a and b, we have $a \cdot b = b \cdot a$.
- *($\cdot 3$) (Associativity)*
  For all real numbers a, b, and c, we have $(a \cdot b) \cdot c = a \cdot (b \cdot c)$.
- *($\cdot 4$) (Identity)*
  For every real number a, $1 \cdot a = a \cdot 1 = a$.
- *($\cdot 5$') (Nonzero Inverse)*
  For every real number a, either $a = 0$ or there exists a real number y such that $a \cdot y = y \cdot a = 1$.
- *($\cdot D+$) (Distributivity)*
  For all real numbers a, b, and c, we have $a \cdot (b + c) = a \cdot b + a \cdot c$ and $(a + b) \cdot c = a \cdot c + b \cdot c$.

Here we treat the terms addition, multiplication, 0, and 1, which are used in the axioms above, as primitives—see the list in Chapter 2. The field axioms do not refer to the other two binary operations of numbers, namely subtraction and division; in fact, these concepts are not on our list of primitives either. This is because these terms can be defined as follows:

Definition 13.4 *The* negative *of a real number a is the real number x for which $a + x = 0$. The negative of a will be denoted by $-a$.*

Definition 13.5 *Given two real numbers a and b, we define the* difference *$a - b$ as the number $a + (-b)$.*

Definition 13.6 *The* reciprocal *of a nonzero real number a is the real number y for which $a \cdot y = 1$. The reciprocal of a will be denoted by $\frac{1}{a}$.*

Definition 13.7 *Given two real numbers a and b, where b is not 0, we define the quotient $\frac{a}{b}$ as the number $a \cdot \frac{1}{b}$.*

Axioms (+5) and (·5') guarantee the existence of the negative of a real number and the reciprocal of a nonzero real number, respectively. It also must be added that these terms are well defined: one can prove that the inverses for addition and multiplication are unique (cf. Problem 11 of Chapter 14).

It is also worth pointing out that the field axioms, as we listed them above, do contain some superfluous statements. For example, it is easy to prove one of the two identities listed under (·D+) using the other one and (·2). Furthermore, we now prove that axiom (+2) follows from the other field axioms.

Theorem 13.8 *Commutativity of addition follows from the other field axioms.*

Proof We assume that the field in question is $\mathbb{R}$ with operations denoted by $+$ and $\cdot$ playing the roles of $*$ and $\diamond$, respectively, and that the corresponding identity elements are 0 and 1. This is just a notational change to make the reading of this proof easier; as long as we do not assume properties of the real numbers beyond those listed in the field axioms, our proof is valid.

We need to prove that for every pair of real numbers a and b, we have $a + b = b + a$. Consider the product

$$(1 + 1) \cdot (a + b).$$

We can use distributivity (·D+) to rewrite it as

$$(1 + 1) \cdot a + (1 + 1) \cdot b,$$

then, using (·D+) again, we further rewrite this as

$$(1 \cdot a + 1 \cdot a) + (1 \cdot b + 1 \cdot b).$$

By (·4), this is equal to

$$(a + a) + (b + b).$$

Starting with the same expression and using the two distributivity axioms in the other order, we get

$$(1 + 1) \cdot (a + b) = (1 \cdot (a + b)) + (1 \cdot (a + b))$$
$$= ((1 \cdot a) + (1 \cdot b)) + ((1 \cdot a) + (1 \cdot b)),$$

from which, using (·4), we get

$$(a + b) + (a + b).$$

Since we started with the same expression, we must have

$$(a + a) + (b + b) = (a + b) + (a + b).$$

Now we add $-a$ on the left as well as $-b$ on the right to both sides (which are available by (+5)), and use (+3) to get

$$(((-a) + a) + a) + (b + (b + (-b))) = (((-a) + a) + b) + (a + (b + (-b))).$$

Four applications of (+5) yield

$$(0 + a) + (b + 0) = (0 + b) + (a + 0),$$

from which the result

$$a + b = b + a$$

follows after applying (+4).   □

Using the terminology of Chapter 4, we can thus say that the field axioms, as listed above, are not independent. We mention, in passing, that the rest of the field axioms are independent. To verify that, one would need to construct structures, for each of the remaining axioms, such that the chosen axiom is false but every other axiom is true.

We will see further examples for rings, integral domains, and fields in the problems at the end of this chapter.

Let us now turn to the axioms that statements and sets satisfy. We introduce the following structure.

Definition 13.9 *Suppose that $\mathcal{B}$ is any collection of objects on which a unary operation, $^-$, and two binary operations, $*$ and $\diamond$, are defined. We say that $\mathcal{B}$ forms a* Boolean algebra *for these operations if all of the following properties hold.*

- *$(^-)^-$ has the closure property;*
- *$(*1)$ $*$ has the closure property;*
- *$(*2)$ $*$ has the commutative property;*
- *$(*3)$ $*$ has the associative property;*
- *$(*4)$ $*$ has the identity property;*
- *$(*C\diamond)$ $*$ has the complementation property with respect to $\diamond$;*
- *$(\diamond 1)$ $\diamond$ has the closure property;*
- *$(\diamond 2)$ $\diamond$ has the commutative property;*
- *$(\diamond 3)$ $\diamond$ has the associative property;*
- *$(\diamond 4)$ $\diamond$ has the identity property;*
- *$(\diamond C*)$ $\diamond$ has the complementation property with respect to $*$;*
- *$(*D\diamond)$ $*$ has the distributive property with respect to $\diamond$; and*
- *$(\diamond D*)$ $\diamond$ has the distributive property with respect to $*$.*

We note that, unlike with rings and fields, the two binary operations in Definition 13.9 can be interchanged: they play perfectly symmetrical roles.

We then have the following theorem about sets.

Theorem 13.10 *Let X be a set. Then $P(X)$, the power set of X, is a Boolean algebra for taking complements in X, and for forming unions and intersections. Namely, we have the following properties.*

- $(^-)$ *(Closure)*
  For all sets $A \in P(X)$, we have $\overline{A} \in P(X)$;
- $(\cup 1)$ *(Closure)*
  For all sets $A, B \in P(X)$, we have $A \cup B \in P(X)$;
- $(\cup 2)$ *(Commutativity)*
  For all sets $A, B \in P(X)$, we have $A \cup B = B \cup A$;
- $(\cup 3)$ *(Associativity)*
  For all sets $A, B, C \in P(X)$, we have $(A \cup B) \cup C = A \cup (B \cup C)$;
- $(\cup 4)$ *(Identity)*
  For all sets $A \in P(X)$, we have $A \cup \emptyset = A$ and $\emptyset \cup A = A$;
- $(\cup C \cap)$ *(Complementation)*
  For all sets $A \in P(X)$, we have $A \cup \overline{A} = X$ and $\overline{A} \cup A = X$;
- $(\cap 1)$ *(Closure)*
  For all sets $A, B \in P(X)$, we have $A \cap B \in P(X)$;
- $(\cap 2)$ *(Commutativity)*
  For all sets $A, B \in P(X)$, we have $A \cap B = B \cap A$;
- $(\cap 3)$ *(Associativity)*
  For all sets $A, B, C \in P(X)$, we have $(A \cap B) \cap C = A \cap (B \cap C)$;
- $(\cap 4)$ *(Identity)*
  For all sets $A \in P(X)$, we have $A \cap X = A$ and $X \cap A = A$;
- $(\cap C \cup)$ *(Complementation)*
  For all sets $A \in P(X)$, we have $A \cap \overline{A} = \emptyset$ and $\overline{A} \cap A = \emptyset$;
- $(\cap D \cup)$ *(Distributivity)*
  For all sets $A, B, C \in P(X)$, we have $A \cap (B \cup C) = (A \cap B) \cup (A \cap C)$ and $(A \cup B) \cap C = (A \cap C) \cup (B \cap C)$;
- $(\cup D \cap)$ *(Distributivity)*
  For all sets $A, B, C \in P(X)$, we have $A \cup (B \cap C) = (A \cup B) \cap (A \cup C)$ and $(A \cap B) \cup C = (A \cup C) \cap (B \cup C)$.

The proof of Theorem 13.10 can be carried out easily using truth tables.

Similarly, we can present the corresponding theorem about statements. Here we need to be a bit more careful: rather than claiming that two statements are equal (which we did not define), we claim that they are equivalent (which was defined by Definition 5.4).

Theorem 13.11 *The collection of all statements forms a Boolean algebra for the operations of negation, disjunction, and conjunction. In particular, we have the following properties for statements.*

- $(\neg)$ *(Closure)*
  For all statements P, $\neg P$ *is a statement;*
- $(\vee 1)$ *(Closure)*
  For all statements P *and* Q, $P \vee Q$ *is a statement;*
- $(\vee 2)$ *(Commutativity)*
  For all statements P *and* Q, *we have* $P \vee Q \Leftrightarrow Q \vee P$;
- $(\vee 3)$ *(Associativity)*
  For all statements P, Q, *and* R, *we have* $(P \vee Q) \vee R \Leftrightarrow P \vee (Q \vee R)$;
- $(\vee 4)$ *(Identity)*
  For all statements P *and all false statements* F, *we have* $P \vee F \Leftrightarrow P$ *and* $F \vee P \Leftrightarrow P$;
- $(\vee C\wedge)$ *(Complementation)*
  For all statements P *and all true statements* T, *we have* $P \vee (\neg P) \Leftrightarrow T$ *and* $(\neg P) \vee P \Leftrightarrow T$;
- $(\wedge 1)$ *(Closure)*
  For all statements P *and* Q, $P \wedge Q$ *is a statement;*
- $(\wedge 2)$ *(Commutativity)*
  For all statements P *and* Q, *we have* $P \wedge Q \Leftrightarrow Q \wedge P$;
- $(\wedge 3)$ *(Associativity)*
  For all statements P, Q, *and* R, *we have* $(P \wedge Q) \wedge R \Leftrightarrow P \wedge (Q \wedge R)$;
- $(\wedge 4)$ *(Identity)*
  For all statements P *and all true statements* T, *we have* $P \wedge T \Leftrightarrow P$ *and* $T \wedge P \Leftrightarrow P$;
- $(\wedge C\vee)$ *(Complementation)*
  For all statements P *and all false statements* F, *we have* $P \wedge (\neg P) \Leftrightarrow F$ *and* $(\neg P) \wedge P \Leftrightarrow F$;
- $(\wedge D\vee)$ *(Distributivity)*
  For all statements P, Q, *and* R, *we have* $P \wedge (Q \vee R) \Leftrightarrow (P \wedge Q) \vee (P \wedge R)$ *and* $(P \vee Q) \wedge R \Leftrightarrow (P \wedge R) \vee (Q \wedge R)$;
- $(\vee D\wedge)$ *(Distributivity)*
  For all statements P, Q, *and* R, *we have* $P \vee (Q \wedge R) \Leftrightarrow (P \vee Q) \wedge (P \vee R)$ *and* $(P \wedge Q) \vee R \Leftrightarrow (P \vee R) \wedge (Q \vee R)$.

Again, each of these statements can be easily proved by constructing a truth table.

Besides sets and statements, there are many other examples for Boolean algebras; we introduce one well-known example next. Let us define B_n to be the set of sequences of a given length n of 0s and 1s; that is, strings of binary digits (called *bits*) of length n. For example,

$$B_3 = \{000, 001, 010, 011, 100, 101, 110, 111\}.$$

We then define the unary operation $^-$ on B_n as the component-wise change of all bits: we turn each 0 bit into a 1 and vice versa. We also define two binary operations: component-wise addition and multiplication specified with the rules in Table 13.5.

Table 13.5 Boolean addition and multiplication

+	0	1
0	0	1
1	1	1

·	0	1
0	0	0
1	0	1

We can verify that B_n is a Boolean algebra by verifying each property in the previous theorems (cf. Problem 2). For example, in B_3 we have

$$101 \cdot \overline{101} = 101 \cdot 010 = 000$$

and

$$(101 + 001) \cdot 011 = 101 \cdot 011 = 001 = 001 + 001 = 101 \cdot 011 + 001 \cdot 011,$$

verifying particular instances of the complementation and distributivity properties. The Boolean algebra B_n is employed frequently in *computer science*.

It is interesting to compare the Boolean algebra properties above to the field axioms. One can make the observation that, in many ways, statements act like numbers with the operations of disjunction, conjunction, and negation acting like addition, multiplication, and taking negatives of numbers, respectively; furthermore, a contradiction plays the role of zero, and a tautology plays the role of 1.

This parallel, however, is not complete; there are differences in the properties above. In particular, there is no inverse to a statement; instead, statements satisfy the complementation property. Also, in a field we have one fewer distributive rule than we do for statements: addition of numbers is not distributive with respect to multiplication. Thus the algebra of statements and sets follows different rules from the algebra of numbers!

Rings, integral domains, fields, and Boolean algebras are just some of the most common structures in abstract mathematics. Others include groups (introduced in Chapter 15), lattices (discussed in Chapter 16), topologies (see Problem 10), and finite geometries (see Problem 11)—each studied extensively in a corresponding branch of mathematics.

Problems

1. For each of the following sets, decide which of the field axioms hold and whether the nonzero product property holds. (Assume that the operations are usual addition and multiplication.) Which are rings? Which are integral domains? Which are fields?

   (a) the set of even integers;
   (b) the set of odd integers;
   (c) the set of positive rational numbers;

(d) the set of rational numbers that can be written in the form $\frac{a}{b}$ where a and b are integers and b is odd;

(e) the real numbers between -1 and 1 (inclusive);

(f) the set of real polynomials (polynomials with real number coefficients) of degree 5;

(g) the set of real polynomials of degree at most 5;

(h) the set of all real polynomials;

(i) the set of integral polynomials (polynomials with integer coefficients);

(j) the set of all functions of the form $\frac{f(x)}{g(x)}$ where f and g are arbitrary real polynomials and $g \neq 0$;

(k) the set of all functions of the form $\frac{f(x)}{g(x)}$ where f and g are arbitrary integral polynomials and $g \neq 0$;

(l) the set of 2-by-2 matrices with real number entries.

(Hints: Recall that a 2-by-2 matrix over the real numbers (or complex numbers) is an array of the form

$$\begin{pmatrix} a & b \\ c & d \end{pmatrix},$$

where $a, b, c, d \in \mathbb{R}$ (or $\mathbb{C}$). Addition and multiplication of such matrices are defined by the rules

$$\begin{pmatrix} a_1 & b_1 \\ c_1 & d_1 \end{pmatrix} + \begin{pmatrix} a_2 & b_2 \\ c_2 & d_2 \end{pmatrix} = \begin{pmatrix} a_1 + a_2 & b_1 + b_2 \\ c_1 + c_2 & d_1 + d_2 \end{pmatrix}$$

and

$$\begin{pmatrix} a_1 & b_1 \\ c_1 & d_1 \end{pmatrix} \cdot \begin{pmatrix} a_2 & b_2 \\ c_2 & d_2 \end{pmatrix} = \begin{pmatrix} a_1 a_2 + b_1 c_2 & a_1 b_2 + b_1 d_2 \\ c_1 a_2 + d_1 c_2 & c_1 b_2 + d_1 d_2 \end{pmatrix},$$

respectively.)

2. Recall that we let B_2 be the collection of all length-2 strings of binary digits; we also defined the operations of negation, addition, and multiplication on B_2 on page 216.

(a) List the elements of B_2.

(b) Construct the operation tables for the three operations.

(c) Verify that B_2 is a Boolean algebra. (Although you are supposed to prove commutativity, associativity, and distributivity for all choices of the elements, it is all right to demonstrate these properties on specific examples.)

3. Suppose that n is a positive integer and let $\mathbb{Z}_n$ consist of the numbers

$$\{0, 1, 2, \ldots, n - 1\}.$$

Define $\oplus$ and $\odot$ to be addition and multiplication of these numbers mod n; that is, let $a \oplus b$ and $a \odot b$ be the remainder of $a + b$ and $a \cdot b$ when divided by n.

(a) Verify that, for these operations, $\mathbb{Z}_n$ is a commutative ring with a multiplicative identity. (It is all right to demonstrate commutativity, associativity, and distributivity on specific examples.)

(b) Find, with proof, all values of n for which $\mathbb{Z}_n$ is an integral domain.

(c) Find some values of n for which $\mathbb{Z}_n$ is a field and some for which it is not a field. Make a conjecture regarding the general case. (You do not need to prove your conjecture—we will do so in Chapter 14.)

4. For each of the following sets, decide if the set forms an integral domain for the usual addition and multiplication operations.

(a) $10\mathbb{Z} = \{10z \mid z \in \mathbb{Z}\}$ (multiples of 10);

(b) $\frac{1}{10}\mathbb{Z} = \{\frac{z}{10} \mid z \in \mathbb{Z}\}$ (multiples of 1/10);

(c) $\mathbb{Z}\left[\frac{1}{10}\right] = \{\frac{z}{10^n} \mid z \in \mathbb{Z}, n \in \mathbb{N}\}$ (finite decimals).

5. Let us define the binary operations $\oplus$ and $\otimes$ on a *finite* set S of real numbers as follows:

$$a \oplus b = \min\{a, b\}$$

and

$$a \otimes b = \max\{a, b\}$$

for every $(a, b) \in S \times S$ (here min and max denote the smaller and the larger—or, in case of equality, either—of the two numbers, respectively). Decide whether these operations satisfy the commutative, associative, distributive, identity, and inverse properties.

6. Let us define the binary operations $*$ and $\diamond$ on the set of real numbers as follows:

$$a * b = a + b + 1 \text{ and } a \diamond b = a \cdot b + a + b$$

for every $(a, b) \in \mathbb{R}^2$. (Here $+$ and $\cdot$ denote ordinary addition and multiplication.)

Decide which of the twelve axioms listed in Definition 10.2 are satisfied. What kind of structure does $\mathbb{R}$ have for these operations?

(Hints: If you claim that a certain axiom holds, make sure you prove the appropriate statement rather than its converse. For example, you cannot prove that the identity property holds by showing that if the identity element exists then it must equal a certain number.)

7. Let S be a nonempty set and let $P(S)$ be the power set of S. Define the operations $*$ and $\diamond$ as follows:

$$A * B = (A \setminus B) \cup (B \setminus A) \text{ and } A \diamond B = A \cap B.$$

(Here $A \setminus B = A \cap \overline{B}$; $A * B$ is called the *symmetric difference* of A and B.)

   (a) Use Venn diagrams to verify that $P(S)$ is a ring for these operations.
   (b) Prove that $P(S)$ is a field if, and only if, S consists of a single element.
       (Hint: To prove that the nonzero inverse property fails when S has more than one element, consider a proper nonempty subset of S, that is, a set A for which $\emptyset \subset A \subset S$.)

8. Let n be a positive integer, and define $D(n)$ to be the set of all positive divisors of n. Define the unary operation $^-$ and the binary operations $*$ and $\diamond$ on $D(n)$ as follows:

$$\overline{a} = \frac{n}{a},$$

$$a * b = \text{lcm}(a, b),$$

and

$$a \diamond b = \gcd(a, b)$$

for every $a, b \in D(n)$. (Here lcm and gcd denote the least common multiple and the greatest common divisor of the two numbers, respectively.)

   (a) Is $D(6)$ a Boolean algebra for these operations?
   (b) Is $D(8)$ a Boolean algebra for these operations?
   (c) Find all values of n under 40 for which $D(n)$ is a Boolean algebra for these operations. Make a conjecture regarding the general case in terms of the prime factorization of n. (You do not need to prove your conjecture.)

9. In this problem we define and study the so-called *order axioms*.

   Definition 13.12 *We say that a ring R (with operations $+$ and $\cdot$) is an* ordered ring *if it contains a subset P with the following properties.*

   - *(O) For every $a \in R$, exactly one of the following holds: $a = 0$, $a \in P$, or $-a \in P$.*
   - *(O+) For all $a \in P$ and $b \in P$, $a + b \in P$.*
   - *(O·) For all $a \in P$ and $b \in P$, $a \cdot b \in P$.*

   Elements in P are called the positive *elements of R. Ordered integral domains (ordered fields) are defined as ordered rings that are also integral domains (fields).*

Here we state as an axiom that, unsurprisingly, $\mathbb{Z}$, $\mathbb{Q}$, and $\mathbb{R}$ satisfy the order axioms.

Axiom 13.13 *The integers form an ordered integral domain and both the rational numbers and the real numbers form ordered fields.*

(a) Define the *negative* elements of an ordered ring R.

   (Hint: Be careful not to confuse the "negative elements" of R with the "negatives of elements" of R. For example, the negative elements of $\mathbb{Z}$ form the set $\{-1, -2, -3, \dots\}$, but every integer (even a positive one) is the negative of some integer!)

(b) Define the order relations *greater than or equal to* and *less than or equal to* and the strict order relations *greater than* and *less than* in an ordered ring

(c) Let $\mathbb{R}[x]$ denote the set of all real polynomials. Verify that $\mathbb{R}[x]$ is an ordered integral domain.

   (Hint: Define a polynomial to be positive if its leading coefficient (the coefficient of the term of highest degree) is positive.)

(d) Let $\mathbb{R}(x)$ denote the set of all functions of the form $\frac{f(x)}{g(x)}$ where f and g are arbitrary real polynomials and $g \neq 0$. Verify that $\mathbb{R}(x)$ is an ordered field.

(e) Prove that $\mathbb{Z}_n$ cannot be an ordered ring for any integer $n \geq 2$.

   (Hints: Suppose, indirectly, that there is a subset P of $\mathbb{Z}_n$ for which the three order axioms hold. By axiom (O), we must have either $1 \in P$ or $n - 1 \in P$, but not both. Show that repeated application of axiom $(O+)$ yields a contradiction.)

10. Let X be a nonempty finite set and let τ be a subset of the power set of X. We say that τ is a *topology* on X if the following three conditions are satisfied:

   - $(\tau 1)$: τ contains $\emptyset$ and X;
   - $(\tau 2)$: if $A \in \tau$ and $B \in \tau$, then $A \cap B \in \tau$;
   - $(\tau 3)$: if $A \in \tau$ and $B \in \tau$, then $A \cup B \in \tau$.

   Suppose that $n \in \mathbb{N}$, $S = \{1, 2, \dots, n\}$, and let

   $$\tau_n = \{U \subseteq S \mid ((a \in U) \wedge (b \in S) \wedge (a|b)) \Rightarrow (b \in U)\}.$$

   (a) Prove that τ_n is a topology for all n.

   (b) Exhibit all elements of τ_n for $n \in \{1, 2, 3, 4, 5, 6\}$.

      (Hints: The sequence $(|\tau_1|, |\tau_2|, |\tau_3|, \dots)$ appears in the On-Line Encyclopedia of Integer Sequences at https://oeis.org as sequence A051026; in particular, $|\tau_1| = 2$, $|\tau_2| = 3$, $|\tau_3| = 5$, $|\tau_4| = 7$, $|\tau_5| = 13$, and $|\tau_6| = 17$.)

11. In this problem we return to Problem 11 of Chapter 4. Given a finite set $\mathcal{P}$ (here called *points*) and a collection $\mathcal{L}$ of subsets of $\mathcal{P}$ (called *lines*), we say that the structure formed by them is a *projective plane* if the following properties hold:

- (P1): $\mathcal{L}$ contains at least two lines.
- (P2): Each line in $\mathcal{L}$ contains at least three points.
- (P3): Every point lies on at least three lines.
- (P4): For each two points, there is exactly one line that contains them.
- (P5): For each two lines, there is exactly one point that lies on both.

Prove that there is a positive integer n (called the *order* of the projective plane) so that every line contains exactly $n + 1$ points, every point is on exactly $n + 1$ lines, and $\mathcal{P}$ and $\mathcal{L}$ both have size $n^2 + n + 1$.

(Hints: First, prove that there exist two lines and a point that is not on either line, and that there exist two points and a line that does not contain either point. Next, prove that, given a line L and a point P not on L, there is a bijection between the set of points on L and the set of lines containing P. Explain how this implies that each line has the same number of points, and each point is on the same number of lines.)

Remarks In Problem 11 of Chapter 4 we verified that projective planes of order 2 and of order 3 exist. It is a very famous open question to decide what the order n of a projective plane can be. It is widely believed that all possible values of n are (nonnegative integer) powers of primes such as $n = 2, 3, 4, 5, 7, 8, 9,$ 11, 13, 16, etc. It has been shown that n cannot be 6, 10, 14, 21, 22, or any of another infinitely many values; of these, $n = 10$ was the one most recently excluded (this was done by a computer program designed by C. W. H. Lam, L. Thiel, and S. Swierz in 1989). Infinitely many others remain undecided as of today, the five smallest of which are $n = 12, 15, 18, 20,$ and 24.

Chapter 14
Working in the Fields (and Other Structures)

In Chapter 13 we got acquainted with mathematical structures such as rings, integral domains, fields, and Boolean algebras. As discussed there, the benefit of abstracting the common properties of various systems into a unifying structure is that, once we prove certain statements about a structure using only the properties that define the structure, they will then be true for each system that models the structure. In this chapter we see examples for such *axiomatic proofs*.

While in theory all theorems of higher mathematics could be proved axiomatically, we will usually not proceed this way; one would need to develop an extensive family of previously proven results, and this would be too tedious and time consuming for us. Nevertheless, the formal axiomatic proofs in this chapter are quite instructive: one can develop a very clear understanding of what constitutes a proof since each step needs to be derived directly from one of the axioms or previously proven statements. Techniques learned in this chapter will be beneficial when studying other axiomatic branches of mathematics such as (abstract) algebra, (real and complex) analysis, topology, and others.

We start by proving some well-known theorems about Boolean algebras. Since we will use only properties that hold for every Boolean algebra, our theorem will be true for the particular Boolean algebras we have seen: statements, sets, and 0-1 sequences.

Theorem 14.1 (The Bound Laws) *Suppose that $\mathcal{B}$ is a Boolean algebra for the unary operation $^{-}$ and the binary operations $*$ and $\diamond$ (as listed in Definition 13.9). Let e_* and $e_\diamond$ denote the identity elements of the operations $*$ and $\diamond$, respectively, and suppose that $a \in \mathcal{B}$. Then we have $a * e_\diamond = e_\diamond$ and $a \diamond e_* = e_*$.*

Proof We start with the first identity. Our proof will consist of a sequence of equations; at each step we indicate the particular property we use (cf. the Boolean algebra properties of Definition 13.9).

$$a * e_\diamond \overset{(\diamond 4)}{=} (a * e_\diamond) \diamond e_\diamond$$

B. Bajnok, *An Invitation to Abstract Mathematics*, Undergraduate Texts in Mathematics, https://doi.org/10.1007/978-3-030-56174-1_14

$$\overset{(*C\diamond)}{=} (a * e_\diamond) \diamond (a * \overline{a})$$

$$\overset{(*D\diamond)}{=} a * (e_\diamond \diamond \overline{a})$$

$$\overset{(\diamond 4)}{=} a * \overline{a}$$

$$\overset{(*C\diamond)}{=} e_\diamond$$

For the second identity, recall that the two binary operations in a Boolean algebra play a symmetrical role, so we can simply interchange $*$ and $\diamond$ (easy with a typesetting program that has a "replace" command).

$$a \diamond e_* \overset{(*4)}{=} (a \diamond e_*) * e_*$$

$$\overset{(\diamond C*)}{=} (a \diamond e_*) * (a \diamond \overline{a})$$

$$\overset{(\diamond D*)}{=} a \diamond (e_* * \overline{a})$$

$$\overset{(*4)}{=} a \diamond \overline{a}$$

$$\overset{(\diamond C*)}{=} e_*$$

This completes our proof.   □

It is worthwhile to state the Bound Laws for our three favorite Boolean algebras.

Corollary 14.2 (The Bound Laws for Statements) *Suppose that T is a true statement, F is a false statement, and P is an arbitrary statement. Then $P \vee T \Leftrightarrow T$ and $P \wedge F \Leftrightarrow F$.*

Corollary 14.3 (The Bound Laws for Sets) *Suppose that A is a set inside a universal set U, and $\emptyset$ is a set with no elements. Then $A \cup U = U$ and $A \cap \emptyset = \emptyset$.*

Corollary 14.4 (The Bound Laws for 0-1 Sequences) *Suppose that $\mathbf{a}$ is a 0-1 sequence of a given length, and $\mathbf{e}$ and $\mathbf{z}$ are the sequences of all 1s and all 0s, respectively, of the same length. Then $\mathbf{a} + \mathbf{e} = \mathbf{e}$ and $\mathbf{a} \cdot \mathbf{z} = \mathbf{z}$.*

Now we turn to proving a theorem about rings (and thus, as a special case, about fields). Our claim will be very similar to one of the claims in Theorem 14.1 above, but it is just a superficial resemblance: we are in a ring, not in a Boolean algebra!

Theorem 14.5 *Suppose that $\mathcal{R}$ is a ring for the binary operations $*$ and $\diamond$ (as listed in Definition 13.1). Let e_* denote the identity element of $*$ and suppose that $a \in \mathcal{R}$. Then we have $e_* \diamond a = a \diamond e_* = e_*$.*

Proof As before, our proof will consist of a sequence of equations; at each step we indicate the particular property we use (cf. the ring properties of Definition 13.1).

$$a \diamond e_* \overset{(*4)}{=} (a \diamond e_*) * e_*$$

$$\overset{(*5)}{=} (a \diamond e_*) * [(a \diamond e_*) * -(a \diamond e_*)]$$

$$\overset{(*3)}{=} [(a \diamond e_*) * (a \diamond e_*)] * -(a \diamond e_*)$$

$$\overset{(\diamond D*)}{=} [a \diamond (e_* * e_*)] * -(a \diamond e_*)$$

$$\overset{(*4)}{=} (a \diamond e_*) * -(a \diamond e_*)$$

$$\overset{(*5)}{=} e_*$$

This proves that $a \diamond e_* = e_*$; the proof of $e_* \diamond a = e_*$ can be completed similarly (see Problem 1).   □

When comparing Theorems 14.1 and 14.5, we see several discrepancies: the properties of Boolean algebras are different from those of rings. In particular, the corresponding dual of the claim of Theorem 14.5 is false; namely, in a ring we generally do not have $a * e_\diamond = e_\diamond$. First of all, the $\diamond$ operation may not even have an identity, and even in a field (when it does), the claim is false! For example, it is not true that for real numbers we have $a + 1 = 1$ (unless, of course, $a = 0$; cf. Problem 4).

Recall from Problem 3 (b) of Chapter 13 that the ring $\mathbb{Z}_n$ is not an integral domain when n is composite: the nonzero product property fails as we have two nonzero elements of $\mathbb{Z}_n$ whose product is zero. Therefore, while by Theorem 14.5 whenever (at least) one of a or b equals e_*, $a \diamond b$ equals e_* as well, the converse is false in general: we may have elements a and b in a ring for which $a \diamond b = e_*$ but $a \neq e_*$ and $b \neq e_*$. However, as we already mentioned in Chapter 13 and as we prove next, the statement holds when our ring is a field.

Theorem 14.6 *Every field is an integral domain. In other words, the field axioms imply the nonzero product property.*

Proof Let $\mathcal{F}$ be a field for the binary operations $*$ and $\diamond$ (as listed in Definition 13.1). Let e_* denote the identity element of $*$, and suppose that $a \in \mathcal{F}$ and $b \in \mathcal{F}$ satisfy the equation $a \diamond b = e_*$. We need to prove that $a = e_*$ or $b = e_*$.

We employ the Law of Case Exclusion and assume that $a \neq e_*$; we then must show that $b = e_*$. Note that by the nonzero inverse property, if $a \neq e_*$, then we have an element $c \in \mathcal{F}$ for which $c \diamond a = e_\diamond$. We can then write

$$b = e_\diamond \diamond b = (c \diamond a) \diamond b = c \diamond (a \diamond b) = c \diamond e_*.$$

By Theorem 14.5, we have $c \diamond e_* = e_*$, and thus $b = e_*$, as claimed.   □

We can now apply Theorems 14.5 and 14.6 to the field of real numbers and combine them as follows:

Corollary 14.7 *Suppose that x and y are real numbers. Then $x \cdot y = 0$ if, and only if, $x = 0$ or $y = 0$.*

According to Theorem 14.6, every field is an integral domain. The converse of this statement is clearly false: the integers, for example, form an integral domain for addition and multiplication but, since only 1 and -1 have multiplicative inverses that are integers, this integral domain is not a field. Yet, our next theorem provides a "partial converse" to Theorem 14.6.

Theorem 14.8 *Every finite integral domain is a field. In other words, a finite integral domain satisfies the nonzero inverse property.*

Proof We employ the method we used in the proof of Proposition 12.3 and elsewhere in Chapter 12. Suppose that our integral domain D, with binary operations $*$ and $\diamond$ (as defined in Definition 13.1), has n elements for some $n \in \mathbb{N}$ and that these elements are $a_1, a_2, \ldots, a_n$. Since D is an integral domain, one of these elements is the identity e_*. Suppose that $a_1 \neq e_*$. We prove that a_1 has an inverse for $\diamond$ in D; since a_1 can be an arbitrary element in D that is different from e_*, this proves the nonzero inverse property.

Consider the set

$$A = \{a_1 \diamond a_1, a_1 \diamond a_2, \ldots, a_1 \diamond a_n\}.$$

Since D is closed for the $\diamond$ operation, $A \subseteq D$. Next, we show that $|A| = n$. Indeed, if

$$a_1 \diamond a_i = a_1 \diamond a_j$$

for some integers i and j with $1 \leq i \leq n$ and $1 \leq j \leq n$, then

$$a_1 \diamond a_i - a_1 \diamond a_j = e_*,$$

which we can rewrite as

$$a_1 \diamond (a_i - a_j) = e_*.$$

(Here, as usual, $-d$ denotes the inverse of $d \in D$ for the $*$ operation.) Since D is an integral domain, it satisfies the nonzero product property, so we must have $a_1 = e_*$ or $a_i - a_j = e_*$, but since we assumed that $a_1 \neq e_*$, we get $a_i - a_j = e_*$ and thus $a_i = a_j$. This proves that the elements listed in A are pairwise distinct, so $|A| = n$.

But if $|A| = |D| = n$, then A can only be a subset of D if $A = D$. Therefore, one of the elements in A equals $e_\diamond$, which means that a_1 has an inverse for $\diamond$ in D, as claimed. $\square$

Recall from Problem 3 (b) of Chapter 13 that for positive prime values of n, $\mathbb{Z}_n$ is an integral domain. Combining this with Theorem 14.8 yields the following well-known result.

Theorem 14.9 *The ring $\mathbb{Z}_p$ is a field whenever p is a positive prime.*

As we mentioned above, in theory every proof can be reduced to axioms and written formally. Doing this for all results in mathematics is not practical, however: it would take much too long to do so. In this chapter we merely attempted to demonstrate formal axiomatic proofs. Generally, we write proofs as most mathematicians do: less formally.

Problems

1. Complete the proof of Theorem 14.5 by showing that $e_* \diamond a = e_*$.
2. Rewrite the proof of Theorem 13.8 in the style of the proofs of Theorems 14.1 and 14.5, that is, as a string of equations starting with $a + b$ and ending with $b + a$. (Be careful not to use the claim itself!)
3. (a) Suppose that $\mathcal{F}$ is a field for the binary operations $*$ and $\diamond$ (as listed in Definition 13.1). Let e_* denote the identity element of $*$ and $e_\diamond$ denote the identity element of $\diamond$. Prove that $e_* \neq e_\diamond$.
    (Hint: Use Theorem 14.5 and the requirement that a field has at least two elements.)
   (b) Consider the following claim.
    Claim. For the real numbers 1 and 0 we have $0 = 1$.
    According to part (a), this claim is false. Find the mistake(s) in the following argument.
    Argument. Let a be an arbitrary real number. By axiom $(+5)$, we must have a real number x such that $a + x = 0$. Multiplying this equation by a yields $(a + x) \cdot a = 0 \cdot a$, or, after using distributivity, $a^2 + x \cdot a = 0 \cdot a$ (denoting $a \cdot a$ by a^2). Axiom $(\cdot 5)$ guarantees an inverse to $a^2 + x \cdot a$; that is, there exists a real number y for which $(a^2 + x \cdot a) \cdot y = 1$. Using this y to multiply our equation $a^2 + x \cdot a = 0 \cdot a$, we get $(a^2 + x \cdot a) \cdot y = (0 \cdot a) \cdot y$. Now the left-hand side equals 1, so we have $1 = (0 \cdot a) \cdot y$. According to Corollary 14.7, 0 times any real number is 0, so $(0 \cdot a) \cdot y = 0 \cdot y = 0$. Therefore, we proved that $0 = 1$.
4. Suppose that $\mathcal{F}$ is a field for the binary operations $*$ and $\diamond$ (as listed in Definition 13.1). Let e_* and $e_\diamond$ denote the identity elements of $*$ and $\diamond$, respectively, and suppose that $a \in \mathcal{F}$.

   (a) Find the mistake(s) in the following argument.
    Claim. We have $a * e_\diamond = e_\diamond$.
    Argument. At each step we refer to a particular property under Definition 13.1.

    $$a * e_\diamond \overset{(\diamond 4)}{=} (a * e_\diamond) \diamond e_\diamond$$

    $$\overset{(\diamond 5')}{=} (a * e_\diamond) \diamond [(a * e_\diamond) \diamond -(a * e_\diamond)]$$

$$\overset{(\diamond 3)}{=} \; [(a * e_\diamond) \diamond (a * e_\diamond)] \diamond -(a * e_\diamond)$$

$$\overset{(*D\diamond)}{=} \; [a * (e_\diamond \diamond e_\diamond)] \diamond -(a * e_\diamond)$$

$$\overset{(\diamond 4)}{=} \; (a * e_\diamond) \diamond -(a * e_\diamond)$$

$$\overset{(\diamond 5')}{=} \; e_\diamond$$

(b) Prove that one cannot have $a * e_\diamond = e_\diamond$ unless $a = e_*$.
   (Hint: Simplify $a * (e_\diamond * -e_\diamond)$ in two different ways. As a corollary, this
   says that if a real number a satisfies $a + 1 = 1$, then we must have $a = 0$.)

5. (a) Let $\mathcal{F}$ be a collection of two or more objects on which two binary
      operations are defined, say $*$ and $\diamond$. Suppose that $*$ satisfies axioms $(*1)$–
      $(*5)$, and that $\diamond$ is distributive with respect to $*$. Prove that $\diamond$ cannot satisfy
      the inverse property.
      (Hint: Prove that the identity element e_* cannot have an inverse for the $\diamond$
      operation.)

   (b) Use the previous part to explain why no real number can be divided by
      zero.

6. In this problem we discuss the following important theorem about Boolean
   algebras.

   Theorem 14.10 (De Morgan's Laws) *Suppose that $\mathcal{B}$ is a Boolean algebra
   for the unary operation $^-$ and the binary operations $*$ and $\diamond$ (as listed in
   Definition 13.9). Let e_* and $e_\diamond$ denote the identity elements of the operations
   $*$ and $\diamond$, respectively, and suppose that $a, b \in \mathcal{B}$.
   Then we have*

   $$\overline{(a * b)} = \overline{a} \diamond \overline{b}$$

   and

   $$\overline{(a \diamond b)} = \overline{a} * \overline{b}.$$

   (a) Rewrite De Morgan's Laws for the Boolean algebra of statements.
   (b) Rewrite De Morgan's Laws for the Boolean algebra of all subsets of a given
      set.
   (c) Consider the following proof of the first identity of Theorem 14.10. For
      each line in the proof, supply all the necessary Boolean algebra properties
      that are used.

   $$\overline{(a * b)} = e_\diamond \diamond e_\diamond \diamond \overline{(a * b)}$$

   $$= (e_\diamond * b) \diamond (e_\diamond * a) \diamond \overline{(a * b)}$$

   $$= [(a * \overline{a}) * b] \diamond [(b * \overline{b}) * a] \diamond \overline{(a * b)}$$

$$= [(a * b) * (\overline{a} \diamond \overline{b})] \diamond \overline{(a * b)}$$

$$= [(a * b) \diamond \overline{(a * b)}] * [(\overline{a} \diamond \overline{b}) \diamond \overline{(a * b)}]$$

$$= e_* * [(\overline{a} \diamond \overline{b}) \diamond \overline{(a * b)}]$$

$$= e_* * e_* * [(\overline{a} \diamond \overline{b}) \diamond \overline{(a * b)}]$$

$$= (e_* \diamond \overline{b}) * (e_* \diamond \overline{a}) * [(\overline{a} \diamond \overline{b}) \diamond \overline{(a * b)}]$$

$$= [(a \diamond \overline{a}) \diamond \overline{b}] * [(b \diamond \overline{b}) \diamond \overline{a}] * [(\overline{a} \diamond \overline{b}) \diamond \overline{(a * b)}]$$

$$= [(\overline{a} \diamond \overline{b}) \diamond (a * b)] * [(\overline{a} \diamond \overline{b}) \diamond \overline{(a * b)}]$$

$$= (\overline{a} \diamond \overline{b}) \diamond [(a * b) * \overline{(a * b)}]$$

$$= (\overline{a} \diamond \overline{b}) \diamond e_\diamond$$

$$= \overline{a} \diamond \overline{b}$$

7. In Problem 5 of Chapter 9 we proved that

$$(A \cup \overline{B} \cup C) \cap \overline{(A \cup C)} = \overline{A} \cap \overline{B} \cap \overline{C}$$

holds for arbitrary sets A, B, and C. This time, prove the identity using the axioms of a Boolean algebra.
(Hints: Starting on the left-hand side, use, in order, the commutative, distributive, complementation, identity, and De Morgan's properties to arrive at the right-hand side.)
8. Use the Boolean algebra properties of sets (or previously proven theorems) to provide an axiomatic proof for the two identities of Problem 7 of Chapter 6.
9. Using the Boolean algebra properties of statements or previously proven theorems, prove that the following properties hold for arbitrary statements P and Q.

   (a) *Idempotent Laws*: $(P \vee P) \Leftrightarrow P$ and $(P \wedge P) \Leftrightarrow P$;
   (b) *Absorption Laws*: $[P \wedge (P \vee Q)] \Leftrightarrow P$ and $[P \vee (P \wedge Q)] \Leftrightarrow P$;
   (c) *Involution Law*: $\neg(\neg P) \Leftrightarrow P$.
   (Hint: Simplify

   $$[\neg P \vee P] \wedge [\neg(\neg P) \vee \neg P] \wedge [\neg(\neg P) \vee P]$$

   in two different ways.)

10. Prove each of the following statements using the field axioms and Definitions 13.4–13.7 (as well as previously proven statements). For parts (e) and (f) you should use Corollary 14.7 as well.

    (a) For every real number a, $-(-a) = a$.
    (b) For every real number a, $(-1) \cdot a = -a$.

(c) For all real numbers a and b, $(-a) \cdot b = -(a \cdot b)$.

(d) For all real numbers a and b, $(-a) \cdot (-b) = a \cdot b$.

(e) If a, b, c, and d are real numbers, $b \neq 0$, and $d \neq 0$, then

$$\frac{a}{b} \cdot \frac{c}{d} = \frac{a \cdot c}{b \cdot d}.$$

(f) If a, b, c, and d are real numbers, $b \neq 0$, and $d \neq 0$, then

$$\frac{a}{b} + \frac{c}{d} = \frac{a \cdot d + b \cdot c}{b \cdot d}.$$

11. Use the field axioms and the order axioms (cf. Definition 13.12) as well as your definitions from parts (a) and (b) of Problem 9 of Chapter 13 to prove the following properties of the real numbers.

(a) *Trichotomy*: For all real numbers a and b, exactly one of the following statements holds: $a < b$, $a = b$, or $a > b$.

(b) *Transitivity*: If a, b, and c are real numbers so that $a > b$ and $b > c$, then $a > c$.

(c) *Addition Law*: If a, b, and c are real numbers so that $a > b$, then $a + c > b + c$.

(d) *Multiplication Law*: If a, b, and c are real numbers so that $a > b$ and $c > 0$, then $ac > bc$.

(e) *Multiplication Law*: If a, b, and c are real numbers so that $a > b$ and $c < 0$, then $ac < bc$.

(f) If a is a real number and $a \neq 0$, then $a^2 > 0$.

(g) If a and b are positive real numbers and $a < b$, then $a^2 < b^2$.

(h) For all real numbers a and b, $a^3 < b^3$ holds if, and only if, $a < b$.

Chapter 15
Group Work

In Chapter 13 we introduced several mathematical structures, and in Chapter 14 we studied them in some detail. In this chapter we discuss an important structure we have not identified yet: the structure of groups.

A group is a set together with a single binary operation, satisfying four rather natural properties. In particular, we make the following definition.

Definition 15.1 *Suppose that G is any set of objects on which a binary operation $*$ is defined. We say that G is a* group *for the operation $*$ if all of the following properties hold.*

- *($*1$) $*$ has the closure property;*
- *($*3$) $*$ has the associative property;*
- *($*4$) $*$ has the identity property; and*
- *($*5$) $*$ has the inverse property.*

If, in addition, we also know that

- *($*2$) $*$ has the commutative property,*

then we say that G is an Abelian group.

(Our labels of these properties are in correspondence with those in Chapter 13.)

Groups are ubiquitous in many parts of mathematics, and they are also used in a large variety of other fields. The term Abelian group is named after the Norwegian mathematician Niels Abel (1802–1829); sometimes the term "commutative group" is used instead.

Recall that the primary number sets $\mathbb{Z}$, $\mathbb{Q}$, $\mathbb{R}$, and $\mathbb{C}$ all form integral domains, so they form Abelian groups for addition as well. Furthermore, the nonzero elements of $\mathbb{Q}$, $\mathbb{R}$, and $\mathbb{C}$ form Abelian groups for multiplication. However, the nonzero elements of $\mathbb{Z}$ do not form a group for multiplication since only 1 and -1 have multiplicative inverses.

© Springer Nature Switzerland AG 2020

B. Bajnok, *An Invitation to Abstract Mathematics*, Undergraduate Texts in Mathematics, https://doi.org/10.1007/978-3-030-56174-1_15

Table 15.1 The operation
table of $\mathbb{Z}_4$

$\oplus$	0	1	2	3
0	0	1	2	3
1	1	2	3	0
2	2	3	0	1
3	3	0	1	2

Table 15.2 The operation
table of $\mathbb{Z}_{10}$

$\oplus$	0	1	2	3	4	5	6	7	8	9
0	0	1	2	3	4	5	6	7	8	9
1	1	2	3	4	5	6	7	8	9	0
2	2	3	4	5	6	7	8	9	0	1
3	3	4	5	6	7	8	9	0	1	2
4	4	5	6	7	8	9	0	1	2	3
5	5	6	7	8	9	0	1	2	3	4
6	6	7	8	9	0	1	2	3	4	5
7	7	8	9	0	1	2	3	4	5	6
8	8	9	0	1	2	3	4	5	6	7
9	9	0	1	2	3	4	5	6	7	8

Besides these groups that have infinitely many elements, there is also a rich variety of finite groups. We will meet several well-known finite groups in the problem set below; here we introduce (revisit) one particular family of finite groups: the *cyclic groups* of order n. (The term *order* refers to the size of the group.)

Suppose that n is a positive integer, and consider

$$\mathbb{Z}_n = \{0, 1, 2, \ldots, n - 1\},$$

the set of nonnegative integers less than n. As before, we define the binary operation $\oplus$ on $\mathbb{Z}_n$ to be "addition mod n"; that is, for elements a and b of $\mathbb{Z}_n$, we set $a \oplus b$ to be the remainder of $a + b$ when divided by n. The operation tables of $\mathbb{Z}_4$ and $\mathbb{Z}_{10}$ are given as Tables 15.1 and 15.2. (Glancing at these tables makes it clear why $\mathbb{Z}_n$ is called a cyclic group.)

One can verify that $\mathbb{Z}_n$ is an Abelian group for every n. In the case of $\mathbb{Z}_{10}$, for example, we can see that all entries in the table remain in the set, thus the set is closed for this operation; 0 is the identity element, as indicated by the first row and the first column; and every element has an inverse, since 0 appears in each row and column. Associativity and commutativity follow from the fact that addition is associative and commutative in $\mathbb{Z}$.

In the abstract treatment of group theory one is not concerned with the particular meaning of the elements and the operation. As we said above, any set of objects on which a binary operation is defined forms a group as long as the four axioms above are satisfied. For example, one can check (cf. Problem 1) that both operation tables in Table 15.3 describe groups.

Table 15.3 Two groups of
order 4

G_1:

*	A	B	C	D
A	A	B	C	D
B	B	C	D	A
C	C	D	A	B
D	D	A	B	C

G_2:

*	A	B	C	D
A	A	B	C	D
B	B	A	D	C
C	C	D	A	B
D	D	C	B	A

Table 15.4 The group of
order 1

*	A
A	A

Table 15.5 The group of
order 2

*	A	B
A	A	B
B	B	A

One may ask the question: How many different groups are there of a certain order n? Clearly, we cannot regard two groups as different if they only differ in the labeling of the elements. We make this notion more precise with the following definition.

Definition 15.2 *We say that two groups are* isomorphic *if there is a bijection from one to the other that preserves the operations. More precisely, suppose that G is a group for the operation $*$ and H is a group for the operation $\diamond$. We say that the two groups are isomorphic to each other if there is a bijection $f : G \to H$ so that*

$$f(g_1 * g_2) = f(g_1) \diamond f(g_2)$$

holds for all elements g_1 and g_2 of G.

The operation-preserving bijection in Definition 15.2 is called a *group isomorphism*.

In Problem 9 we prove that if f is a group isomorphism from G to H, then its inverse is a group isomorphism from H to G, hence our phrase that the groups are isomorphic to each other is appropriate. When two groups are isomorphic, we consider them to be the same, although the proper terminology is to say that they belong to the same *isomorphism class*.

Let us now try to see how many different groups there are of some small orders. It is quite obvious that there is only one group of order 1, the one given in Table 15.4 (whether we call the single element A or something else does not matter; the same holds for the notation for the operation). More precisely, using Definition 15.2 we say that any two groups of order 1 are isomorphic.

Similarly, it only takes a moment to verify that there is only one group of order two, given in Table 15.5. Indeed, we can decide to label the identity element A, yielding the values $A * A = A$ and $A * B = B * A = B$; we then must have $B * B = A$ in order for B to have an inverse. Thus there is a single isomorphism class for groups of order 2 (which could be represented by $\mathbb{Z}_2$).

Table 15.6 The group of
order 3

*	A	B	C
A	A	B	C
B	B	C	A
C	C	A	B

Table 15.7 Some gnu values

n	1	2	3	4	5	6	7	8	9	10
gnu(n)	1	1	1	2	1	2	1	5	2	2

n	11	12	13	14	15	16	17	18	19	20
gnu(n)	1	5	1	2	1	14	1	5	1	5

It is a bit more work to prove that there is only one group of order 3 as well—
see Problem 1. Therefore, all groups of order 3 are isomorphic to one another (the
isomorphism class could be represented by $\mathbb{Z}_3$).

Thus far we have described groups of order 4 in three different ways. Are these
three groups all isomorphic too? Consider first the groups $\mathbb{Z}_4$ and G_1 above. We can
see that these two groups are isomorphic by mapping 0, 1, 2, and 3 to A, B, C, and
D, respectively. Indeed, one can easily check that this map is operation preserving;
for example, in $\mathbb{Z}_4$ we have $2 \oplus 3 = 1$, and in G_1 we have the corresponding equation
$C * D = B$.

Let us turn now to the group G_2 above. We argue that this group is not isomorphic
to $\mathbb{Z}_4$, as follows. Consider the element A. The operation table indicates that every
element "squared" in the group is equal to A (that is, $x * x = A$ for every element
x). Since none of 0, 1, 2, or 3 in $\mathbb{Z}_4$ has this property (the main diagonal of the
table for $\mathbb{Z}_4$ contains more than one element), we see that none of the elements of
$\mathbb{Z}_4$ can be mapped to A. Therefore, the two groups are not isomorphic. The group
G_2 is referred to as the *Klein 4-group*, named after the German mathematician Felix
Klein (1849–1925), who was one of the pioneers of group theory. In Problem 1 we
prove that every group of order 4 is isomorphic to either $\mathbb{Z}_4$ or the Klein 4-group.

We should mention that, in spite of the examples listed thus far, not every group
is Abelian. In fact, for every even order 6 or higher, at least one group of that order
is nonabelian (see also Problems 2, 3, and 4 below).

The classification problem of all finite groups can be stated as follows: For every
positive integer n, find the number of pairwise non-isomorphic groups of order n,
called the *group number of n* and denoted by gnu(n), and find a representative of
each isomorphism class. Today we are quite far from a full solution to this problem;
however, the number of pairwise non-isomorphic groups of order up to 20 is given
in Table 15.7.

We see that the sequence of group numbers is quite peculiar: It contains
occasional values that are much higher than any previous value (e.g., gnu(8) = 5
and gnu(16) = 14) but, quite often, we have gnu(n) = 1. Regarding orders for
which the gnu is 1, we will prove in Problem 8 (c) below that gnu(p) = 1 for all
positive primes p (but, as gnu(15) = 1 shows, gnu(n) may be 1 for composite n
as well). As for high gnu values, it was recently shown by John Horton Conway

(1937–2020) and his collaborators that when n is a power of 2, gnu(n) is large. (The term and notation for group numbers comes from them also.) For example, of the approximately fifty billion (pairwise non-isomorphic) groups of order less than 2048 (= 2^{11}), more than 99 percent have order exactly 1024 (= 2^{10}). In spite of this and other results achieved mainly by computational techniques, we are quite far today from being able to characterize all finite groups.

While the classification of all groups remains hopelessly out of reach, we now believe we have a complete classification of all finite *simple groups*. We will not define simple groups here (these concepts are defined and studied in detail in group theory or abstract algebra); let us just say that, much like integers factor into primes, groups can be built from smaller simple groups. For example, it turns out that the group $\mathbb{Z}_n$ is simple exactly when n is a positive prime number.

The enormous task of classifying all finite simple groups was announced as completed in 2004 after the contributions of many mathematicians on thousands of pages of journal articles spanning over one hundred years. The result can be summarized as follows:

Theorem 15.3 (Classification of Finite Simple Groups) *Every finite simple group is isomorphic to one member of three infinite families or is isomorphic to one of twenty-six other "sporadic" groups.*

One of the three infinite families mentioned in Theorem 15.3 consists of the cyclic groups $\mathbb{Z}_n$ for n prime. In Problem 3 below we introduce another family, the *alternating groups* A_n; as it turns out, A_n is simple for all $n \geq 5$. Beyond these two infinite families, simple groups become rapidly more complex; the third infinite family is formed by groups of *Lie type*, which you may encounter in advanced studies in algebra. Beyond that, the twenty-six sporadic groups are also the subject of considerable recent research; the largest one, the so-called *Monster group*, has order precisely

$$808, 017, 424, 794, 512, 875, 886, 459, 904, 961, 710, 757, 005, 754, 368, 000, 000, 000.$$

The main result of the classification is that the list is complete; every finite simple group is accounted for. Simple groups have been studied extensively and possess many applications.

At the present time, no one person has read the proof of the classification in its entirety, though attempts are currently under way. One might naturally wonder if such a lengthy work contains any errors. In fact, quite a few gaps in the proof have been discovered, but each of these could be fixed relatively easily. While no one knows for certain, most mathematicians believe that the proof—and certainly the result—is now essentially complete.

The concept of a group is one of the most fundamental ones in algebra, and the methods of group theory have been employed in many different branches of mathematics. Groups play important roles in a variety of other fields as well, including physics, chemistry, and cryptography; recently they have also been providing convenient frameworks in certain areas of the social sciences. The branch of mathematics that studies groups and their properties is called *group theory*. The problems below serve as an introduction to this important and far-reaching subject.

Problems

1. (a) Let G be a group for the binary operation $*$. For a given element a of G, we let

$$a * G = \{a * g \mid g \in G\}$$

and

$$G * a = \{g * a \mid g \in G\}.$$

Prove that $a * G = G * a = G$ for all a.

Remark We can interpret our claims to say that in the operation table of G, each row and each column contains each element of G exactly once.

(b) Prove that all groups of order 3 are isomorphic to each other. (In other words, every group of order 3 is isomorphic to the cyclic group $\mathbb{Z}_3$.)
(Hints: Suppose that the three elements of G are A, B, and C; without loss of generality, assume that A is the identity element. Use part (a) to conclude that the operation table of G must be the one on page 234.)

(c) Prove that every group of order 4 is isomorphic to either G_1 or G_2, given in Table 15.3. (In other words, every group of order 4 is isomorphic to the cyclic group $\mathbb{Z}_4$ or to the Klein 4-group.)
(Hints: Let the four elements of G be A, B, C and D, with A being the identity element. We have two choices for $B * B$: it is one of C or D (without loss of generality, $B * B = C$) or it is the identity element A. Use part (a) above to conclude that in the first case G is isomorphic to the cyclic group G_1. The operation table in the second case can be completed in two ways, with $C * C = A$ or $C * C = B$. Prove that these two subcases lead to G being isomorphic to G_1 or G_2, respectively.)

(d) Consider the operation table given in Table 15.8.

Table 15.8 Is this a group?

$*$	A	B	C	D	E
A	A	B	C	D	E
B	B	A	D	E	C
C	C	E	A	B	D
D	D	C	E	A	B
E	E	D	B	C	A

The conditions of part (a) are satisfied: each row and each column contains the five elements of G. Does the table belong to a group?

2. Suppose that n is an integer with $n \geq 3$. The dihedral group D_n is defined as the group of rigid motions taking a regular n-gon back to itself, where the operation is composition. As an example, consider the regular hexagon in Figure 15.1.

Fig. 15.1 A hexagon with labeled vertices

If f denotes rotation around the center by $180°$ (in this case it makes no difference if we rotate clockwise or counterclockwise) and g denotes reflection with respect to the vertical axis (the line connecting vertices 1 and 4 in the figure above), then the effects of f, g, and $f \circ g$ are shown, in order, in Figure 15.2.

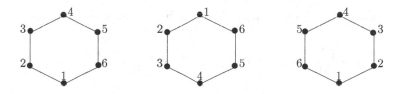

Fig. 15.2 The hexagon after transformations f, g, and $f \circ g$, respectively

We see that, in this case, $f \circ g$ has the same effect on our original hexagon as reflection with respect to the horizontal axis (the line connecting the midpoint of side 23 with the midpoint of side 56); denoting this transformation by h, we thus have $f \circ g = h$.

(a) Prove that $|D_n| = 2n$.
   (Hints: To show that $|D_n| \geq 2n$, identify n rotations and n reflections that are in D_n. To prove that $|D_n| \leq 2n$, consider the possibilities for how a rigid motion may operate on two adjacent vertices of the n-gon.)

(b) Exhibit the operation table for the dihedral group D_3. Verify that D_3 is a nonabelian group.

(c) Exhibit the operation table for the dihedral group D_4. Verify that D_4 is a nonabelian group.

3. Recall from Problem 11 of Chapter 8 that S_n is defined as the set of bijections on $\{1, 2, \ldots, n\}$. It can be shown that S_n forms a group for the operation of composition; this important group is called the *symmetric group of degree n*. We also described the cycle representation for the elements of S_n. Using that description, we define A_n, called the *alternating group* of degree n, as the collection of those elements of S_n whose cycle representation contains an even number of cycles that have even length. For instance, the example $f = (1528)(37)(4)(6) \in S_8$ that we presented there has a cycle of length 4, another of length 2, and two cycles of length 1; therefore, f is an element of A_8.

(a) List all elements of A_4 in cycle notation.
(b) Exhibit the operation table of A_4. (Note that the operation is function composition.)
(c) Verify that A_4 is a group.
    (Hint: Note that associativity follows from the fact that the composition of functions is an associative operation; cf. page 130.)
(d) Is A_4 an Abelian group?

4. In this problem we present two groups of order 8, both formed by 2-by-2 matrices for multiplication. (Recall that multiplication of 2-by-2 matrices was defined in Problem 1 (l) of Chapter 13.)

(a) Consider the set S_1 consisting of the following eight 2-by-2 matrices:

$$A_1 = \begin{pmatrix} 1 & 0 \\ 0 & 1 \end{pmatrix}, \ A_2 = \begin{pmatrix} 0 & -1 \\ 1 & 0 \end{pmatrix}, \ A_3 = \begin{pmatrix} -1 & 0 \\ 0 & -1 \end{pmatrix}, \ A_4 = \begin{pmatrix} 0 & 1 \\ -1 & 0 \end{pmatrix},$$

$$A_5 = \begin{pmatrix} 1 & 0 \\ 0 & -1 \end{pmatrix}, \ A_6 = \begin{pmatrix} 0 & 1 \\ 1 & 0 \end{pmatrix}, \ A_7 = \begin{pmatrix} -1 & 0 \\ 0 & 1 \end{pmatrix}, \ A_8 = \begin{pmatrix} 0 & -1 \\ -1 & 0 \end{pmatrix}.$$

Construct the multiplication table for S_1, and verify that S_1 is a nonabelian group for multiplication.

(b) Consider the set S_2 consisting of the following eight 2-by-2 matrices: A_1, A_2, A_3, and A_4 from part (a) above, as well as i times A_5, A_6, A_7, and A_8, namely,

$$i A_5 = \begin{pmatrix} i & 0 \\ 0 & -i \end{pmatrix}, \ i A_6 = \begin{pmatrix} 0 & i \\ i & 0 \end{pmatrix}, \ i A_7 = \begin{pmatrix} -i & 0 \\ 0 & i \end{pmatrix}, \ i A_8 = \begin{pmatrix} 0 & -i \\ -i & 0 \end{pmatrix},$$

where i is the imaginary number with $i^2 = -1$. Construct the multiplication table for S_2, and verify that S_2 is a nonabelian group for multiplication.

5. Suppose that n is a positive integer with $n \geq 2$, and let $\mathcal{U}_n$ consist of those elements from the set

$$\{0, 1, 2, \ldots, n - 1\}$$

that are relatively prime to n. Let $\odot$ denote multiplication of these numbers mod n; that is, $a \odot b$ is the remainder of $a \cdot b$ when divided by n.

(a) Exhibit the operation table for $\mathcal{U}_9$, and verify that $\mathcal{U}_9$ is a group for the operation.
(b) Prove that $\mathcal{U}_n$ is an Abelian group for every integer $n \geq 2$.
   (Hints: The associative and commutative properties follow from the same properties of the integers, and the identity property holds since $1 \in \mathcal{U}_n$ for all $n \geq 2$.
   To verify the closure property, we need to show that if a and b are elements of $\mathcal{U}_n$, then $a \odot b$ is also an element of $\mathcal{U}_n$; that is, if a and b are relatively prime to n and $ab = qn + r$ for some integers q and r with $0 \leq r \leq n - 1$, then r is also relatively prime to n. Argue indirectly, and assume that n and r have a common divisor $d \geq 2$. By the Factorization Theorem (see Theorem 11.4), d has a prime divisor p. Show that a or b then must also be divisible by p, which would imply that they are not both relatively prime to n, a contradiction.
   Finally, to verify the inverse property, we must show that if $a \in \mathcal{U}_n$, then there is an element $x \in \mathcal{U}_n$ for which $a \odot x = 1$, that is, that $ax = qn + 1$ for some integer q. Use Theorem 12.3.)

6. Let G be a group for the binary operation $*$, with identity element e_*. A subset H of G is called a *subgroup* of G if it is a group for the same operation; that is,

   • for any two elements h_1, h_2 of H, $h_1 * h_2$ is also in H;
   • the identity element e_* is in H;
   • for any element h of H, the inverse h^{-1} is also in H.

(Associativity in H follows from associativity in G.)

(a) Find all four subgroups of $\mathbb{Z}_{10}$. (You do not need to prove that there are no others.)
(b) Find all ten subgroups of D_4. (Cf. Problem 2 (c) above. You do not need to prove that there are no others.)
(c) Given an element $a \in G$, the *centralizer* $C(a)$ of a in G is defined as the collection of elements of G that commute with a; that is,

$$C(a) = \{c \in G \mid c * a = a * c\}.$$

Prove that for every $a \in G$, the centralizer $C(a)$ of a is a subgroup of G.

Remark Observe that the centralizer of the identity element is G.

(d) The *center* C of G is defined as the collection of elements of G that commute with all elements of the group; that is,

$$C = \{c \in G \mid c * a = a * c \text{ for all } a \in G\}.$$

Prove that the center of a group is a subgroup of the centralizer of every element of G. (In particular, the center is a subgroup of G.)

(e) Given a subgroup H of G and an element a of G, the *left coset $a * H$* is defined as

$$a * H = \{a * h \mid h \in H\}.$$

Prove that for any two elements $a_1, a_2 \in G$, the left cosets $a_1 * H$ and $a_2 * H$ are either equal or disjoint.

(Hints: Let a_1^{-1} denote the inverse of a_1 in G. Prove that if $a_1^{-1} * a_2 \in H$ then $a_1 * H = a_2 * H$, and if $a_1^{-1} * a_2 \notin H$ then $(a_1 * H) \cap (a_2 * H) = \emptyset$.)

(f) Prove the following well-known result.

Theorem 15.4 (Lagrange's Theorem) *Let H be a subgroup of a finite group G. Then the order of H divides the order of G.*

(Hint: Use part (e).)

7. Let G be a finite nonabelian group with $*$ as the operation, and let C denote its center (see Problem 6 above).

(a) Prove that C has order at most $|G|/4$.
   (Hints: Since G is nonabelian, one can find an element $a \in G$ whose centralizer $C(a)$ is not all of G. Recall from Problem 6 that C is a subgroup of $C(a)$ and that $C(a)$ is a subgroup of G, then use Lagrange's Theorem.)

(b) Find a group G whose center has order exactly $|G|/4$.

(c) One way to measure the degree to which G is Abelian is to count the percentage of ordered pairs of elements that commute; namely, we may determine the quantity

$$A(G) = \frac{|\{(a, b) \in G^2 \mid a * b = b * a\}|}{|G|^2}.$$

Prove that $A(G) \leq 5/8$.
(Hints: Show that

$$|\{(a, b) \in G^2 \mid a * b = b * a\}| = \sum_{g \in C} |C(g)| + \sum_{g \notin C} |C(g)|,$$

then use part (a).)

(d) Find a group G with $A(G) = 5/8$.

8. Let G be a group for the binary operation $*$, with identity element e_*. For an element $a \in G$, we set $a^0 = e_*$, and for a positive integer n, we define a^n recursively as $a^{n-1} * a$.

   (a) Prove that in a finite group G, every element a has a positive exponent d for which $a^d = e_*$. The smallest such exponent is called the *order* of a in G. (Cf. Problem 11 (d) of Chapter 8.)
   (Hint: Start by observing that, by the Pigeonhole Principle, there are distinct positive integers k and l for which $a^k = a^l$.)
   (b) Prove that the order of any element in a finite group is a divisor of the order (size) of the group.
   (Hint: Use Lagrange's Theorem from above.)
   (c) Let p be a positive prime. Prove that all groups of order p are isomorphic to each other and thus $\mathrm{gnu}(p) = 1$.
   (Hint: Use part (b).)

9. Let G be a group for the binary operation $*$ (with identity element e_*), and H be a group for the binary operation $\diamond$ (with identity element $e_\diamond$). Suppose that $f : G \to H$ is a group isomorphism. Prove each of the following statements.

   (a) The inverse function f^{-1} of f is a group isomorphism from H to G.
   (b) We have $f(e_*) = e_\diamond$.
   (c) For each $a \in G$, the order of a in G equals the order of $f(a)$ in H.
   (d) If G is cyclic, then so is H.
   (e) If G is Abelian, then so is H.

10. (a) Recall that $\mathbb{Z}_6$ is the group on $\{0, 1, 2, 3, 4, 5\}$ for addition mod 6; $\mathcal{U}_9$ is the group on $\{1, 2, 4, 5, 7, 8\}$ for multiplication mod 9 (see Problem 5 above); we also let $\mathbb{Z}_7^*$ be the group on $\{1, 2, 3, 4, 5, 6\}$ for multiplication mod 7. (Note that $\mathbb{Z}_7$ is a field by Theorem 14.9 and thus its nonzero elements form a group for multiplication.) Prove that $\mathbb{Z}_6$, $\mathbb{Z}_7^*$, and $\mathcal{U}_9$ are all isomorphic to each other.

   Remarks It is easy to see that $\mathbb{Z}_n^*$ is only a group when n is prime, and one can also prove that for prime values of n, $\mathbb{Z}_n^*$ is cyclic and thus isomorphic to $\mathbb{Z}_{n-1}$. It is a famous result of Carl Friedrich Gauss (1777–1855) that $\mathcal{U}_n$ is cyclic if, and only if, $n \in \{1, 2, 4\}$ or $n = p^k$ or $n = 2p^k$ for an odd prime p and positive integer k.

   (b) We introduced the group D_4 in Problem 2 (c) above, and we also defined two groups of order 8 in Problem 4. Are any two of these three groups isomorphic to each other?
   (c) Are D_6 and A_4 (see Problems 2 and 3 above) isomorphic to each other?
   (d) Prove that the infinite groups $\mathbb{Z}$ and $\mathbb{Q}$ (for addition) are not isomorphic to each other.
   (Hints: Suppose, indirectly, that there is an isomorphism $f : \mathbb{Z} \to \mathbb{Q}$. By Problem 9 (b) above, we have $f(0) = 0$; let $f(1) = q$. Consider the element $f^{-1}(q/2)$ in $\mathbb{Z}$.)

11. (a) Suppose that $\mathcal{X}$ is a set on which a binary operation $*$ is defined that satisfies the closure, associative, and identity properties (with identity element e). Suppose further that some element $a \in \mathcal{X}$ has a *left inverse*, that is, an element $x \in \mathcal{X}$ for which $x * a = e$, and that it also has a *right inverse*, that is, an element $y \in \mathcal{X}$ for which $a * y = e$. Prove that x is then the unique left inverse of a, y is the unique right inverse of a, and $x = y$.
   (Hint: We proved a similar statement for functions in Theorem 8.5.)

   (b) Suppose that $\mathcal{X}$ is a set on which a binary operation $*$ is defined that satisfies the closure and associative properties. Suppose further that there is an element e in $\mathcal{X}$ that acts as a *left identity*; that is, for all $a \in \mathcal{X}$, we have $e * a = a$. Suppose furthermore that every element of $\mathcal{X}$ has a *left inverse* with respect to e; that is, for all $a \in \mathcal{X}$, we have an $x \in \mathcal{X}$ for which $x * a = e$. Prove that the operation then also has the identity and inverse properties and thus $\mathcal{X}$ is a group for $*$.
   (Hints: Prove first that if x is a left inverse of a with respect to e, then it is also a right inverse of a with respect to e. Then use this to prove that the left identity is also a right identity.)

   Remark According to this result, Definition 15.1 for groups has some redundancy.

12. A group G is called an *elementary Abelian 2-group* if $a * a = e$ holds for every $a \in G$ ($*$ is the operation in G and e is the identity element).

    (a) Prove that, as the name suggests, an elementary Abelian 2-group is indeed Abelian.
    (b) Find all values of n up to 40 for which the group $\mathcal{U}_n$ (cf. Problem 5 above) is an elementary Abelian 2-group.

    Remark It can be proven using some more advanced methods that there are no values of n beyond 40 for which $\mathcal{U}_n$ is an elementary Abelian 2-group.

13. Let m and n be positive integers, and let us define the *direct product* of groups $\mathbb{Z}_m$ and $\mathbb{Z}_n$, denoted by $\mathbb{Z}_m \times \mathbb{Z}_n$, to consist of all *ordered pairs* (a, b) where a is an element of $\mathbb{Z}_m$ and b is an element of $\mathbb{Z}_n$; with addition in $\mathbb{Z}_m \times \mathbb{Z}_n$ performed component-wise; that is, $(a_1, b_1) + (a_2, b_2)$ is the ordered pair (a_3, b_3) where a_3 and b_3 equal $a_1 + a_2$ mod m and $b_1 + b_2$ mod n, respectively. The direct product of three or more groups can be defined analogously.

    (a) Verify that $\mathbb{Z}_m \times \mathbb{Z}_n$ is indeed a group for this operation.
    (b) Find the order of $(1, 1)$ in $\mathbb{Z}_m \times \mathbb{Z}_n$.
    (c) Prove that $\mathbb{Z}_m \times \mathbb{Z}_n$ is isomorphic to $\mathbb{Z}_{mn}$ if, and only if, m and n are relatively prime.
    (Hint: Use part (b).)
    (d) Let p be a positive prime. Prove that no two of the groups

    $$\mathbb{Z}_{p^4}, \; \mathbb{Z}_{p^3} \times \mathbb{Z}_p, \; \mathbb{Z}_{p^2} \times \mathbb{Z}_{p^2}, \; \mathbb{Z}_{p^2} \times \mathbb{Z}_p \times \mathbb{Z}_p, \; \mathbb{Z}_p \times \mathbb{Z}_p \times \mathbb{Z}_p \times \mathbb{Z}_p$$

    are isomorphic to each other.

(Hint: By Problem 9 (c) above, considering the highest possible order of an element in each group distinguishes between any two groups listed, with one exception.)

Remark According to this claim, there are exactly five pairwise nonisomorphic groups of order p^4 that are direct products of cyclic groups.

(e) We have the following result on the classification of finite Abelian groups:

Theorem 15.5 (Fundamental Theorem of Finite Abelian Groups)
Each finite Abelian group is isomorphic to an essentially unique direct product of cyclic groups of prime-power order.

Similarly to the Fundamental Theorem of Arithmetic (see Theorem 11.8), the phrase "essentially unique" refers to the fact that we don't consider two direct products different if they differ only in the order of factors. Based on this theorem, we have a complete classification of finite Abelian groups; in particular, for each positive integer n, we know the number of isomorphism classes of Abelian groups of order n, which we denote by agnu(n), analogously to gnu(n), introduced on page 234.

Verify each of the agnu values in Table 15.9.

Table 15.9 Some Abelian gnu values

n	1	2	3	4	5	6	7	8	9	10
agnu(n)	1	1	1	2	1	1	1	3	2	1

n	11	12	13	14	15	16	17	18	19	20
agnu(n)	1	2	1	1	1	5	1	2	1	2

Chapter 16
Good Relations

We continue our adventure into advanced mathematics with the study of one of the most fundamental objects: relations.

We have already seen several instances in which two objects in a collection have a specific relationship. For example, two real numbers may satisfy an inequality, an integer might be divisible by another, two statements might be equivalent, a set might be a subset of another, and so on. In this chapter we study relationships like these in a general—and, therefore, more abstract—framework.

Definition 16.1 *Suppose that A and B are two sets. A relation R from A to B is any subset of $A \times B$; here A is called the* domain *of R and B is called the* codomain *of R. If the domain and the codomain of a relation are the same set S, then we say that R is a* relation on S.

If elements $a \in A$ and $b \in B$ are such that $(a, b) \in R$, then we say that a is in (this) relationship *with b, and denote this by $a \sim_R b$ or simply $a \sim b$. If $(a, b) \notin R$, then we write $a \nsim b$.*

Note that in our terminology a *relation* is a collection of *relationships*. Furthermore, a relation can be an arbitrary subset of the direct product and does not have to satisfy any particular characteristics; there are as many relations from one set to another as there are subsets of their direct product. For example, the number of different relations on $S = \{1, 2, 3\}$ can be determined by calculating the size of the power set of $S \times S$; since $S \times S$ consists of 9 elements, its power set has size $2^9 = 512$. Among these 512 relations on S are

- equality: $R_= = \{(1, 1), (2, 2), (3, 3)\}$ (the two components are equal);
- order: $R_< = \{(1, 2), (1, 3), (2, 3)\}$ (the first component is less than the second); and
- divisibility: $R_| = \{(1, 1), (1, 2), (1, 3), (2, 2), (3, 3)\}$ (the first component divides the second).

© Springer Nature Switzerland AG 2020

B. Bajnok, *An Invitation to Abstract Mathematics*, Undergraduate Texts in Mathematics, https://doi.org/10.1007/978-3-030-56174-1_16

However, most of the other 509 relations on S have no particular distinguishing features. In this chapter we discuss arbitrary relations; we are, however, most interested in the "good relations" that satisfy certain properties—we will identify what we mean by this soon.

There are several ways to specify and describe relations. The simplest way to describe a relation, particularly if the relation consists of only a few relationships, is the *list* notation. For example,

$$R = \{(1, \{1\}), (1, \{1, 2\}), (2, \{2\}), (2, \{1, 2\})\}$$

is the relation from the set $A = \{1, 2\}$ to the set $P(A) = \{\emptyset, \{1\}, \{2\}, \{1, 2\}\}$ (which is the power set of A); it describes the relation of "being an element of." We should note that, just as is the case for sets in general, relations cannot always be put in list notation!

Another option for describing a relation might be the *formula* or *conditional* description; the relation above, for example, can be written as

$$R = \{(x, X) \mid x \in A, X \subseteq A, x \in X\}$$

or

$$R = \{(x, X) \in A \times P(A) \mid x \in X\}.$$

The *matrix* description of a relation R from a domain A to a codomain B is a table in which rows are indexed by the elements of A, columns are indexed by the elements of B, and the entry in row a ($a \in A$) and column b ($b \in B$) is 1 or 0 depending on whether $(a, b) \in R$ or not. For example, if $A = \{1, 2, 3\}$, $B = P(A)$, and R is the relation of "being an element of," then R can be given by Table 16.1.

Relations may also be represented in several ways graphically. If the domain and the codomain of the relation are both sets of real numbers, then the relation can be graphed in the Descartes coordinate plane: let the "horizontal" axis represent the domain and the "vertical" axis represent the codomain; mark the point (a, b) if the relation contains the ordered pair (a, b). The resulting *graphs* are often quite helpful in visualizing and understanding the relation. For example, the relations

$$R_1 = \{(x, y) \in \mathbb{R}^2 \mid x^2 + y^2 \leq 9\},$$

$$R_2 = \{(x, y) \in \mathbb{R}^2 \mid |x| + |y| \leq 9\},$$

Table 16.1 An example for the matrix description of a relation

	$\emptyset$	$\{1\}$	$\{2\}$	$\{3\}$	$\{1,2\}$	$\{1,3\}$	$\{2,3\}$	$\{1,2,3\}$
1	0	1	0	0	1	1	0	1
2	0	0	1	0	1	0	1	1
3	0	0	0	1	0	1	1	1

Fig. 16.1 An example for the
digraph representation of a
relation

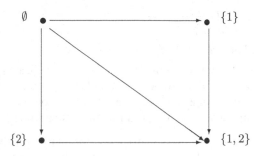

and

$$R_3 = \{(x, y) \in \mathbb{R}^2 \mid |x - y| \le 9\}$$

can be viewed in the usual Descartes coordinate plane, in order, as a "full" disk of
radius 3 centered at the origin, a "full" square with vertices $(\pm 9, 0)$ and $(0, \pm 9)$,
and a "diagonal" band bounded by the lines $y = x \pm 9$.

Another graphical representation is primarily used when describing a relation on
a finite set A. Here the elements of A are represented by different points in the plane
(usually it does not matter how these points are arranged), and we draw an arrow
from a to b if $a \sim b$ (if we also have $b \sim a$, then we draw another arrow from b to
a). The resulting diagram is called the *directed graph* or *digraph* of the relation. For
example, the digraph representation of the $\subsetneq$ ("strict subset") relation on $P(\{1, 2\})$,
the power set of the two-element set $\{1, 2\}$, is given in Figure 16.1.

If the relation is *symmetric*, that is, if for any two points in its digraph, either there
is no arrow between the points or there is an arrow in both directions, then we might
simplify the diagram by replacing each of these double arrows by a single line; the
resulting picture is called the (undirected) *graph* of the relation. For example, the
graph in Figure 16.2 illustrates a group of five people of which certain pairs are
friends.

In Chapter 8 we introduced functions, but we did not quite define them precisely.
We can now do so using the terminology of relations:

Fig. 16.2 Five amigos

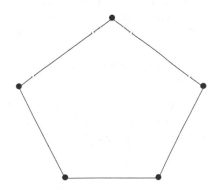

Definition 16.2 *Suppose that A and B are nonempty sets, and let f be a relation from A to B. We say that f is a* function *from A to B if for every a ∈ A there is a unique b ∈ B with (a, b) ∈ f.*

As before, we write $f : A \to B$ to indicate that f is a function from A to B; if $(a, b) \in f$ for some $a \in A$ and $b \in B$, we may also write $a \mapsto b$ or $f(a) = b$.

We can easily see from the matrix or graph representation of a relation if it is a function. A relation is a function whenever

- each row in its matrix description has exactly one 1 in it;
- its graph crosses every vertical line (over its domain) exactly once (this is called the *vertical line test*).

Note that, while the vertical line test tells us whether the relation is a function, the horizontal line test, introduced on page 130, determines if the function is injective.

In our title we promised some "good relations." Indeed, there are several relations that deserve special mention; as is the case with any good relations, they are "visited" often (and we promise to do so in this book). In particular, we make the following definitions.

Definition 16.3 *A relation R on the set A is said to be*

- reflexive *if for every a ∈ A, a ∼ a;*
- irreflexive *if for every a ∈ A, a ≁ a;*
- symmetric *if for every a ∈ A and b ∈ A, a ∼ b implies that b ∼ a;*
- asymmetric *if for every a ∈ A and b ∈ A, a ∼ b implies that b ≁ a;*
- antisymmetric *if for every a ∈ A and b ∈ A, a ∼ b and b ∼ a imply that a = b; and*
- transitive *if for every a ∈ A, b ∈ A, and c ∈ A, a ∼ b and b ∼ c imply that a ∼ c.*

For example, it is easy to see that

- < (less than) is an irreflexive, asymmetric, and transitive relation on the set of real numbers;
- ≤ (less than or equal to) is a reflexive, antisymmetric, and transitive relation on the set of real numbers;
- | (divisibility) is a reflexive, antisymmetric, and transitive relation on the set of positive integers;
- ⊆ (subset) is a reflexive, antisymmetric, and transitive relation on a set of sets; and
- ⇔ (equivalence) is a reflexive, symmetric, and transitive relation on a set of statements.

Relations with several of the characteristics we defined, such as the ones above, play important roles. The following three types of relations are particularly ubiquitous in mathematics.

Definition 16.4 *Suppose that R is a relation on a set S. We say that R is*

- *an equivalence relation if R is reflexive, symmetric, and transitive;*
- *an order relation if R is reflexive, antisymmetric, and transitive; and*
- *a graph relation if R is irreflexive and symmetric.*

In the rest of this chapter we focus on equivalence relations, we study order relations in Chapter 17, and we discuss graph relations in Appendix E.

Suppose that R is an equivalence relation on a set A. Note that, as a consequence of R being an equivalence relation on A, any element of A is in the relationship with itself; furthermore, for any two distinct elements x and y of A, either both $x \sim y$ and $y \sim x$ or neither relationship holds. For a given element $x \in A$, the collection of all elements of A that are in the relationship with x, denoted by

$$[x]_R = \{y \in A \mid x \sim y\},$$

is called the *equivalence class* of x determined by R. As we just observed, every element of A is an element of its own equivalence class, and if x is in the equivalence class of y, then y is also in the equivalence class of x.

Let us consider some examples. First let $A = P(\{1, 2, 3, 4\})$, the power set of $\{1, 2, 3, 4\}$, and let R be the relation on A of "having the same number of elements." Clearly, R is an equivalence relation on A with five different equivalence classes:

$$[\{1, 2, 3, 4\}]_R = \qquad \{\{1, 2, 3, 4\}\}$$

$$[\{1, 2, 3\}]_R = \{\{1, 2, 3\}, \{1, 2, 4\}, \{1, 3, 4\}, \{2, 3, 4\}\}$$

$$[\{1, 2\}]_R = \{\{1, 2\}, \{1, 3\}, \{1, 4\}, \{2, 3\}, \{2, 4\}, \{3, 4\}\}$$

$$[\{1\}]_R = \qquad \{\{1\}, \{2\}, \{3\}, \{4\}\}$$

$$[\emptyset]_R = \qquad \{\emptyset\}$$

The sizes of these equivalence classes are $1, 4, 6, 4,$ and 1—which we recognize as row 4 in Pascal's Triangle (cf. Figure 11.1). Observe also that the five equivalence classes are pairwise disjoint and their union gives all of $A = P(\{1, 2, 3, 4\})$.

As another example, consider the relation R on $\mathbb{N}$ of "having the same number of positive divisors." It is easy to see that R is an equivalence relation, and we find that $[1]_R = \{1\}$ (only 1 has one positive divisor); $[2]_R$ and $[3]_R$ are both equal to the set of all positive primes (cf. Definition 2.1);

$$[4]_R = \{4, 9, 25, 49, \ldots\}$$

is the set of prime squares, and so on. It is not difficult to prove that there are infinitely many equivalence classes, and that each one, except for $[1]_R$, consists of infinitely many elements. (This provides another approach to Problem 10 (e) of Chapter 6.)

Perhaps the most important example of equivalence relations is the congruence relation among integers. As an extension of Definition 3.2, we make the following definition.

Definition 16.5 *Suppose that m is a positive integer. We say that integers a and b are* congruent mod *m if a and b leave the same remainder when divided by m; this is denoted by $a \equiv b$ mod m. We call m the* modulus *of the congruence.*

It is obvious that a congruence is an equivalence relation. The equivalence class of an integer $x \in \mathbb{Z}$ for a modulus m is the *congruence class*

$$[x]_m = \{x + km \mid k \in \mathbb{Z}\} = \{x, x + m, x - m, x + 2m, x - 2m, \dots\};$$

for example,

$$[3]_{10} = \{3, 13, -7, 23, -17, \dots\}.$$

It is easy to see that there are exactly m congruence classes mod m; for example, the two congruence classes mod 2 are the set of even integers and the set of odd integers:

$$[0]_2 = \{0, 2, -2, 4, -4, 6, -6, \dots\}$$

and

$$[1]_2 = \{1, -1, 3, -3, 5, -5, \dots\}.$$

As each of our examples demonstrates, the equivalence classes of a given relation on a set partition the set into pairwise disjoint and nonempty parts with each element of the underlying set belonging to an equivalence class. We make the following general definition.

Definition 16.6 *A* partition *of the nonempty set A is a set Π of its subsets with the following properties.*

- *No subset in Π is empty.*
- *Any two distinct subsets in Π are disjoint.*
- *The union of all the subsets in Π is the entire set A.*

Furthermore, we also introduce the following terminology.

Definition 16.7 *Given a partition Π of the set A, a subset T of A that intersects every member of Π in a single element is called a* set of representatives *for Π.*

For instance, the set $T = \{0, 1, 2, \dots, m - 1\}$ is a set of representatives for the partition

$$\Pi = \{[0]_m, [1]_m, [2]_m, \ldots, [m-1]_m\}$$

of the set of integers, but this partition has (infinitely) many other sets of representatives as well:

$$\{0, 1, 2, 3, 4, 5, 6\},$$

$$\{10, 11, 12, 13, 14, 15, 16\},$$

and

$$\{10, 20, 30, 40, 50, 60, 70\}$$

are all sets of representatives for the congruence partition of the integers mod 7. As a less explicit example, consider a partition

$$\Pi = \{S_1, S_2, \ldots, S_n\}$$

of the set of natural numbers $\mathbb{N}$. Since, by definition, each part S_i ($i = 1, \ldots, n$) is nonempty, we can (for example) invoke Theorem 10.6 to select the minimum element m_i of S_i; the set

$$T = \{m_1, m_2, \ldots, m_n\}$$

is then a set of representatives for Π.

One of the most natural assumptions about set partitions is that one can always choose a set of representatives for the partition: this only entails selecting an element from each part of the partition. (Note that each part is nonempty.) It might thus be quite surprising that, if Π is infinite and no particular rule is available for choosing all the representative elements for T, then the existence of T cannot be proven from the basic properties of set theory and must be assumed as an axiom.

Axiom 16.8 (The Axiom of Choice) *Any partition of any set has a set of representatives.*

To shed some more light on the Axiom of Choice, consider the following three examples:

- let Π_1 be an arbitrary partition of the set of natural numbers $\mathbb{N}$;
- let Π_2 be a partition of the set of real numbers $\mathbb{R}$ into intervals; and
- let Π_3 be an arbitrary partition of the set of real numbers $\mathbb{R}$.

One can easily choose a set of representatives for Π_1: for example, we can choose the minimum element of each part in the partition (the minimum element exists by Theorem 10.6). We can also find a set of representatives for Π_2, by choosing the midpoint of each interval, for example. But no one has ever been able to find a set of representatives for Π_3, and the fact that there is one cannot be proven from the other axioms of the real numbers. Thus, as is (almost) customary, we add the Axiom of Choice to our collection of axioms. However, there are reasons to be skeptical about the validity of this axiom; see Appendix C for more on this subject.

As a slight detour, let us fulfill an earlier promise that the Axiom of Choice allows us to do. In Chapter 6, we defined the direct product of a finite number of sets $A_1, \ldots, A_n$ as

$$\prod_{i=1}^{n} A_i = A_1 \times A_2 \times \cdots \times A_n$$

$$= \{(x_1, x_2, \ldots, x_n) \mid (x_1 \in A_1) \wedge (x_2 \in A_2) \wedge \cdots \wedge (x_n \in A_n)\},$$

and said that we would define the product of an infinite number of sets in Chapter 16. We proceed to do so now.

Note first that there is a one-to-one correspondence between the direct product $\prod_{i=1}^{n} A_i$ above and the set of functions

$$\left\{ f : \{1, 2, \ldots, n\} \to \cup_{i=1}^{n} A_i \mid (f(1) \in A_1) \wedge (f(2) \in A_2) \wedge \cdots \wedge (f(n) \in A_n) \right\}.$$

Indeed, the element

$$(x_1, x_2, \ldots, x_n) \in \prod_{i=1}^{n} A_i$$

can be identified with the function f given by the following table representation.

i	1	2	...	n
$f(i)$	x_1	x_2	...	x_n

This gives us the suggestion for the following definition.

Definition 16.9 *Let I be a set, and suppose that X_i is a set for each $i \in I$. Then the* Cartesian *or* direct product *of the set of sets X_i is defined as the set of functions*

$$\prod_{i \in I} X_i = \{f : I \to \cup_{i \in I} X_i \mid \forall i \in I, f(i) \in X_i\}.$$

Note that the definition we gave in Chapter 6 for the case in which I is nonempty and finite is a special case of Definition 16.9. Of course, if I or any X_i is the empty-set, then so is $\prod_{i \in I} X_i$. Conversely, we have the following theorem.

Theorem 16.10 *Let I be a set, and let X_i be a set for each $i \in I$. Suppose that $I \neq \emptyset$ and $X_i \neq \emptyset$ for each $i \in I$. Then $\prod_{i \in I} X_i \neq \emptyset$.*

We prove Theorem 16.10 in Problem 11; in fact, we prove that it is equivalent to the Axiom of Choice.

When $I = \mathbb{N}$, the infinite direct product is also denoted by

$$\prod_{i=1}^{\infty} X_i = X_1 \times X_2 \times \cdots = \{(x_1, x_2, x_3, \ldots) \mid \forall i \in \mathbb{N}, x_i \in X_i\};$$

furthermore, when $X_i = X$ for all $i \in \mathbb{N}$, then

$$\prod_{i=1}^{\infty} X = \{(x_1, x_2, x_3, \ldots) \mid \forall i \in \mathbb{N}, x_i \in X\}$$

is simply denoted by X^{∞}. The elements of X^{∞} are called *infinite sequences*. We have already discussed sequences occasionally, and we will take these discussions to the limit in Chapter 18 (pun intended).

We now return to partitions. If our set A is small, we can easily list all of its partitions. For example, the set $A = \{1, 2, 3\}$ has five partitions, namely

$$\Pi_1 = \{\{1\}, \{2\}, \{3\}\},$$

$$\Pi_2 = \{\{1, 2\}, \{3\}\},$$

$$\Pi_3 = \{\{1\}, \{2, 3\}\},$$

$$\Pi_4 = \{\{1, 3\}, \{2\}\}, \text{ and}$$

$$\Pi_5 = \{\{1, 2, 3\}\}.$$

As the size of the set increases, however, the number of partitions grows in a rapid, but not entirely transparent, manner. Denoting the number of distinct partitions of a set with n elements by B_n—named the n-th *Bell number* after the Scottish mathematician and science fiction author Eric Temple Bell (1883–1960)—one can determine the following values:

n	1	2	3	4	5	6	7	8	9	10
B_n	1	2	5	15	52	203	877	4, 140	21, 147	115, 975

For example, as we have seen above, $B_3 = 5$. (For more on this fascinating sequence, see A000110 in the On-Line Encyclopedia of Integer Sequences at https://oeis.org.)

As our examples above show, there is a strong correspondence between equivalence relations and partitions, as stated in the following important theorem.

Theorem 16.11 (The Fundamental Theorem of Equivalence Relations) *Let A be a set.*

1. *If R is an equivalence relation on A, then the equivalence classes on A determined by R form a partition of A.*
2. *Conversely, let Π be a partition of A, and define a relation R by making two elements have the relationship whenever they are in the same set in the partition Π. Then R is an equivalence relation on A.*

We leave the rather easy proof for Problem 8.

The concept of a relation is one of the most fundamental ones in mathematics, and "good" relations—those that satisfy some desired properties—are used in many areas. In particular, the special relations of functions and of equivalence relations appear in virtually all parts of mathematics.

Problems

1. Decide if the following claim is true or false. If the claim is false, provide a counterexample; in either case, analyze the argument given. Be as specific as possible.
   Claim. In the definition of an equivalence relation, it is enough to require that the relation be symmetric and transitive. That is, if the relation R on a set A is both symmetric and transitive, then it is also reflexive.
   Argument. Since R is symmetric, $a \sim b$ implies $b \sim a$. But R is also transitive, so $a \sim b$ and $b \sim a$ imply that $a \sim a$. Therefore, R is reflexive.
2. Describe the (Cartesian) graph of each of the following relations, and decide whether the relation is reflexive, symmetric, and/or transitive. For those that are equivalence relations, describe the equivalence classes.

   (a) $R = \{(x, y) \in \mathbb{R}^2 \mid x \leq y + 1\}$
   (b) $R = \{(x, y) \in \mathbb{R}^2 \mid xy > 0\}$
   (c) $R = \{(x, y) \in \mathbb{R}^2 \mid xy \geq 0\}$
   (d) $R = \{(x, y) \in \mathbb{R}^2 \mid xy \neq 0\}$
   (e) $R = \{(x, y) \in (\mathbb{R} \setminus \{0\})^2 \mid xy > 0\}$
   (f) $R = \{(x, y) \in \mathbb{R}^2 \mid |x - y| \leq 1\}$
   (g) $R = \{(a, b) \in \mathbb{Z}^2 \mid a - b \text{ is divisible by } 2\}$
   (h) $R = \{(a, b) \in \mathbb{Z}^2 \mid a + b \text{ is divisible by } 2\}$
   (i) $R = \{(a, b) \in \mathbb{Z}^2 \mid a - b \text{ is divisible by } 5\}$
   (j) $R = \{(a, b) \in \mathbb{Z}^2 \mid a + b \text{ is divisible by } 5\}$

3. (a) Which of the following collections of congruence classes form partitions of the set of integers $\mathbb{Z}$?

   i. $\{[1234]_4, [4123]_4, [3412]_4, [2341]_4\}$
   ii. $\{[0]_2, [1]_4, [2]_8, [6]_8\}$

    iii. $\{[0]_2, [1]_4, [3]_8, [7]_8\}$
    iv. $\{[0]_2, [0]_3, [1]_6, [5]_6\}$

(b) Find some examples for partitions of $\mathbb{Z}$ into exactly five parts in which each part is a congruence class.

Remarks It has been known that no such partition exists with all moduli distinct. In fact, if m is the largest modulus, then it was shown by Davenport, Mirsky, Newman, and Rado that a congruence class partition of $\mathbb{Z}$ will contain at least p distinct congruence classes mod m where p is the smallest positive prime divisor of m. Each example in part (a) satisfies this condition (though not all are partitions).

4. Verify the values of the first four Bell numbers given on page 253.
5. Suppose that S is a set with 4 elements. Find the number of

  (a) relations
  (b) reflexive relations
  (c) irreflexive relations
  (d) symmetric relations
  (e) asymmetric relations
  (f) antisymmetric relations
  (g) graph relations
  (h) equivalence relations

on the set S. (Note that we did not ask for the number of transitive relations as it is difficult to determine.)

    (Hints: Except for the last part, count the number of relations using matrix representations; that is, count the number of choices we have for choosing which of the 16 matrix entries are 1s. Use the Fundamental Theorem of Equivalence Relations for the last part.)

6. Let S be a nonempty set, and suppose that R_1 and R_2 are relations on S. Prove or disprove each of the following statements.

  (a) If R_1 and R_2 are reflexive, then $R_1 \cup R_2$ is reflexive.
  (b) If R_1 and R_2 are symmetric, then $R_1 \cup R_2$ is symmetric.
  (c) If R_1 and R_2 are transitive, then $R_1 \cup R_2$ is transitive.
  (d) If R_1 or R_2 is reflexive, then $R_1 \cup R_2$ is reflexive.
  (e) If R_1 or R_2 is symmetric, then $R_1 \cup R_2$ is symmetric.
  (f) If R_1 or R_2 is transitive, then $R_1 \cup R_2$ is transitive.
  (g) If R_1 and R_2 are reflexive, then $R_1 \cap R_2$ is reflexive.
  (h) If R_1 and R_2 are symmetric, then $R_1 \cap R_2$ is symmetric.
  (i) If R_1 and R_2 are transitive, then $R_1 \cap R_2$ is transitive.
  (j) If R_1 or R_2 is reflexive, then $R_1 \cap R_2$ is reflexive.
  (k) If R_1 or R_2 is symmetric, then $R_1 \cap R_2$ is symmetric.
  (l) If R_1 or R_2 is transitive, then $R_1 \cap R_2$ is transitive.

7. Prove that each of the following relations is an equivalence relation. For each relation R, describe explicitly the equivalence classes $[(3, 1)]_R$, $[(2, 2)]_R$, and $[(1, 3)]_R$, then find an explicit bijection between the set of equivalence classes and a certain familiar number set.

(a) $R = \{((a, b), (c, d)) \in (\mathbb{N}^2)^2 \mid a + d = b + c\}$

(b) $R = \{((a, b), (c, d)) \in (\mathbb{Z} \times (\mathbb{Z} \setminus \{0\}))^2 \mid a \cdot d = b \cdot c\}$

8. Prove the Fundamental Theorem of Equivalence Relations (Theorem 16.11).

9. (a) Sketch the (combinatorial) graph of the following relations:

   i. $R_1 = \{(a, b) \in \{1, 2, 3, 4, 5\}^2 \mid |a - b| \in \{1, 4\}\}$

   ii. $R_2 = \{(a, b) \in \{1, 2, \ldots, 8\}^2 \mid |a - b| \in \{1, 4, 7\}\}$

   iii. $R_3 = \{(a, b) \in \{1, 2, \ldots, 13\}^2 \mid |a - b| \in \{1, 5, 8, 12\}\}$

   (b) What does each of these graphs have to do with the Ramsey numbers? (See Definition 12.7 and also Proposition 12.6, Problem 10 (b) in Chapter 3, and Problem 7 in Chapter 12.)

10. Prove that isomorphism among groups is an equivalence relation. (To be precise, we should say that, while reflexivity, symmetry, and transitivity hold, group isomorphism is not a relation as the collection of all groups is not a set— there are too many of them!)

11. Prove that Theorem 16.10 is equivalent to the Axiom of Choice.

    (Hints: To prove that the Axiom of Choice implies Theorem 16.10, define, for each $i \in I$,

$$Y_i = \{(i, x_i) \mid x_i \in X_i\};$$

then consider the set $Y = \cup_{i \in I} Y_i$.)

Chapter 17
Order, Please!

As promised, in this chapter we discuss an important and highly applicable type of relations: order relations. Since the usual "less than or equal to" relation on $\mathbb{R}$ (or a subset of $\mathbb{R}$ such as $\mathbb{N}$, $\mathbb{Z}$, or $\mathbb{Q}$) is a primary example of an order relation, it is useful to use notation resembling the $\leq$ sign; however, since our discussion applies to other orderings as well, we choose the symbol $\preceq$ that is similar, but not identical, to the symbol $\leq$. Throughout this chapter (and beyond), $a \preceq b$ denotes the fact that $a \sim_R b$ holds for some elements a and b (of a given set) and order relation R (on the same set).

We now restate the definition of an order relation. Before doing so, however, we should point out that we here talk about order relations that do not necessarily satisfy the property that any two elements can be compared; that is, given elements a and b in our set, it is not necessarily the case that we have $a \preceq b$ or $b \preceq a$. For this reason, order relations are often referred to as *partial order relations*—in contrast to *total order relations* where any two elements are comparable.

Definition 17.1 *A relation $\preceq$ on a set P is said to be a* partial order relation *(or simply* order relation*) if it is reflexive, antisymmetric, and transitive; in this case we say that P forms a* partially ordered set *or* poset *for the relation.*

Posets, therefore, are mathematical structures in the sense of Chapter 13—matching the importance of groups, rings, fields, and Boolean algebras. Since a set P may be a poset for a variety of different partial order relations, when referring to a particular relation $\preceq$, we will talk about the *poset system* $(P, \preceq)$. (If the relation is already clear from the context, then we will simply talk about the poset P.)

The following three examples are important poset systems, and we will return to them throughout this chapter.

- Any set X of real numbers is a poset for the "less than or equal to" relation; this poset system is denoted by $(X, \leq)$. In fact, $(X, \leq)$ is a total order system since any two real numbers are comparable.

© Springer Nature Switzerland AG 2020
B. Bajnok, *An Invitation to Abstract Mathematics*, Undergraduate Texts in
Mathematics, https://doi.org/10.1007/978-3-030-56174-1_17

- The power set $P(U)$ of any set U is a poset for the "subset" relation; this system is called the *Boolean poset* and is denoted by $(P(U), \subseteq)$. As a special case, if $U = \{1, 2, \ldots, n\}$ for some positive integer n, then we simply write $\mathcal{B}_n$ for $(P(U), \subseteq)$.
- Any set of nonnegative integers S is a poset for the "divisibility" relation; this system is called the *divisor poset* and is denoted by $(S, |)$. In particular, if S is the set of all positive divisors of a positive integer n, we let $\mathcal{D}_n$ denote $(S, |)$.

We should note that the set of integers is not a poset for divisibility as it fails antisymmetry; for example, we have $7|(-7)$ and $(-7)|7$, but, of course, $7 \neq -7$.

Each partial order relation has a corresponding "strict" version, as follows:

Definition 17.2 *Suppose that $\preceq$ is a partial order relation on a set P. The* strict order *relation corresponding to $\preceq$ is the relation $\prec$ on P consisting of all ordered pairs $(a, b) \in P^2$ for which $a \preceq b$ but $a \neq b$.*

It is then not hard to prove that $\prec$ is an irreflexive, asymmetric, and transitive relation on P—see Problem 1.

Partially ordered sets may be visualized via diagrams constructed as follows. Suppose that P is a set with partial order relation $\preceq$ (and corresponding strict order $\prec$). For two distinct elements $a \in P$ and $b \in P$ we say that b is a (consecutive) *successor* of a if $a \prec b$ and there is no $c \in P$ for which $a \prec c$ and $c \prec b$. For example, in the divisor poset $\mathcal{D}_{18}$, 18 is a successor of 6, since $6|18$ and there is no $c \in \mathbb{N}$ different from 6 and 18 for which $6|c$ and $c|18$. However, 18 is not a successor of 3, as we have $3|9$ and $9|18$.

The *Hasse diagram* of a poset P—so named after the German mathematician Helmut Hasse (1898–1979)—is a graph G in the plane (in the sense of Chapter 16) so that the vertices of G correspond to the elements of P; two vertices of G are adjacent if, and only if, one is a successor of the other; and the vertices are positioned in the plane in such a fashion that if $b \in P$ is a successor of $a \in P$, then the vertex corresponding to a is "lower" than the vertex corresponding to b. Note that, since the partial order relation is antisymmetric, we cannot have two vertices so that one of them is both "lower" and "higher" than the other. On the other hand, if two elements of P are such that neither is a successor of the other, then we may choose to put either one "lower" than the other, or we may put them at the same "height." Of course, a poset has many different Hasse diagrams, and it is not always easy to decide if two diagrams correspond to the same poset.

Hasse diagrams are particularly convenient for finite posets. The Hasse diagram of $\mathcal{D}_{108}$, for example, is given in Figure 17.1.

The most well-known (and used) Hasse diagram is probably that of $\mathcal{B}_n$; due to its structure, it is referred to as the n-dimensional *hypercube*. For example, the Hasse diagram of $\mathcal{B}_3$ is the (not very "hyper") 3-dimensional cube of Figure 17.2.

Hasse diagrams of infinite posets can sometimes also be clear. For example, the Hasse diagrams of the total orders $(\mathbb{N}, \leq)$ and $(\mathbb{Z}, \leq)$ are vertical paths, infinite in both directions for $\mathbb{Z}$, but infinite only upward for $\mathbb{N}$. The Hasse diagrams for $(\mathbb{Q}, \leq)$

Fig. 17.1 The Hasse diagram of $\mathcal{D}_{108}$

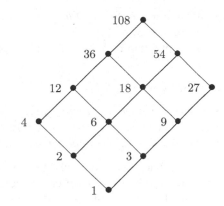

Fig. 17.2 The Hasse diagram of $\mathcal{B}_3$

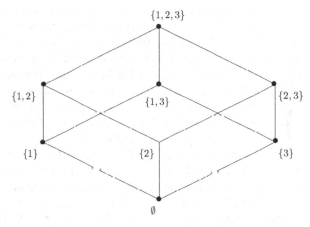

or $(\mathbb{R}, \leq)$, however, are not feasible: there are no successors as $\mathbb{Q}$ and $\mathbb{R}$ are dense (cf. Problem 2 in Chapter 12).

Posets and their properties are discussed in almost all branches of mathematics. Here we examine some of their basic properties. Let us start by introducing some terminology.

Definition 17.3 *Suppose that $\preceq$ is a partial order relation on a set P, and let S be a subset of P.*
We say that an element $a \in S$ is

- *a minimal element of S if there is no $b \in S$ different from a for which $b \preceq a$;*
- *a maximal element of S if there is no $b \in S$ different from a for which $a \preceq b$;*
- *the minimum element of S if $a \preceq b$ holds for every $b \in S$;*
- *the maximum element of S if $b \preceq a$ holds for every $b \in S$.*

Furthermore, we say that an element $a \in P$ is

- *a lower bound for S in P if $a \preceq b$ holds for every $b \in S$;*
- *an upper bound for S in P if $b \preceq a$ holds for every $b \in S$;*

- *the* infimum *of S in P if it is a lower bound of S and c $\preceq$ a holds for every lower bound c of S; and*
- *the* supremum *of S in P if it is an upper bound of S and a $\preceq$ c holds for every upper bound c of S.*

Finally, we say that S is

- bounded below *in P if it has at least one lower bound in P;*
- bounded above *in P if it has at least one upper bound in P; and*
- bounded *in P if it is both bounded below and bounded above in P.*

It is probably wise to briefly discuss these rather delicate terms. We start with the first four definitions. The difference between *a minimal* element and *the minimum* element of a subset S of a poset P is that, while both capture the fact that no element of S can be less than them, the minimum element has the additional property that it must be comparable to (and, therefore, less than) every other element of S. Clearly (cf. Problem 4), if the minimum element of S exists at all, then it must be unique, and if S has more than one minimal element, then it cannot have a minimum; the analogous facts hold for maximal and maximum elements. If they exist, then we will use the notations min S and max S for the minimum and maximum elements of S, respectively. For example, in the poset P given by its Hasse diagram in Figure 17.3, the elements a, b, c, and g are minimal elements of P; there is no minimum element, f is the only maximal element, and max $P = f$.

Let us now examine the rest of the terms in Definition 17.3. First note that, unlike the minimal, maximal, minimum, and maximum elements of the subset S of P, lower and upper bounds and the infimum and supremum elements do not need to be in S. The set of lower bounds for S in P is denoted by $S^{\downarrow}$ (pronounced "S lower"), and the set of upper bounds for S in P is denoted by $S^{\uparrow}$ (pronounced "S upper"). If neither $S^{\downarrow}$ nor $S^{\uparrow}$ is empty, then S is bounded in P. Clearly, a subset S in a poset P

Fig. 17.3 A charging horse

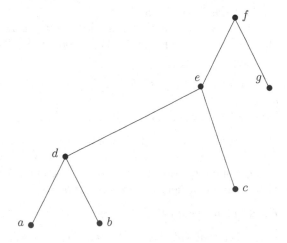

may have many upper or lower bounds, but the infimum and supremum of S in P, if they exist, must be unique; they will be denoted by $\inf S$ and $\sup S$, respectively.

Let us further illuminate these concepts through some examples. Consider first the set $S = \{2, 6, 9\}$ in the poset $(\mathbb{N}, \leq)$. Then $x = 1$ is a lower bound for S since $1 \leq y$ holds for every $y \in S$. However, $x = 1$ is not the infimum of S as $x \leq 1$ does not hold for every lower bound x of S: clearly, $x = 2$ is a lower bound and $2 \not\leq 1$. However, $\inf S = 2$ as 1 and 2 are the only lower bounds of S in P and $1 \leq 2$. We can also verify that $\sup S = 9$. It is easy to see that every nonempty subset of $\mathbb{N}$ has an infimum (the minimum element of the set, which exists by Theorem 10.6), but an infinite subset does not have a supremum.

Next, consider again the set $S = \{2, 6, 9\}$, but now in the divisor poset $(\mathbb{N}, |)$. This time, we have $\inf S = 1$ and $\sup S = 18$. For example, 18 is the supremum as it is a common multiple of 2, 6, and 9, and every common multiple of 2, 6, and 9 is a multiple of 18 (so 18 is the *least common multiple* of 2, 6, and 9). We find that every nonempty subset of $\mathbb{N}$ has an infimum (the *greatest common divisor*), but an infinite subset does not have a supremum; cf. Theorem 17.8 below.

Partially ordered sets in which every subset (or at least every finite subset) has an infimum and a supremum play important roles in many parts of mathematics.

Definition 17.4 *Suppose that $\preceq$ is a partial order relation on a set P.*

If every two-element subset of P has an infimum and a supremum, then $(P, \preceq)$ is called a lattice.

If every subset of P has an infimum and a supremum, then $(P, \preceq)$ is called a complete lattice.

It is obvious that, in every poset, all one-element subsets have an infimum and a supremum. Using induction, it is easy to prove that, if every two-element subset in P has a supremum and an infimum (i.e., P is a lattice), then every *nonempty finite* subset has an infimum and a supremum—see Theorem 17.11.

It is worth examining the case of the empty-set: when does $\emptyset$ have an infimum and a supremum in a poset P? Note that, since (trivially) every element of P serves both as an upper bound and a lower bound for $\emptyset$, in order for $\emptyset$ to have a supremum or infimum, P itself must have a minimum or maximum element, respectively.

Proposition 17.5 *In a poset P, $\inf \emptyset$ exists if, and only if, $\max P$ does, in which case*

$$\inf \emptyset = \max P;$$

and $\sup \emptyset$ exists if, and only if, $\min P$ does, in which case

$$\sup \emptyset = \min P.$$

Therefore, since in a complete lattice even $\emptyset$ must have an infimum and a supremum, we get the following:

Corollary 17.6 *A complete lattice must have both a minimum and a maximum element. In particular, every complete lattice is bounded.*

Let us now examine our three main examples for posets on page 257. Clearly, $(\mathbb{N}, \leq), (\mathbb{Z}, \leq), (\mathbb{Q}, \leq)$, and $(\mathbb{R}, \leq)$ are all lattices, with the minimum and maximum of the two numbers in question serving as infimum and supremum, respectively; for example, $\inf\{3, 8\} = 3$ and $\sup\{3, 8\} = 8$. However, none of these posets are complete lattices, since none of them have maximum elements (and only $\mathbb{N}$ has a minimum element). Thus, in these posets, only nonempty sets may possibly have infima and suprema. Indeed, in both $(\mathbb{N}, \leq)$ and $(\mathbb{Z}, \leq)$, every nonempty and bounded subset has both a supremum and an infimum; for example, if S is the set of all integers strictly between 3 and 8, then $\inf S = 4$ and $\sup S = 7$.

This is not the case, however, in $(\mathbb{Q}, \leq)$. Consider the set

$$S = \{x \in \mathbb{Q} \mid x^2 < 2\}$$

in this poset. (We see that

$$S = \{x \in \mathbb{Q} \mid -\sqrt{2} < x < \sqrt{2}\},$$

but we wanted to define S in terms of rational numbers only.) Clearly, S is nonempty and bounded in $\mathbb{Q}$. However, we find that S has neither an infimum nor a supremum; indeed, the set of lower bounds of S and the set of upper bounds of S in $\mathbb{Q}$ are given by

$$S^{\downarrow} = \{y \in \mathbb{Q} \mid y \leq -\sqrt{2}\}$$

and

$$S^{\uparrow} = \{y \in \mathbb{Q} \mid y \geq \sqrt{2}\},$$

respectively, but $S^{\downarrow}$ has no maximum element and $S^{\uparrow}$ has no minimum element among the rational numbers. Thus, in $(\mathbb{Q}, \leq)$, even a nonempty bounded set may have no supremum and infimum. We will return to this example in Chapter 21.

At the same time, we find that the set S above has both an infimum and a supremum in $(\mathbb{R}, \leq)$: clearly, $\inf S = -\sqrt{2}$ and $\sup S = \sqrt{2}$. It turns out that, in fact, every nonempty and bounded subset of $\mathbb{R}$ has an infimum and a supremum. Since, clearly, this fact cannot be proven from the ordered field axioms—otherwise it would apply to the ordered field of rational numbers as well (see also Problem 7)— we state this as an axiom.

Axiom 17.7 (The Completeness Axiom) *If S is a nonempty and bounded subset of $\mathbb{R}$, then S has both a supremum and an infimum in the poset $(\mathbb{R}, \leq)$.*

(Here we state this statement as an axiom, as it is customary, but we should note that it is possible to use a different axiom system for the set of real numbers from which the Completeness Axiom can be derived; cf. Problem 9 in Chapter 21.) The fact that the ordered field of the real numbers satisfies the Completeness Axiom is often stated in the condensed form that $\mathbb{R}$ is a *complete ordered field*. We will examine some of the important and well-known consequences of the Completeness Axiom in Chapter 18.

Regarding the Boolean poset $(P(U), \subseteq)$, we can easily verify that it is a complete lattice for every set U. For any set $S \subseteq P(U)$, the infimum and supremum of S are given by the intersection and the union of all sets in S, respectively; that is,

$$\inf \, S = \bigcap_{X \in S} X \quad \text{and} \quad \sup \, S = \bigcup_{X \in S} X.$$

(We should note that, by the axioms of set theory, these intersections and unions are themselves subsets of U—see Appendix C.) Clearly, we have $\min P(U) = \emptyset$ and $\max P(U) = U$.

Finally, we examine the divisor posets. As our earlier example shows, $(\{2, 6, 9\}, |)$ is not a lattice: for example, $S = \{6, 9\}$ has neither an infimum nor a supremum in the set. On the other hand, we have the following result.

Theorem 17.8 *The divisor poset* $(\mathbb{N}, |)$ *is a lattice. That is, for all pairs of positive integers a and b,*

- *there is a unique positive integer d such that*

  - *d is a common divisor of a and b, and*
  - *d is a multiple of all common divisors of a and b;*

- *there is a unique positive integer l such that*

  - *l is a common multiple of a and b, and*
  - *l is a divisor of all common multiples of a and b.*

Proof We use the Fundamental Theorem of Arithmetic (cf. Theorem 11.8) to show that $\gcd(a, b)$ and $\operatorname{lcm}(a, b)$ satisfy the respective requirements, and it will also be clear from our proof that no other integers do. Since our claim clearly holds if $a = 1$ or $b = 1$, we will assume that $a \geq 2$ and $b \geq 2$.

Let P_a and P_b be the set of positive primes that divide a and b, respectively, and set $P = P_a \cup P_b$. For each $p \in P$, let α_p be the number of times p appears in the prime factorization of a; define β_p similarly. (Note that α_p or β_p may be 0.) With these notations, we can write

$$a = \prod_{p \in P} p^{\alpha_p} \quad \text{and} \quad b = \prod_{p \in P} p^{\beta_p}.$$

It is easy to see that a positive integer c is a common divisor of a and b if, and only if, c is of the form

$$c = \prod_{p \in P} p^{\gamma_p}$$

where

$$0 \le \gamma_p \le \min\{\alpha_p, \beta_p\}$$

holds for each prime $p \in P$. In particular, setting

$$\gamma_p = \min\{\alpha_p, \beta_p\}$$

for each $p \in P$, we get the greatest common divisor, which then is clearly a multiple of every c in the form given above.

Similarly, we see that any common multiple of a and b will have a prime factorization that contains every $p \in P$ with exponent at least

$$\delta_p = \max\{\alpha_p, \beta_p\};$$

and thus the least common multiple of a and b is

$$\mathrm{lcm}(a, b) = \prod_{p \in P} p^{\delta_p},$$

and this number is then a divisor of all common multiples of a and b.   □

Theorem 17.8 also implies that the finite divisor poset $\mathcal{D}_n$ is a lattice for every $n \in \mathbb{N}$ and, since it's finite, it is a complete lattice (cf. Theorem 17.12). In contrast, $(\mathbb{N}, |)$ is not a complete lattice: there is no $n \in \mathbb{N}$ that is divisible by all positive integers. However, as we show in Problem 9, $(\mathbb{N} \cup \{0\}, |)$ is a complete lattice!

Before we close this chapter on order relations, we must mention a special type of order relation that plays a crucial role in the development of set theory. Recall Theorem 10.6, which states that in the poset $(\mathbb{N}, \le)$, every nonempty set contains a minimum element. We make the following definition.

Definition 17.9 *Suppose that $\preceq$ is a partial order relation on a set P. If every nonempty subset of P contains a minimum element, then $\preceq$ is said to be a* well-order *on P, and $(P, \preceq)$ is called a* well-ordered set *or* woset.

Here we have to restrict the requirement of having a minimum element to nonempty subsets of P; the empty-set, of course, has no minimum element. An interesting question—with a surprising answer—is: Which sets can be well-ordered; that is, on which sets can one define a well-ordering relation?

We already noted that, by Theorem 10.6, the poset $(\mathbb{N}, \le)$ is a woset. It may be worth pointing out that there are many ways to well-order $\mathbb{N}$ besides the usual "less than or equal to" relation. Here are some examples (with $m \prec n$ denoting the strict order that $m \preceq n$ but $m \ne n$):

$$1 \prec 2 \prec 3 \prec 4 \prec 5 \prec 6 \prec \cdots$$

$$2 \prec 1 \prec 4 \prec 3 \prec 6 \prec 5 \prec \cdots$$

$$4 \prec 5 \prec 6 \prec 7 \prec \cdots \prec 1 \prec 2 \prec 3$$

$$1 \prec 3 \prec 5 \prec 7 \prec \cdots \prec 2 \prec 4 \prec 6 \prec 8 \prec \cdots$$

$$1 \prec 2 \prec 4 \prec 8 \prec \cdots \prec 3 \prec 6 \prec 12 \prec 24 \prec \cdots \prec 5 \prec 10 \prec 20 \prec 40 \prec \cdots$$

We can easily verify that each order is a well-order.

Clearly, $(\mathbb{Z}, \leq)$, $(\mathbb{Q}, \leq)$, and $(\mathbb{R}, \leq)$ are not wosets as there is no minimum integer, rational number, or real number. It is not too surprising, however, that the set of integers and even the set of rational numbers can be well-ordered; after all, each set can even be put in a list (cf. Problem 4 in Chapter 6). For example, the integers can be well-ordered as

$$0 \prec 1 \prec -1 \prec 2 \prec -2 \prec 3 \prec -3 \prec 4 \prec -4 \prec \cdots$$

or

$$0 \prec 1 \prec 2 \prec 3 \prec 4 \prec \cdots \prec -1 \prec -2 \prec -3 \prec -4 \prec \cdots.$$

It was one of the most shocking events in the history of mathematics when in 1904 the German mathematician Ernst Zermelo announced his proof of the following theorem.

Theorem 17.10 (The Well-Ordering Theorem) *Every set has a well-ordering; that is, for every set X, there exists an order $\preceq$, so that $(X, \preceq)$ is a woset.*

While Zermelo's proof is not too complicated, the statement itself is hard to grasp. No one, to this day, knows, for example, how to create a well-ordering of the real numbers (Zermelo's proof is not constructive). Any proof of the Well-Ordering Theorem uses the Axiom of Choice (cf. page 251); in fact, it can be proven that they are equivalent: each one implies the other. So we have two equivalent statements— the Axiom of Choice and the Well-Ordering Theorem—where one statement (the Axiom of Choice) agrees with our intuition and the other seems impossible to imagine. The majority of mathematicians have already made peace with the Well-Ordering Theorem, but having to accept consequences like this made some argue for dropping the Axiom of Choice from our axiom system. (Another, perhaps even more striking, consequence of the Axiom of Choice is the so-called Banach–Tarski Paradox, cf. Theorem C.9.)

Lattices, especially complete lattices, are important mathematical structures that appear in algebra, linear algebra, analysis, combinatorics, and topology—just about all branches of mathematics.

Problems

1. (a) Suppose that $\preceq$ is a partial order relation on a set P. Prove that the strict order relation $\prec$ corresponding to $\preceq$ (defined by Definition 17.2) is an irreflexive, asymmetric, and transitive relation on P.
   (b) Suppose that $\prec$ is an irreflexive, asymmetric, and transitive relation on a set P. Prove that the relation $\preceq$ corresponding to $\prec$ (defined by the union of $\prec$ and $=$) is a partial order relation on P.

2. Recall that in Definition 13.12 we defined ordered rings, ordered integral domains, and ordered fields via respective subsets P of "positives" for which axioms (O), $(O+)$, and $(O\cdot)$ held. (We then used P to define the usual "less than or equal to" relation on such structures.) In this problem we prove that the set P is unique in $\mathbb{Z}$, $\mathbb{Q}$, and $\mathbb{R}$; furthermore, $\mathbb{C}$ contains no subset P that makes it an ordered field.

   (a) Prove that if P is a subset of $\mathbb{Z}$ for which the order axioms hold, then $P = \mathbb{N}$.
      (Hint: First prove that $1 \in P$.)
   (b) Prove that if P is a subset of $\mathbb{Q}$ for which the order axioms hold, then P is the usual set of positive rationals.
   (c) Prove that if P is a subset of $\mathbb{R}$ for which the order axioms hold, then $P = (0, \infty)$.
      (Hint: Use the fact that every element of $(0, \infty)$ has a square root in $\mathbb{R}$.)
   (d) Prove that $\mathbb{C}$ contains no subset P that makes it an ordered field.
      (Hint: $i^2 = -1$.)

3. For a positive integer n, the *partition lattice* $\mathcal{P}_n$ of order n consists of all partitions of $\{1, 2, \ldots, n\}$; with the partial ordering $\preceq$ defined as "refinement": for $\Pi_1 \in \mathcal{P}_n$ and $\Pi_2 \in \mathcal{P}_n$, we say that $\Pi_1 \preceq \Pi_2$ if every element of Π_1 is a subset of some element of Π_2. For example, among the partitions of the set $\{1, 2, 3\}$ listed on page 253, we have $\Pi_1 \preceq \Pi_2$ and $\Pi_3 \preceq \Pi_5$, but $\Pi_2 \not\preceq \Pi_3$ since $\{1, 2\} \not\subseteq \{1\}$ and $\{1, 2\} \not\subseteq \{2, 3\}$.

   (a) List all elements of the partition lattice of order 4.
   (b) Draw the Hasse diagram of the partition lattice of order n for every $n \leq 4$.

4. Suppose that P is a partially ordered set and that S is a subset of P. For each statement below, decide (by providing a proof or a counterexample) if the statement is true or false.

   (a) If the minimum element of S exists at all, then it must be unique.
   (b) If S has more than one minimal element, then it cannot have a minimum element.
   (c) If S has only one minimal element, then this element is its minimum element.
   (d) If the infimum of S in P exists at all, then it must be unique.
   (e) If S has more than one lower bound, then it cannot have an infimum in P.

(f) If S has only one lower bound, then this element is its infimum.

(g) If S has a minimum element, then this element is also the infimum of S in P.

(h) If S has an infimum in P, then this element is also the minimum element of S.

(i) If S has an infimum in P, then this element is also the maximum element of $S^{\downarrow}$.

(j) If S has a minimum element, then this element is also the supremum element of $S^{\downarrow}$ in P.

5. In Definition 17.4, we defined a lattice as a poset in which every two-element subset has an infimum and a supremum and a complete lattice as a poset in which every subset has an infimum and a supremum. The following theorems state alternative criteria. Prove each result.

(a) **Theorem 17.11** *A partially ordered set P is a lattice if, and only if, every nonempty finite subset of P has an infimum and a supremum.*

(b) **Theorem 17.12** *A finite partially ordered set P is a complete lattice if, and only if, P is a lattice.*
(Hint: Use Theorem 17.11, but don't forget about the empty-set!)

(c) **Theorem 17.13** *A partially ordered set P is a complete lattice if, and only if, every subset of P has an infimum.*
(Hints: Let $S \subseteq P$; by assumption, $a = \inf S^{\uparrow}$ exists. Prove that $a = \sup S$.)

Remarks According to Theorem 17.13, it is enough to verify "half" of what Definition 17.4 requires. A similar result can be attained by replacing "infimum" by "supremum."

6. For each of the following posets, decide whether it is a lattice and, if so, if it is a complete lattice.

(a) P is given by its Hasse diagram in Figure 17.4.

Fig. 17.4 A hexagon with a chord

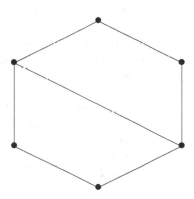

Fig. 17.5 A hexagon with
two chords

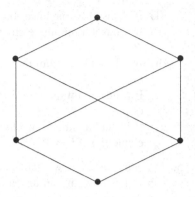

(b) P is given by its Hasse diagram in Figure 17.5.

(c) $P = \{1, 2, 4, 5, 6, 12, 20, 30, 60\}$ with the partial order being divisibility.

(d) $P = \{1, 2, 5, 15, 20, 60\}$ with the partial order being divisibility.

(e) $P = (6, 7]$ with the partial order being $\leq$.

(f) $P = [6, 7]$ with the partial order being $\leq$.

    (Hints: Let $S \subseteq P$. Note that, by the Completeness Axiom, $i = \inf S$ and $s = \sup S$ exist in $\mathbb{R}$. Decide whether $i \in P$ and $s \in P$ must hold. Consider the cases $S = \emptyset$ and $S \neq \emptyset$ separately.)

(g) $P = \{5\} \cup (6, 7]$ with the partial order being $\leq$.

    (Hints: Let $S \subseteq P$, and consider the cases (i) $S = \emptyset$; (ii) $S = \{5\}$; (iii) $5 \in S$ and $S \cap (6, 7] \neq \emptyset$; and (iv) $5 \notin S$ and $S \cap (6, 7] \neq \emptyset$ separately.)

7. Recall that the set of rational functions $\mathbb{R}(x)$ (cf. Problem 9 (d) of Chapter 13) is an ordered field. Prove that $\mathbb{R}(x)$ is not a complete ordered field.

    (Hint: Recall that in $\mathbb{R}(x)$ a polynomial is positive whenever its leading coefficient is positive, and a rational function f/g (with real polynomials f and $g \neq 0$) is positive whenever the leading coefficients of f and g have the same sign. Prove that the Completeness Axiom fails in $\mathbb{R}(x)$ by showing that the subset $\mathbb{R}$ of $\mathbb{R}(x)$ is bounded but has no supremum.)

8. (a) Consider the following claim.

    Claim 1. The poset $(\mathbb{N}, \leq)$ is a complete lattice.

    We have already pointed out that Claim 1 is false; for example, it violates Corollary 17.6. Therefore, the following argument must be incorrect; explain why.

    Argument. According to Theorem 17.13, it is enough to verify that every set of positive integers has an infimum for $\leq$. But, by Theorem 10.6, every subset of $\mathbb{N}$ has a minimum element; this element then is clearly the infimum of the set.

  (b) Consider the following claim.

    Claim 2. The poset $(\mathbb{N}, \mid)$ is a lattice.

In Theorem 17.8 we proved that Claim 2 was true. The following argument seems considerably simpler—unfortunately, it is not correct. Explain what is wrong with it.

Argument. We need to prove that for all pairs of positive integers a and b, there are integers $i = \inf\{a, b\}$ and $s = \sup\{a, b\}$. We will show that $i = \gcd(a, b)$ and $s = \text{lcm}(a, b)$ satisfy the definition of infimum and supremum, respectively. We will only do this here for i; the argument for s is similar.

Note that the partial order $\preceq$ here is divisibility; therefore, we need to prove that i is a common divisor of a and b and that, if c is any common divisor of a and b, then i is greater than or equal to c. But both of these claims follow trivially from the definition of the greatest common divisor. This proves that $\mathbb{N}$ is a lattice for the divisibility relation.

9. Prove that the set of nonnegative integers is a complete lattice for the divisibility relation. Explain why this does not violate Corollary 17.6.

10. Consider the following proposition.

Proposition 17.14 *Suppose that L is a lattice with partial order $\preceq$, and let $a, b, c \in L$. If $a \preceq b$, then $\inf\{a, c\} \preceq \inf\{b, c\}$.*

(a) Restate Proposition 17.14 for the lattice $(\mathbb{R}, \leq)$.
(b) Restate Proposition 17.14 for the Boolean lattice $(P(U), \subseteq)$. (As usual, $P(U)$ denotes the power set of a set U.)
(c) Restate Proposition 17.14 for the divisor lattice $(\mathbb{N}, |)$.
(d) Prove Proposition 17.14.
(e) Is the converse of Proposition 17.14 true?

11. (a) Consider the following theorem.

Theorem 17.15 *Suppose that L is a lattice with partial order $\preceq$, and let $a, b, c \in L$. Then*

$$\sup\{\inf\{a, b\}, \inf\{a, c\}\} \preceq \inf\{a, \sup\{b, c\}\}.$$

i. Restate Theorem 17.15 for the Boolean lattice $(P(U), \subseteq)$. ($P(U)$ denotes the power set of a set U.)
ii. What is wrong with the following "proof"? Wlog we can assume that $b \prec c$; so $\sup\{b, c\} = c$. By Proposition 17.14, we also have $\inf\{a, b\} \prec \inf\{a, c\}$, and thus $\sup\{\inf\{a, b\}, \inf\{a, c\}\} = \inf\{a, c\}$. Therefore, the claim of Theorem 17.15 simplifies to

$$\inf\{a, c\} \preceq \inf\{a, c\},$$

which obviously holds.
iii. Prove Theorem 17.15.
(Hint: Prove first that $\inf\{a, b\} \preceq a$, $\inf\{a, c\} \preceq a$, $\inf\{a, b\} \preceq \sup\{b, c\}$, and $\inf\{a, c\} \preceq \sup\{b, c\}$.)

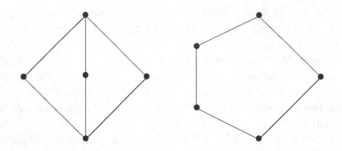

Fig. 17.6 The lattices M_3 and N_5

(b) Lattices for which equality holds in Theorem 17.15 play an important role in the algebra of lattices.

Definition 17.16 *A lattice L with partial order $\preceq$ is called a* distributive lattice *if for every $a, b, c \in L$,*

$$\sup\{\inf\{a, b\}, \inf\{a, c\}\} = \inf\{a, \sup\{b, c\}\}.$$

(It can be shown that requiring the condition above is equivalent to requiring the dual condition in which inf and sup are interchanged.)

  i. Verify that $(\mathbb{R}, \leq)$ is a distributive lattice.
  ii. Verify that the Boolean lattice $(P(U), \subseteq)$ is a distributive lattice.
 iii. Verify that the divisor lattice $(\mathbb{N}, |)$ is a distributive lattice.
     (Hint: Use prime factorizations.)
 iv. Show that neither of the two lattices, given by their Hasse diagrams in Figure 17.6, is a distributive lattice.

Remarks The two lattices above are usually denoted by M_3 and N_5, respectively. We have the following famous result, reminiscent of Kuratowski's Theorem (see Appendix E).

Theorem 17.17 *A lattice L is distributive if, and only if, it does not contain M_3 or N_5 as a sublattice.*

12. (a) Prove the following proposition.

Proposition 17.18 *Every well-order is a total order; that is, if $\preceq$ is a well-order on a set X, then for any $a, b \in X$ we have $a \preceq b$ or $b \preceq a$.*

(Hint: Consider the set $\{a, b\}$.)
    (b) Prove the following proposition.

Proposition 17.19 *In every woset, at most one element has no successor.*

(Hints: Let $\preceq$ be a well-order on a set X; we need to prove that, given any pair of distinct elements a and b of X, at least one has a successor. By Proposition 17.18, we can assume, wlog, that $a \preceq b$. Consider the set of *strict* upper bounds of $\{a\}$; that is, $\{a\}^{\Uparrow} = \{a\}^{\uparrow} \setminus \{a\}$.)

(c) One can easily think of examples for totally ordered sets where no element has a successor (for example, $(\mathbb{R}, \leq)$), and total orders where exactly one element does not have one (for example, the well-order

$$4 \prec 5 \prec 6 \prec 7 \prec \cdots \prec 1 \prec 2 \prec 3$$

of $\mathbb{N}$ that we have seen earlier). Find an example of a total order on a set where exactly two elements have no successors.

13. Let S be a set with a partial order relation $\preceq$, and let f be a bijection on S. Suppose further that f is *order-preserving* (or *increasing*); that is, for any $a, b \in S$, $a \preceq b$ implies that $f(a) \preceq f(b)$. Prove or disprove each of the following statements.

(a) $f = id_S$.
(b) If $\preceq$ is a total order, then $f = id_S$.
(c) If $\preceq$ is a well-order, then $f = id_S$.

14. In this problem we investigate a certain equivalence relation among partially ordered sets. We make the following definition.

Definition 17.20 *Let A and B be sets with partial orders $\preceq_A$ and $\preceq_B$, respectively. We say that the two posets are* isomorphic, *and write*

$$(A, \preceq_A) \cong (B, \preceq_B),$$

if there is an order-preserving bijection from A to B; that is, a bijection $f : A \to B$ with the property that for any pair of elements a_1 and a_2 of A, we have $a_1 \preceq_A a_2$ if, and only if, $f(a_1) \preceq_B f(a_2)$. When two posets are isomorphic, we also say that they have the same order type.

(a) Prove that having the same order type is an equivalence relation. (To be precise, we should say that, while reflexivity, symmetry, and transitivity hold, having the same order type is not a relation as the collection of all posets is not a set—there are too many of them!)

(b) Prove each of the following statements.

   i. Every total order on a finite set determines the same order type.
   ii. The intervals $(0, 1)$, $(0, 1]$, $[0, 1)$, and $[0, 1]$, with the usual order $\leq$, all have different order types.
   iii. The number sets $\mathbb{N}$, $\mathbb{Z}$, and $\mathbb{Q}$, with the usual order $\leq$, all have different order types.
   iv. The rational numbers in the interval $(0, 1)$ and the set $\mathbb{Q}$ of all rational numbers, with the usual order $\leq$, have the same order types.

(c) Suppose that posets $(A, \preceq_A)$ and $(B, \preceq_B)$ are isomorphic. Prove that if one poset is a woset then so is the other.

(d) On page 264 we listed the following well-orders of $\mathbb{N}$:

$$1 < 2 < 3 < 4 < 5 < 6 < \cdots$$

$$4 \prec 5 \prec 6 \prec 7 \prec \cdots \prec 1 \prec 2 \prec 3$$

$$1 \prec 3 \prec 5 \prec 7 \prec \cdots \prec 2 \prec 4 \prec 6 \prec 8 \prec \cdots$$

$$1 \prec 2 \prec 4 \prec 8 \prec \cdots \prec 3 \prec 6 \prec 12 \prec 24 \prec \cdots \prec 5 \prec 10 \prec 20 \prec 40 \prec \cdots$$

Prove that each of these well-orders provides a different order type.

15. Prove the following theorem.

Theorem 17.21 (The Principle of Transfinite Induction) *Let X be a nonempty set, $\preceq$ be a well-order on X with corresponding strict order $\prec$, m be the (unique) minimum element of X, and $P(x)$ be a predicate that becomes a statement for every $x \in X$. If*

- *$P(m)$ and*
- *$\forall a \in X, \wedge_{x \prec a} P(x) \Rightarrow P(a)$*

both hold, then $P(x)$ is true for every $x \in X$.

(Hints: Proceed indirectly, and assume that the truth set X_P of the predicate is a proper subset of X. Then $X \setminus X_P$ is nonempty; consider its minimum element.)

Remarks We included the assumption that $P(m)$ holds, even though it vacuously follows from the inductive assumption since there are no elements in X for which $x \prec m$.

The Principle of Transfinite Induction is a far-reaching generalization of the Principle of Induction; it is powerful in proving statements involving "very large" sets (cf. Chapter 20). For example, it helps us prove strikingly simple-sounding statements, such as the following result.

Theorem 17.22 *There is a subset of the Euclidean plane that intersects every line exactly twice.*

Recall from Problem 2 of Chapter 4 that a subset of the plane that intersects every line exactly once does not exist! Theorem 17.22 can be proven by a recursive construction using transfinite induction.

Chapter 18
Now That's the Limit!

Recall from Chapter 8 that infinite sequences can be defined as functions whose domain is the set of natural numbers $\mathbb{N}$ or, equivalently, as the elements of the infinite Cartesian product X^∞ for some set X. We have seen examples for infinite sequences throughout this book; in this chapter we study the most important attribute of some sequences: their limits.

Limits are frequently discussed in mathematics and are widely used in many applications. A precise definition of limits, however, is usually not given at an elementary level. This is no surprise as the limit concept is a rather difficult one; it took mathematicians many centuries to come up with the precise definition. The concept was finally crystallized by the French mathematician Augustine-Louis Cauchy (1789–1857) and others in the nineteenth century—relatively late in the development of "basic" mathematics! In this chapter we formulate the precise definition using triple quantifiers (cf. Chapter 7). (Here we deal only with limits of sequences; functions and their limits can be treated analogously and can be found in books on analysis.)

We have already discussed statements involving one or two quantifiers: In Chapter 9 we considered statements of the form

$$\forall x \in U, P(x);$$

and in Chapter 12 we saw examples of the form

$$\forall x \in U, \exists y \in V, P(x, y).$$

To study limits, we now add one more quantifier, and analyze statements of the form

$$\forall x \in U, \exists y \in V, \forall z \in W, P(x, y, z).$$

But let's develop the definition one step at a time.

© Springer Nature Switzerland AG 2020

B. Bajnok, *An Invitation to Abstract Mathematics*, Undergraduate Texts in Mathematics, https://doi.org/10.1007/978-3-030-56174-1_18

What, exactly, do we mean when we say that the limit of the sequence

$$\left(\frac{1}{n}\right)_{n=1}^{\infty} = \left(\frac{1}{1}, \frac{1}{2}, \frac{1}{3}, \frac{1}{4}, \frac{1}{5}, \cdots\right)$$

is zero?

In a first approach, one might say that

- the limit of a sequence is zero if the terms get smaller and smaller.

This cannot possibly be the precise definition, however. We certainly wish to say that the sequence

$$\left(\frac{(-1)^n}{n}\right)_{n=1}^{\infty} = \left(-\frac{1}{1}, \frac{1}{2}, -\frac{1}{3}, \frac{1}{4}, -\frac{1}{5}, \cdots\right)$$

has limit zero, but it is not true that "the terms get smaller and smaller." In view of this, we may try to say that

- the limit of a sequence is zero if the terms get closer and closer to zero, or, somewhat more precisely, that
- the limit of a sequence is zero if the absolute value of the terms gets smaller and smaller.

One quickly finds that this will not work either. Consider a sequence such as

$$\left(\frac{\sqrt[n]{2}}{100}\right)_{n=1}^{\infty} \approx (0.02, 0.0141, 0.0126, 0.0119, 0.0115, \dots).$$

The terms seem to be getting "closer and closer to zero" (indeed, they are!). However, the limit of this sequence is not zero! The value of $\sqrt[n]{2}$ is always greater than 1, so the terms in our sequence will never get below 0.01. (It can be shown, in fact, that the limit is 0.01.) So a correct definition should require that the terms become *arbitrarily close* to zero; that is, closer than any tiny bound (we will make this notion more precise soon).

On the other hand, how about the sequence

$$\left(\frac{\sin n}{n}\right)_{n=1}^{\infty} \approx (0.841, 0.455, 0.047, -0.189, -0.192, \dots)?$$

Does it have limit 0?

This time, the terms do not "get closer and closer to zero" (e.g., the third term is much closer to zero than the next two terms are), but the limit of the sequence is zero! To convince yourself that the limit is indeed zero, note that the value of $\sin n$ is always between -1 and 1; thus our sequence lies somewhere between

$$\left(\frac{1}{n}\right)_{n=1}^{\infty} \text{ and } \left(-\frac{1}{n}\right)_{n=1}^{\infty},$$

both of which have limit zero (this argument is based on the *Squeeze Theorem*—see Problem 6). This example shows that a correct definition should allow for the terms to become arbitrarily close to zero *eventually*, but the terms do not need to get closer to zero in a *monotone* way. For example, since $|\sin n| \leq 1$, the terms of $(\sin n/n)_{n=1}^{\infty}$ will all be less than 0.1 away from zero when $n > 10$, although, as we have seen, some of the first ten terms might also be less than 0.1 away from zero.

Before stating the definition correctly, let us consider one other example:

$$\left(1 + (-1)^n\right)_{n=1}^{\infty} = (0, 2, 0, 2, 0, \dots).$$

Do we want to say that this sequence has limit zero? No; the limit should not be zero since, although there are infinitely many zero terms, the sequence will never become "arbitrarily close" to zero, not even "eventually." (Similarly, the limit of this sequence is not 2 either—in fact, one can easily see that this sequence has no limit!)

In summary, what we wish to say is something like this:
- the limit of a sequence is zero if the absolute values of the terms eventually become arbitrarily close to zero.

What do the terms "eventually" and "arbitrarily close" mean? We say that the terms become *eventually* "whatever," if there is an index K so that all the terms after the K-th term (and perhaps some of the ones even before) are "whatever." That is, the sequence $a = (a_1, a_2, a_3, \dots)$ is eventually "whatever," if

$$\exists K \in \mathbb{N}, \forall n \subset \mathbb{N} \cap (K, \infty), a_n \text{ is "whatever."}$$

And by *arbitrarily close* to zero we mean that, no matter how small a distance we specify, the terms become closer than the specified distance to zero.

Now we are ready for the formal definition.

Definition 18.1 *Let* $(a_1, a_2, a_3, \dots)$ *be an infinite sequence of real numbers. We say that the* limit *of the sequence is zero if*

$$\forall \epsilon \in (0, \infty), \exists K \in \mathbb{N}, \forall n \in \mathbb{N} \cap (K, \infty), |a_n| < \epsilon.$$

It is important to note that the order of the quantifiers cannot be switched: first we fix ϵ, then K, and finally n. We can understand this better by playing the following two-person game that we call the *Zero-Limit game*. It goes as follows:

1. Player I chooses a sequence of real numbers $(a_1, a_2, a_3, \dots)$, then
2. Player II chooses a positive number ϵ, then
3. Player I chooses a positive integer K, and finally
4. Player II chooses an integer n such that $n > K$.

The game is won by Player I if $|a_n| < \epsilon$, and Player II wins if $|a_n| \geq \epsilon$.

We then see that Player I has a winning strategy for this game, namely to choose a sequence that has limit zero. For example, suppose Player I chooses the sequence $(1/n)_{n=1}^{\infty}$. Let's say Player II chooses $\epsilon = 0.0001$. Then Player I can respond by $K = 10,000$; now Player II will be unable to pick an $n > 10,000$ for which $|1/n| \geq 0.0001$. We will see this more precisely in Proposition 18.4 below.

On the other hand, suppose that Player I chooses the sequence $\left(\sqrt[n]{2}/100 \right)_{n=1}^{\infty}$. If Player II chooses $\epsilon = 0.005$ (or any value under 0.01), Player I will not be able to pick a K so that $|\sqrt[n]{2}/100| < 0.005$ for every (actually, for any!) $n > K$—this is because the terms of this sequence are all greater than 0.01.

Similarly, if Player I chooses the sequence $(0, 2, 0, 2, 0, \ldots)$, then Player II can again win by selecting $\epsilon = 0.5$ (indeed, any positive ϵ less than 2). No matter which index K Player I responds with, Player II will be able to find an $n > K$ (making sure to choose an even value for n), so that $|a_n| \geq \epsilon$.

Of course, one is interested in limit values other than zero as well. More generally, we can define a sequence having a limit of any real number as follows:

Definition 18.2 *Let* $\mathbf{a} = (a_1, a_2, a_3, \ldots)$ *be an infinite sequence of real numbers, and let L be a real number. We say that the* limit *of the sequence* $\mathbf{a}$ *is L if*

$$\forall \epsilon \in (0, \infty), \exists K \in \mathbb{N}, \forall n \in \mathbb{N} \cap (K, \infty), |a_n - L| < \epsilon.$$

If the limit of $\mathbf{a}$ *is L, then we write* $\lim_{n \to \infty} a_n = L$, *or simply* $\lim a_n = L$ *or* $\lim \mathbf{a} = L$.

Furthermore, if $\lim \mathbf{a} = L$ *for some real number L, then we say that the sequence* $\mathbf{a}$ *is* convergent; *otherwise we say it is* divergent.

We are almost ready to return to our first sequence, $(1/n)_{n=1}^{\infty}$, and prove that, according to our definition, its limit is indeed zero. Before we do this, however, we state and prove an important, though quite obvious and unsurprising, fact that we will need.

Theorem 18.3 (The Archimedean Property of $\mathbb{R}$) *For every real number x, there is a natural number n such that $n > x$.*

The Archimedean Property is named after the ancient Greek mathematician Archimedes of Syracuse (cca. 287–212 BCE). The statement is usually taken for granted; after all, one can easily "construct" the desired natural number—for example, $n = \lceil x \rceil + 1$ will do. But such "constructions," of course, take the Archimedean Property itself for granted, thus a proof is called for. We should also note that the order of the two quantifiers here is not interchangeable: there is no natural number that is greater than every real number!

Proof We prove the equivalent claim that, for any $x \in \mathbb{R}$,

$$A = \{n \in \mathbb{N} \mid n \leq x\}$$

is a proper subset of $\mathbb{N}$.

Suppose, indirectly, that $A = \mathbb{N}$. Then x is an upper bound of $\mathbb{N}$, so, by the Completeness Axiom, $s = \sup \mathbb{N}$ exists. Since $s - 1 < s$, $s - 1$ is not an upper bound for $\mathbb{N}$; therefore, there exists an $n \in \mathbb{N}$ for which $s - 1 < n$. Thus, $s < n + 1$, which contradicts the fact s is an upper bound of $\mathbb{N}$.   □

With the Archimedean Property in our repertoire, we can now evaluate our first limit.

Proposition 18.4 $\lim \frac{1}{n} = 0$.

Proof The predicate in Definition 18.2, $|a_n - L| < \epsilon$, can be written in our case as $|\frac{1}{n}| < \epsilon$ and, since $\frac{1}{n} > 0$, we can remove the absolute value sign and write $\frac{1}{n} < \epsilon$. Since both n and ϵ are to be positive, we can further rewrite this inequality as $n > \frac{1}{\epsilon}$. Thus, rewriting the definition gives

$$\forall \epsilon \in (0, \infty), \exists K \in \mathbb{N}, \forall n \in \mathbb{N} \cap (K, \infty), n > \frac{1}{\epsilon}.$$

We now explain why this last statement is true. Let us try to unwrap it from the "inside." The predicate

$$\forall n \in \mathbb{N} \cap (K, \infty), n > \frac{1}{\epsilon}$$

means that, for a given ϵ and K (fixed by the first two quantifiers), we have $K + 1 > \frac{1}{\epsilon}$, $K + 2 > \frac{1}{\epsilon}$, $K + 3 > \frac{1}{\epsilon}$, etc. Though we have infinitely many inequalities that we need to satisfy, it is clearly enough to make sure that the first one holds: if $K + 1 > \frac{1}{\epsilon}$, then we will have $K + 2 > \frac{1}{\epsilon}$, $K + 3 > \frac{1}{\epsilon}$, etc. as well. Thus our statement is true as long as we prove

$$\forall \epsilon \in (0, \infty), \exists K \in \mathbb{N}, K + 1 > \frac{1}{\epsilon}$$

or, equivalently,

$$\forall \epsilon \in (0, \infty), \exists K \in \mathbb{N}, K > \frac{1}{\epsilon} - 1.$$

This last statement asks: Once we fix an arbitrary positive real number ϵ, is there a positive integer K that is bigger than $\frac{1}{\epsilon} - 1$? The answer to this question is "yes" by the Archimedean Property of $\mathbb{R}$: the quantity $\frac{1}{\epsilon} - 1$ is a real number (negative if $\epsilon > 1$ and nonnegative if $0 < \epsilon \leq 1$), and, no matter how much it is, there will be a positive integer K that is larger.   □

We can also define the concept of a sequence approaching infinity or negative infinity, as follows:

Definition 18.5 *Let* $\mathbf{a} = (a_1, a_2, a_3, \dots)$ *be an infinite sequence of real numbers. We define the* limit *of the sequence to be* infinity *if*

$$\forall B \in (0, \infty), \exists K \in \mathbb{N}, \forall n \in \mathbb{N} \cap (K, \infty), a_n > B.$$

If the limit of $\mathbf{a}$ *is infinity, then we write* $\lim_{n \to \infty} a_n = \infty$, *or simply* $\lim a_n = \infty$ *or* $\lim \mathbf{a} = \infty$.
Similarly, we define the limit *of the sequence to be* negative infinity *if*

$$\forall B \in (0, \infty), \exists K \in \mathbb{N}, \forall n \in \mathbb{N} \cap (K, \infty), a_n < -B.$$

If the limit of $\mathbf{a}$ *is negative infinity, then we write* $\lim_{n \to \infty} a_n = -\infty$, *or simply* $\lim a_n = -\infty$ *or* $\lim \mathbf{a} = -\infty$.

Note that we changed ϵ to B in the definitions above; this was done purely to emphasize that, while both ϵ and B can be arbitrary positive real numbers, we prefer to think of ϵ as very small ("almost zero") and B as very large ("almost infinity").

Let us see an example of a sequence with limit infinity.

Proposition 18.6 $\lim(2n + 3) = \infty$.

Proof The proof is similar to the one given for Proposition 18.4 above, but here we will be more concise.

Let B be an arbitrary positive real number; by the Archimedean Property, we can choose a positive integer K for which

$$K \geq \frac{B - 3}{2}.$$

With our choice for K, we have

$$\forall n \in \mathbb{N} \cap (K, \infty), n > \frac{B - 3}{2},$$

or, equivalently,

$$\forall n \in \mathbb{N} \cap (K, \infty), 2n + 3 > B.$$

Therefore, we proved that

$$\forall B \in (0, \infty), \exists K \in \mathbb{N}, \forall n \in \mathbb{N} \cap (K, \infty), 2n + 3 > B,$$

and thus $\lim(2n + 3) = \infty$. □

Admittedly, our two examples, Propositions 18.4 and 18.6, feature sequences whose limits are quite obvious and are included here only to facilitate a better understanding of Definitions 18.2 and 18.5. Our next example is much less apparent.

Proposition 18.7 *Define the sequence* $\mathbf{a} = (a_1, a_2, a_3, \dots)$ *recursively by* $a_1 = 1$
and

$$a_{n+1} = \frac{3a_n + 4}{2a_n + 3}$$

for $n \geq 1$. *Then* $\mathbf{a}$ *is convergent and* $\lim \mathbf{a} = \sqrt{2}$.

Our proof of Proposition 18.7 will rely on a far-reaching method. First a couple
of definitions.

Definition 18.8 *The sequence* $\mathbf{a} = (a_1, a_2, \dots)$ *of real numbers is said to be*
bounded *if the set* $\{a_1, a_2, \dots\}$ *is bounded in* $\mathbb{R}$.

Definition 18.9 *The sequence* $\mathbf{a} = (a_1, a_2, \dots)$ *of real numbers is said to be*
increasing *if* $a_n \leq a_{n+1}$ *for all* $n \in \mathbb{N}$; *the sequence is* decreasing *if* $a_n \geq a_{n+1}$
for all $n \in \mathbb{N}$. *A sequence is called* monotone *if it is increasing or decreasing.*

Of course, every sequence $\mathbf{a} = (a_1, a_2, \dots)$ satisfies the trivial condition

$$\forall n \in \mathbb{N}, (a_n \leq a_{n+1} \vee a_n \geq a_{n+1});$$

the sequence is called monotone only if

$$(\forall n \in \mathbb{N}, a_n \leq a_{n+1}) \vee (\forall n \in \mathbb{N}, a_n \geq a_{n+1}).$$

We then have the following useful result.

Theorem 18.10 (The Monotone Convergence Theorem) *Every monotone and*
bounded sequence of real numbers is convergent. In particular, if $\mathbf{a}$ *is increasing,*
then $\lim \mathbf{a} = \sup\{a_1, a_2, \dots\}$, *and if* $\mathbf{a}$ *is decreasing, then* $\lim \mathbf{a} = \inf\{a_1, a_2, \dots\}$.

For a proof, see Problem 7. For instance, the sequence $(\frac{1}{n})_{n=1}^{\infty}$ is decreasing
and has infimum 0, thus has limit 0 in accordance with Proposition 18.4. (The
sequence $(2n + 3)_{n=1}^{\infty}$ is increasing and is unbounded, and thus has limit infinity—
see Theorem 18.21 in Problem 7.)

The beauty of the Monotone Convergence Theorem is that it enables us to find the
limit of a monotone sequence even without knowing what the infimum or supremum
of it is. In the case of Proposition 18.7, we can proceed as follows:

Proof of Proposition 18.7 First we show that the sequence $\mathbf{a}$ is bounded and
increasing—see Problem 5. Therefore, by the Monotone Convergence Theorem, it
converges to a finite limit L.

We can find L as follows. First we use some elementary properties of convergent
sequences (see, for example, the Addition Theorem in Problem 6) to calculate that
if $\lim a_n = L$, then

$$\lim a_{n+1} = \lim \frac{3a_n + 4}{2a_n + 3} = \frac{3 \lim a_n + 4}{2 \lim a_n + 3} = \frac{3L + 4}{2L + 3}.$$

(Since the terms of the sequence are positive, none of the denominators above equal 0.) Next, we note that the limit of a sequence is not altered by omitting (or changing) the first (or any finite number of) terms, so

$$\lim a_{n+1} = \lim(a_2, a_3, a_4, \dots) = \lim(a_1, a_2, a_3, \dots) = L.$$

But the sequence $(a_{n+1})_{n=1}^{\infty}$ has a unique limit (see Theorem 18.16), so we must have

$$L = \frac{3L + 4}{2L + 3},$$

from which a simple calculation, noting also that we must have $L > 0$, yields $L = \sqrt{2}$.   □

Each infinite sequence of real numbers gives rise to a related sequence, called an *infinite series*, defined as follows:

Definition 18.11 *Given an infinite sequence of real numbers* $\mathbf{a} = (a_1, a_2, \dots)$, *we define the* infinite series *(or sequence of partial sums)* $\mathbf{s} = (s_1, s_2, \dots)$ *of* $\mathbf{a}$ *recursively by* $s_1 = a_1$ *and* $s_n = s_{n-1} + a_n$ *for* $n \geq 2$.
If the limit of $\mathbf{s}$ *exists (is a finite number, infinity, or negative infinity), then we let* $\sum_{n=1}^{\infty} a_n$ *or simply* $\sum a_n$ *or* $\sum \mathbf{a}$ *denote* $\lim \mathbf{s}$.
Furthermore, if $\lim \mathbf{s}$ *exists and is a finite number, then we say that the series* $\sum \mathbf{a}$ *is* convergent; *otherwise we say that it is* divergent.

We will examine some well-known infinite series in Problem 11; here we treat only one famous series: the series of Euler's number e.

We first prove the following lemma.

Lemma 18.12 *Define the infinite sequence* $(a_n)_{n=1}^{\infty}$ *by*

$$a_n = \frac{1}{(n-1)!}.$$

(Note that 0! *is defined to be 1.) Let* $(s_n)_{n=1}^{\infty}$ *be the corresponding sequence of partial sums. Then for all positive integers n and m we have*

$$s_n - s_m < \frac{m+1}{m \cdot m!}.$$

In particular, we have $1 \leq s_n < 3$ *for all* $n \in \mathbb{N}$.

Proof Let us start with our first claim. If $n \leq m$, then $s_n - s_m \leq 0$ since the terms of the sequence $(a_n)_{n=1}^{\infty}$ are positive, so our claim obviously holds.

Assume then that $n > m$, in which case we have

$$s_n - s_m = a_{m+1} + a_{m+2} + a_{m+3} + \cdots + a_n$$

$$= \frac{1}{m!} + \frac{1}{(m+1)!} + \frac{1}{(m+2)!} + \cdots + \frac{1}{(n-1)!}$$

$$= \frac{1}{m!}\left(1 + \frac{1}{m+1} + \frac{1}{(m+1)(m+2)} + \cdots + \frac{1}{(m+1)\cdots(n-1)}\right).$$

Note that

$$\frac{1}{(m+1)(m+2)} \leq \left(\frac{1}{m+1}\right)^2,$$

$$\frac{1}{(m+1)(m+2)(m+3)} \leq \left(\frac{1}{m+1}\right)^3,$$

and so on, so we have

$$s_n - s_m \leq \frac{1}{m!}\left(1 + \frac{1}{m+1} + \left(\frac{1}{m+1}\right)^2 + \cdots + \left(\frac{1}{m+1}\right)^{n-m-1}\right).$$

We can find a closed form for the sum above using Lemma 4.2, namely

$$1 + \frac{1}{m+1} + \left(\frac{1}{m+1}\right)^2 + \cdots + \left(\frac{1}{m+1}\right)^{n-m-1} = \frac{1 - \left(\frac{1}{m+1}\right)^{n-m}}{1 - \frac{1}{m+1}};$$

this quantity is clearly less than

$$\frac{1}{1 - \frac{1}{m+1}} = \frac{m+1}{m}.$$

Therefore,

$$s_n - s_m < \frac{1}{m!} \cdot \frac{m+1}{m},$$

as claimed.

Clearly, $s_n \geq s_1 = a_1 = 1$ for all $n \in \mathbb{N}$; substituting $m = 1$ into our inequality, we get $s_n < s_1 + 2 = 3$, completing our proof. $\square$

As an immediate corollary, we get the following:

Theorem 18.13 *The series*

$$\sum_{n=1}^{\infty} \frac{1}{(n-1)!} = \frac{1}{0!} + \frac{1}{1!} + \frac{1}{2!} + \frac{1}{3!} + \cdots$$

converges to a real number.

Proof According to Lemma 18.12, the sequence $(s_n)_{n=1}^{\infty}$ of partial sums is bounded (between 1 and 3); since the series is clearly increasing, our claim follows from the Monotone Convergence Theorem. □

The real number in Theorem 18.13 is called *Euler's number* and is denoted by e, thus

$$e = \sum_{n=1}^{\infty} \frac{1}{(n-1)!} = \frac{1}{0!} + \frac{1}{1!} + \frac{1}{2!} + \frac{1}{3!} + \cdots .$$

As a consequence of Lemma 18.12 and the Monotone Convergence Theorem, we have the following:

Corollary 18.14 *Let m be any positive integer, and set*

$$s_m = \sum_{i=1}^{m} \frac{1}{(i-1)!} = \frac{1}{0!} + \frac{1}{1!} + \frac{1}{2!} + \cdots + \frac{1}{(i-1)!}.$$

We then have

$$s_m < e \le s_m + \frac{m+1}{m \cdot m!}.$$

Using Corollary 18.14 for $m = 3$, we get the bounds

$$2.5 = s_3 < e \le s_3 + \frac{3+1}{3 \cdot 3!} = 2.777\ldots,$$

while an even better estimate would give e $\approx$ 2.71828. The number e plays a crucial role in many parts of mathematics and other fields. Of its many fascinating properties, let us mention the identity

$$\frac{1}{e} = \sum_{n=1}^{\infty} (-1)^{n-1} \frac{1}{(n-1)!} = \frac{1}{0!} - \frac{1}{1!} + \frac{1}{2!} - \frac{1}{3!} + - \cdots ,$$

to which we will return later in the book. Thus, interestingly, the infinite series of e and its reciprocal differ only in the signs of every other term.

In closing, we prove the following result.

Theorem 18.15 *The number e is irrational.*

Proof Let us assume indirectly that $e = \frac{a}{b}$ for some integers a and b; we may also assume that $b \in \mathbb{N}$.

By Corollary 18.14 (with $m = b + 1$), we have

$$0 < e - s_{b+1} \le \frac{b+2}{(b+1) \cdot (b+1)!};$$

multiplying by $b!$ we get

$$0 < b! \cdot e - b! \cdot s_{b+1} \le \frac{b+2}{(b+1)^2}.$$

Note that

$$\frac{b+2}{(b+1)^2} < \frac{b+2}{b^2 + 2b} = \frac{1}{b} \le 1,$$

and thus

$$0 < b! \cdot e - b! \cdot s_{b+1} < 1;$$

in particular, $b! \cdot e - b! \cdot s_{b+1}$ cannot be an integer. We show, however, that both $b! \cdot e$ and $b! \cdot s_{b+1}$ are integers.

Indeed, by our assumption,

$$b! \cdot e = b! \cdot \frac{a}{b} = (b-1)! \cdot a$$

is an integer, and so is

$$b! \cdot s_{b+1} = b! \cdot \left(\frac{1}{0!} + \frac{1}{1!} + \frac{1}{2!} + \cdots + \frac{1}{b!} \right).$$

This is a contradiction. □

The first proof for the irrationality of e was given by Leonhard Euler in the 1730s, though the proof we presented above is due to Joseph Fourier from the end of the eighteenth century. The other famous number that is known to be irrational is π, for which many different proofs exist. It is interesting to note that we now know that e^π is irrational, but we still don't have a proof for $e + \pi$, $e \cdot \pi$, or π^e being irrational.

Problems

1. (a) Suppose that you are playing the *Zero-Limit* game, you are Player II, and
   Player I started the game by choosing the sequence

$$\mathbf{a} = \left(\frac{n+4}{3n}\right)_{n=1}^{\infty}.$$

Can you win? What is a good choice for your ϵ?

   (b) Suppose that you are playing the *Zero-Limit* game, you are Player I, and
   you made the mistake of starting the game by choosing the sequence

$$\mathbf{a} = \left(\frac{n+4}{3n}\right)_{n=1}^{\infty}.$$

Fortunately (?), your opponent responded with $\epsilon = 1/2$. Can you then win?
What is a good choice for your K?

2. Suppose that $\mathbf{a} = (a_1, a_2, a_3, \dots)$ is a given infinite sequence of real numbers,
   and consider the following statements.

   (i) $\exists \epsilon \in (0, \infty), \exists K \in \mathbb{N}, \forall n \in \mathbb{N} \cap (K, \infty), |a_n - 1| < \epsilon$
   (ii) $\exists \epsilon \in (0, \infty), \forall K \in \mathbb{N}, \forall n \in \mathbb{N} \cap (K, \infty), |a_n - 1| < \epsilon$
   (iii) $\forall K \in \mathbb{N}, \exists \epsilon \in (0, \infty), \forall n \in \mathbb{N} \cap (K, \infty), |a_n - 1| < \epsilon$
   (iv) $\forall \epsilon \in (0, \infty), \exists K \in \mathbb{N}, \forall n \in \mathbb{N} \cap (K, \infty), |a_n - 1| < \epsilon$
   (v) $\exists K \in \mathbb{N}, \forall \epsilon \in (0, \infty), \forall n \in \mathbb{N} \cap (K, \infty), |a_n - 1| < \epsilon$
   (vi) $\forall \epsilon \in (0, \infty), \forall K \in \mathbb{N}, \forall n \in \mathbb{N} \cap (K, \infty), |a_n - 1| < \epsilon$

   For each of the sequences below, decide which of the statements above are true
   and which are false.

   (a) $(1, 1, 1, 1, 1, \dots)$
   (b) $(4, 3, 2, 1, 1, 1, 1, \dots)$
   (c) $(1, 0, 1, 0, 1, 0, \dots)$
   (d) $(1, 2, 1, 3, 1, 4, 1, 5, \dots)$
   (e) $(1.1, 1.01, 1.001, 1.0001, 1.00001, \dots)$
   (f) $(1, 1.1, 1, 1.01, 1, 1.001, 1, 1.0001, \dots)$

3. Suppose that $\mathbf{a} = (a_1, a_2, a_3, \dots)$ is a given infinite sequence of real numbers,
   and consider the following statements.

   (i) $\exists c \in (0, \infty), \{n \in \mathbb{N} \mid a_n > c\}$ is finite,
   (ii) $\forall c \in (0, \infty), \{n \in \mathbb{N} \mid a_n > c\}$ is finite,
   (iii) $\exists c \in (0, \infty), \{n \in \mathbb{N} \mid a_n > c\}$ is infinite,
   (iv) $\forall c \in (0, \infty), \{n \in \mathbb{N} \mid a_n > c\}$ is infinite,
   (v) $\exists c \in (0, \infty), \{n \in \mathbb{N} \mid a_n < c\}$ is finite,
   (vi) $\forall c \in (0, \infty), \{n \in \mathbb{N} \mid a_n < c\}$ is finite,
   (vii) $\exists c \in (0, \infty), \{n \in \mathbb{N} \mid a_n < c\}$ is infinite,

(viii) $\forall c \in (0, \infty)$, $\{n \in \mathbb{N} \mid a_n < c\}$ is infinite,

  (ix) $\exists c \in (0, \infty)$, $\{n \in \mathbb{N} \mid |a_n| > c\}$ is finite,

   (x) $\forall c \in (0, \infty)$, $\{n \in \mathbb{N} \mid |a_n| > c\}$ is finite,

  (xi) $\exists c \in (0, \infty)$, $\{n \in \mathbb{N} \mid |a_n| > c\}$ is infinite,

 (xii) $\forall c \in (0, \infty)$, $\{n \in \mathbb{N} \mid |a_n| > c\}$ is infinite,

(xiii) $\exists c \in (0, \infty)$, $\{n \in \mathbb{N} \mid |a_n| < c\}$ is finite,

(xiv) $\forall c \in (0, \infty)$, $\{n \in \mathbb{N} \mid |a_n| < c\}$ is finite,

 (xv) $\exists c \in (0, \infty)$, $\{n \in \mathbb{N} \mid |a_n| < c\}$ is infinite,

(xvi) $\forall c \in (0, \infty)$, $\{n \in \mathbb{N} \mid |a_n| < c\}$ is infinite.

(By saying that a set is finite or infinite, we mean that the set has finitely many or infinitely many elements, respectively.)

(a) Which of the statements above are necessary for $\lim \mathbf{a} = \infty$?

(b) Which of the statements above are sufficient for $\lim \mathbf{a} = \infty$?

(c) Which of the statements above are equivalent to $\lim \mathbf{a} = \infty$?

(d) Which of the statements above are necessary for $\lim \mathbf{a} = 0$?

(e) Which of the statements above are sufficient for $\lim \mathbf{a} = 0$?

(f) Which of the statements above are equivalent to $\lim \mathbf{a} = 0$?

4. Decide whether the following limits exist. If a limit exists, find it. Prove your answers using the definitions.

  (a) $\lim \frac{2}{5n+3}$

  (b) $\lim \frac{2n}{n+3}$

  (c) $\lim \frac{1}{\sqrt{n}}$

  (d) $\lim \frac{1}{\ln n}$

  (e) $\lim n^2$

  (f) $\lim 1.1^n$

     (Hint: Use Proposition 10.8.)

  (g) $\lim r^n$ where r is a real number with $0 < r < 1$.

     (Hint: Use the Monotone Convergence Theorem and the method of Proposition 18.7.)

  (h) $\lim(n + (-1)^n n)$

  (i) $\lim(2n + (-1)^n n)$

5. (a) Prove that the sequence in Proposition 18.7 is bounded by $\sqrt{2}$ from above.
    (Hint: Use induction.)

  (b) Prove that the sequence in Proposition 18.7 is increasing.
    (Hint: Use part (a).)

6. Prove the following well-known theorems.

  (a) **Theorem 18.16 (The Uniqueness of Limits)** *A sequence of real numbers can have at most one limit.*
    (Hints: You need to prove that for any sequence $\mathbf{a}$ and for any two distinct real numbers L_1 and L_2, we can have at most one of $\lim \mathbf{a} = L_1$, $\lim \mathbf{a} = L_2$, $\lim \mathbf{a} = \infty$, or $\lim \mathbf{a} = -\infty$.)

(b) **Theorem 18.17 (The Comparison Theorem)** *Let* $\mathbf{a} = (a_1, a_2, \dots)$ *and* $\mathbf{b} = (b_1, b_2, \dots)$ *be sequences for which*

$$\forall n \in \mathbb{N}, a_n \leq b_n,$$

and suppose that $\lim \mathbf{a} = \infty$. *Then* $\lim \mathbf{b} = \infty$.

Remark Similarly, if $\forall n \in \mathbb{N}, a_n \leq b_n$ and $\lim \mathbf{b} = -\infty$, then $\lim \mathbf{a} = -\infty$.

(c) **Theorem 18.18 (The Squeeze Theorem)** *Let* L *be a real number and* $\mathbf{a} = (a_1, a_2, \dots)$, $\mathbf{b} = (b_1, b_2, \dots)$, *and* $\mathbf{c} = (c_1, c_2, \dots)$ *be sequences for which*

$$\forall n \in \mathbb{N}, a_n \leq b_n \leq c_n,$$

and suppose that $\lim \mathbf{a} = L$ *and* $\lim \mathbf{c} = L$. *Then* $\mathbf{b}$ *is convergent, and* $\lim \mathbf{b} = L$.

(d) **Theorem 18.19 (The Addition Theorem)** *Let* L_1 *and* L_2 *be real numbers and* $\mathbf{a} = (a_1, a_2, \dots)$ *and* $\mathbf{b} = (b_1, b_2, \dots)$ *be sequences for which* $\lim \mathbf{a} = L_1$ *and* $\lim \mathbf{b} = L_2$. *Define* $\mathbf{c} = (c_1, c_2, \dots)$ *by*

$$\forall n \in \mathbb{N}, c_n = a_n + b_n.$$

Then $\mathbf{c}$ *is convergent, and* $\lim \mathbf{c} = L_1 + L_2$.
(Hint: Use the Triangle Inequality: Theorem 9.6.)

7. Prove each of the following theorems.

(a) **Theorem 18.20** *Every convergent sequence of real numbers is bounded.*
(b) The Monotone Convergence Theorem, as stated on page 279.
(c) **Theorem 18.21** *Suppose that the sequence* $\mathbf{a}$ *is increasing but not bounded. Then* $\lim \mathbf{a} = \infty$.
(d) **Theorem 18.22** *Suppose that the sequence* $\mathbf{a}$ *is decreasing but not bounded. Then* $\lim \mathbf{a} = -\infty$.

8. Suppose that the infinite series $\sum \mathbf{a}$ corresponding to the sequence $\mathbf{a}$ is convergent. Prove that the sequence $\mathbf{a}$ is also convergent and $\lim \mathbf{a} = 0$.
(Hint: Use the Addition Theorem.)

9. Let F be an ordered field (cf. Definition 13.12). We say that a subset S of F is *dense in* F if for any two distinct elements of F there is an element of S that is strictly between them. By Problem 2 of Chapter 12, $\mathbb{Q}$ is dense in itself and $\mathbb{R}$ is also dense in itself.

(a) Prove that $\mathbb{Q}$ is dense in $\mathbb{R}$.
(Hint: Use the Archimedean Property twice: first to select the denominator and then to select the numerator of the rational number to be constructed.)
(b) Prove that $\mathbb{R} \setminus \mathbb{Q}$ is dense in $\mathbb{R}$.
(Hint: Use part (a).)

10. Suppose that A is a set of positive integers and, for a positive integer n, let $A(n)$ denote the number of elements of A that are between 1 and n (inclusive). The *natural density* of A is defined as

$$d(A) = \lim \frac{A(n)}{n},$$

if this limit exists.

Find, with precise proof, the natural density of each of the following sets, or prove that the natural density does not exist. (Try to guess the answers first. Cf. Problem 6 of Appendix B.)

(a) The set of positive integers that are divisible by a million.
(b) The set of perfect squares.
(c) The set of positive primes.
    (Hint: Use the Prime Number Theorem.)
(d) The set of positive integers whose decimal contains six consecutive 0s.
    (Hint: What is the natural density of positive integers that have a 0 digit in base 10^6?)

    Remark Note that these numbers include those in part (a) above.

(e) The set of positive integers that have an odd number of decimal digits.
    (Hint: Prove that this set has no natural density.)

11. (a) Prove that

$$\sum_{n=1}^{\infty} \frac{1}{2^n} = \frac{1}{2} + \frac{1}{4} + \frac{1}{8} + \frac{1}{16} + \cdots = 1.$$

(Hint: Use Lemma 4.2.)

(b) Prove that if r is a real number with $0 < r < 1$, then the so-called *geometric series* $\sum_{n=1}^{\infty} r^n$ converges to $\frac{r}{1-r}$.
    (Hint: Use part (g) of Problem 4 above.)

(c) Evaluate

$$\sum_{n=1}^{\infty} \frac{1}{2^{\lceil \log_2 n \rceil}} = \frac{1}{1} + \frac{1}{2} + \frac{1}{4} + \frac{1}{4} + \frac{1}{8} + \cdots.$$

(d) Prove the following classical result.

Proposition 18.23 *The* harmonic series, *defined as the series*

$$\sum_{n=1}^{\infty} \frac{1}{n} = \frac{1}{1} + \frac{1}{2} + \frac{1}{3} + \frac{1}{4} + \frac{1}{5} + \cdots,$$

is divergent.

(Hint: Use part (c) and the Comparison Theorem.)

(e) Evaluate

$$\sum_{n=1}^{\infty} \frac{1}{n(n+1)} = \frac{1}{2} + \frac{1}{6} + \frac{1}{12} + \frac{1}{20} + \cdots .$$

(Hint: Use part (c) of Problem 6 in Chapter 10.)

(f) Evaluate

$$\sum_{n=1}^{\infty} \frac{n}{(n+1)!} = \frac{1}{2} + \frac{1}{3} + \frac{1}{8} + \frac{1}{30} + \cdots .$$

(Hint: Use part (d) of Problem 6 in Chapter 10.)

12. As an extension of Theorem 12.10, one can prove the following:

Theorem 18.24 *Suppose that a is a positive real number and that $b > 1$ is a positive integer. Then there exist integers $m \geq 0$; $r_m, r_{m-1}, \ldots, r_1, r_0$ with $0 \leq r_i \leq b - 1$ for each $i = 0, 1, \ldots, m$; and $n_1, n_2, n_3, \ldots$ with $0 \leq n_j \leq b - 1$ for each $j \in \mathbb{N}$, so that*

$$a = r_m b^m + r_{m-1} b^{m-1} + \cdots + r_1 b + r_0 + \frac{n_1}{b} + \frac{n_2}{b^2} + \cdots .$$

Furthermore, if $m > 0$, then we may also assume that $r_m > 0$.
The expression of a in this form is unique with one exception: If, for some index $k \in \mathbb{N}$, $n_k > 0$ but $n_i = 0$ for all $i > k$, then this number can also be written so that $n'_k = n_k - 1$ and $n'_i = b - 1$ for $i > k$.

(A similar result can be stated for negative real numbers.)
Note that, by the Monotone Convergence Theorem, the expression above determines a unique real number. When the base b is clear, we usually write simply

$$a = \pm r_m r_{m-1} \ldots r_1 r_0 . n_1 n_2 n_3 \ldots ;$$

this is called the *base b representation of a*.

(a) Table 18.1 shows (the beginnings of) the *binary* ($b = 2$), *ternary* ($b = 3$), and *decimal* ($b = 10$) representations of some numbers; if the number has two representations, both are given.
Find the *septenary* (base 7) representation of the same four real numbers. (The first few "septenary digits" will suffice.)

(b) We say that the decimal representation

$$a = \pm r_m r_{m-1} \ldots r_1 r_0 . n_1 n_2 n_3 \ldots$$

Table 18.1 Binary, ternary, and decimal representations of some numbers

	Binary	Ternary	Decimal
1/2	0.10000000000000...	0.11111111111111...	0.50000000000000...
	0.01111111111111...		0.49999999999999...
1/3	0.01010101010101...	0.10000000000000...	0.33333333333333...
		0.02222222222222...	
$\sqrt{2}$	1.01101010000010...	1.10201122122200...	1.41421356237309...
π	11.00100100001111...	10.01021101222201...	3.14159265358979...

is *periodic* if there are positive integers k and p so that $n_{k+i} = n_{k+p+i}$ holds for every nonnegative integer i. (Periodic representations for other bases can be defined analogously.) For example, the table above shows that 1/2 and 1/3 have repeating decimal forms (1/2 has two), but $\sqrt{2}$ and π do not have repeating decimal forms.

Prove that a real number has a periodic decimal representation if, and only if, it is rational.

(Hints: To show that every rational number has a periodic decimal representation, use the Pigeonhole Principle. Use part (b) of Problem 11 above to prove the converse.)

(c) We say that the binary representation

$$a = \pm r_m r_{m-1} \ldots r_1 r_0 . n_1 n_2 n_3 \ldots$$

is *finite* if there is a positive integer k so that $n_{k+i} = 0$ for every positive integer i, in which case we may further assume that $n_k = 1$, or if $n_k = 0$ for all positive integers k (that is, a is an integer). (Finite representations for other bases can be defined similarly.) We can, of course, just ignore the zeros after n_k and write

$$\pm r_m r_{m-1} \ldots r_1 r_0 . n_1 n_2 n_3 \ldots n_k$$

(and write $\pm r_m r_{m-1} \ldots r_1 r_0$ is a is an integer). The set of all real numbers that have a finite binary representation is denoted by $\mathbb{Z}[\frac{1}{2}]$ (cf. Problem 4 (c) of Chapter 13).

For each of the following statements, decide if the statement is true or false.

i. Every real number with a finite binary form can be written as $\frac{z}{2^k}$ for some integer z and positive integer k.

ii. Every real number with a finite binary form can be written uniquely as $\frac{z}{2^k}$ for some integer z and positive integer k.

iii. Every real number with a finite binary form is either an integer or can be written as $\pm \left(n + \frac{a}{2^k} \right)$ for some nonnegative integer n, positive integer k, and odd integer a with $1 \leq a < 2^k$.

iv. Every real number with a finite binary form is either an integer or can be written uniquely as $\pm \left(n + \frac{a}{2^k} \right)$ for some nonnegative integer n, positive integer k, and odd integer a with $1 \leq a < 2^k$.

v. Every real number that can be written as $\pm \left(n + \frac{a}{2^k} \right)$ for some nonnegative integer n, positive integer k, and odd integer a with $1 \leq a < 2^k$ has a finite binary form.

vi. Every real number that can be written as $\pm \left(n + \frac{a}{2^k} \right)$ for some nonnegative integer n, positive integer k, and odd integer a with $1 \leq a < 2^k$ has a unique finite binary form.

vii. $\mathbb{Z}[\frac{1}{2}]$ is an integral domain.

viii. $\mathbb{Z}[\frac{1}{2}]$ is a field.

13. The famous Cantor set—named after the German mathematician Georg Cantor (1845–1918)—is defined recursively, as follows:

We let I_1 denote the interval $[0, 1]$; for each integer $n \geq 1$, we define

$$I_{n+1} = \left\{ \frac{x}{3} \mid x \in I_n \right\} \cup \left\{ \frac{x+2}{3} \mid x \in I_n \right\}.$$

So, we have

$$I_2 = [0, 1/3] \cup [2/3, 1],$$

$$I_3 = [0, 1/9] \cup [2/9, 1/3] \cup [2/3, 7/9] \cup [8/9, 1],$$

and so on; I_{n+1} is what is left after we remove the open middle 1/3 of each interval that makes up I_n. We illustrate the first few iterations in Figure 18.1. We then define the *Cantor set* as

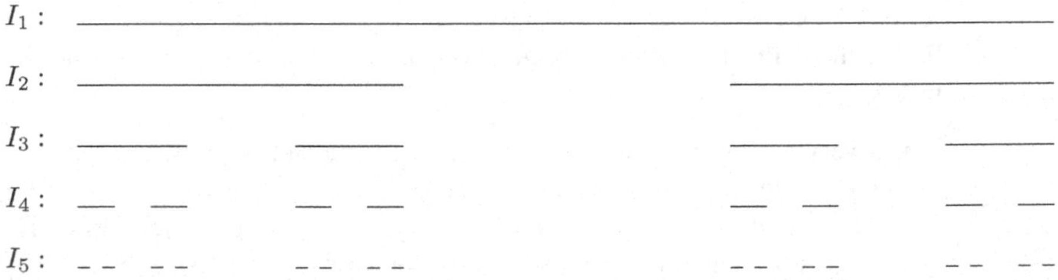

Fig. 18.1 An illustration of the Cantor set

$$C = \bigcap_{i=1}^{\infty} I_n.$$

(a) Let a_n denote the total length of all the intervals that make up I_n. Prove that $\lim a_n = 0$.

    (Hint: You can do this in two different ways: directly, or by adding up the lengths of all the intervals that got removed and subtracting that from 1.)

(b) Characterize the elements of the Cantor set using their ternary representations. In particular, show that the Cantor set has infinitely many elements.

Remarks A vague question one may ask is: How big is the Cantor set? Depending on how one makes this question precise, one gets entirely different answers. As we have just seen, in a certain sense, C is small as it has *measure zero* but, in another sense, it is large, since it has infinitely many elements. In Problem 4 of Chapter 20 we examine further how large the Cantor set really is.

The Cantor set has many other interesting properties as well; it has been an influential object of study in various branches of mathematics. One such property is that it is a prime example of a *self-similar set*, popularly known as a *fractal*. What we mean by self-similarity is that, if instead of starting with the interval $I_1 = [0, 1]$, we start with the interval $[0, 1/3]$ (which is the left half of I_2), as the middle thirds get repeatedly removed, one sees exactly the same picture as one did with the original interval—except that everything looks 1/3 as big. Similarly, zooming in on $[2/3, 1]$ generates the same smaller-scale version of the original set. One can make these ideas more precise, and show that the *fractional dimension* of the Cantor set is $\log_3 2 \approx 0.68$—a value between 0 (the dimension of "isolated" points) and 1 (the dimension of a "continuous" interval).

Chapter 19
Sizing It Up

Counting is probably one of our earliest intellectual pursuits, and it is a ubiquitous task in everyday life. The principles of counting are also what several branches of mathematics are based on, especially combinatorics, probability theory, and statistics. In this chapter we discuss elementary counting in a systematic and precise—shall we say abstract—manner.

A typical counting problem asks us to determine the *size* of a set: the size of a set A, denoted by $|A|$, is the number of elements in A. Clearly, each set has either finite or infinite size. In this chapter we focus on finite sets (sets with finite size) only; we will discuss infinite sets in Chapter 20.

We have already seen several counting questions. For example, in Problem 9 of Chapter 6, we studied the sizes of unions, intersections, Cartesian products, and power sets of given finite sets. The answers to these and other simple questions are based on two fundamental principles of counting: the Addition Rule and the Multiplication Rule.

Lemma 19.1 (The Addition Rule) *If A and B are disjoint finite sets, then we have*

$$|A \cup B| = |A| + |B|.$$

More generally, if A_1, A_2, ..., A_n are pairwise disjoint finite sets ($n \in \mathbb{N}$), then we have

$$|A_1 \cup \cdots \cup A_n| = |A_1| + \cdots + |A_n|.$$

Proof The identity is quite clear for $n = 2$: by the definition of union, if $|A| = k$, $|B| = l$, $A = \{a_1, \ldots, a_k\}$, and $B = \{b_1, \ldots, b_l\}$, then $A \cup B = \{a_1, \ldots, a_k, b_1, \ldots, b_l\}$. Since A and B are disjoint, these $k+l$ elements are distinct, hence $|A \cup B| = k + l$, as claimed. The identity for n sets can be established by induction. $\qquad\square$

© Springer Nature Switzerland AG 2020
B. Bajnok, *An Invitation to Abstract Mathematics*, Undergraduate Texts in Mathematics, https://doi.org/10.1007/978-3-030-56174-1_19

Lemma 19.2 (The Multiplication Rule) *For finite sets A and B, we have*

$$|A \times B| = |A| \cdot |B|.$$

More generally, for finite sets $A_1, A_2, \ldots, A_n$ ($n \in \mathbb{N}$), we have

$$|A_1 \times \cdots \times A_n| = |A_1| \cdot \cdots \cdot |A_n|.$$

Proof We can reduce the identity for $n = 2$ to the Addition Rule, as follows. First note that if $|A| = k$ and $A = \{a_1, \ldots, a_k\}$, then

$$A \times B = \{(a_1, b) \mid b \in B\} \cup \cdots \cup \{(a_k, b) \mid b \in B\};$$

if $a_1, \ldots, a_k$ are distinct, then the k sets on the right-hand side above are pairwise disjoint. Furthermore, these sets all have the same size as B, and, therefore, by the Addition Rule, we get

$$|A \times B| = k \cdot |B|,$$

as claimed. The identity for n sets can then be proved by induction.   □

Observe that the Addition Rule—unlike the Multiplication Rule—requires that the sets be pairwise disjoint. If we don't know whether $A_1, A_2, \ldots, A_n$ are pairwise disjoint, then we can only claim that

$$|A_1 \cup \cdots \cup A_n| \le |A_1| + \cdots + |A_n|.$$

It is worth pointing out that the Pigeonhole Principle and the Generalized Pigeonhole Principle (cf. Theorems 12.4 and 12.5) are easy corollaries of this inequality. Namely, if we have

$$|A_1 \cup \cdots \cup A_n| > kn$$

for some nonnegative integer k, then there must be an index $i \in \{1, \ldots, n\}$ for which $|A_i| \ge k + 1$, otherwise the inequality above would fail.

Later in this chapter we state a more precise result for the size of the union of n (not necessarily pairwise disjoint) sets.

We now turn to the size of the power set of a given set. (The formula explains the name "power" set; cf. Chapter 6.)

Proposition 19.3 *For a finite set A, we have*

$$|P(A)| = 2^{|A|}.$$

Proof Observe that our claim holds when $A = \emptyset$, so we may assume that $|A| = n$ for some $n \in \mathbb{N}$. Let $A = \{a_1, \ldots, a_n\}$. First we find a bijection from $P(A)$ to the set

$$\{0, 1\}^n = \{(x_1, \ldots, x_n) \mid x_1 \in \{0, 1\}, \ldots, x_n \in \{0, 1\}\},$$

as follows.

Let X be an arbitrary subset of A. Define $f(X)$ to be the element

$$(f_1(X), \ldots, f_n(X)) \in \{0, 1\}^n$$

where, for $i \in \{1, \ldots, n\}$,

$$f_i(X) = \begin{cases} 1 & \text{if } a_i \in X; \\ 0 & \text{if } a_i \notin X. \end{cases}$$

Then f is both injective and surjective, so it is a bijection.

So, there is a bijection from $P(A)$ to the set $\{0, 1\}^n$, and, therefore, these two sets have the same size (cf. Problem 3 in Chapter 8). Since for the size of $\{0, 1\}^n$ we can apply the Multiplication Rule, we immediately get our result that $|P(A)| = 2^n$. $\square$

Before we move on to the four basic counting questions in mathematics, we review some familiar terminology and notations and introduce some new ones. Recall that, for a given set A and positive integer m, an element $(a_1, a_2, \ldots, a_m)$ of A^m is called a sequence of length m. The order of the terms in the sequence matters; for example, the sequence $(2, 3, 4, 5)$ of integers is different from $(3, 2, 4, 5)$. On the other hand, a subset of A of size m is simply a collection of m of its elements where two subsets are considered equal without regard to the order in which the terms are listed; for example, $\{2, 3, 4, 5\}$ and $\{3, 2, 4, 5\}$ are equal subsets of the set of integers. Recall also that a set remains unchanged if we choose to list some of its elements more than once (cf. Definition 6.5); for example, the sets $\{2, 3, 3, 5\}$, $\{2, 3, 5, 5\}$, and $\{2, 3, 5\}$ are all equal, while the sequences $(2, 3, 3, 5)$, $(2, 3, 5, 5)$, and $(2, 3, 5)$ are all different. Thus, we can consider sets as two-fold relaxations of sequences: we don't care about the order in which the elements are listed, nor do we care how many times the elements are listed.

It will be useful for us to introduce two other objects. First we say that a sequence $(a_1, a_2, \ldots, a_m)$ of elements of a set A is a *list*, if the m terms are pairwise distinct. Thus, in a list, the order of the elements still matters, but each element is only allowed to appear once. For example, the sequence $(2, 3, 4, 5)$ is a list, but $(2, 3, 3, 5)$ is not. Conversely, in a so-called *multiset* of size m, denoted by $[a_1, a_2, \ldots, a_m]$, the order of the elements $a_1, a_2, \ldots, a_m$ of A does not matter (as is the case with sets), but elements may appear repeatedly (as they may in sequences). For example, the multisets $[2, 3, 3, 5]$, $[2, 3, 5, 5]$, and $[2, 3, 5]$ are all different, but $[2, 3, 3, 5]$ is still the same as $[2, 5, 3, 3]$.

Table 19.1 Lists and
sequences of length m, sets
and multisets of size m

	Order matters	Order does not matter
Elements distinct	m-lists	m-sets
Elements may repeat	m-sequences	m-multisets

Given a set A and a positive integer m, we are interested in counting the number of m-sequences (sequences of length m), m-lists (lists of length m), m-multisubsets (multisubsets of size m), and m-subsets (subsets of size m) of A. The schematic summary of these four terms is given in Table 19.1.

Obviously, if $|A| < m$, then A has neither m-lists nor m-subsets. If $|A| = m$, then the (only) m-subset of A is A itself, while, as we will soon see, if $|A| = m$, then A has $m!$ m-lists. For other situations, we introduce the following notations.

Suppose that n is a nonnegative integer and m is a positive integer. We define the *rising factorial m-th power* and the *falling factorial m-th power* of n to be

$$n^{\overline{m}} = n(n+1)\cdots(n+m-1)$$

and

$$n^{\underline{m}} = n(n-1)\cdots(n-m+1),$$

respectively. For example, we have $10^{\overline{3}} = 10\cdot 11\cdot 12 = 1320$ and $10^{\underline{3}} = 10\cdot 9\cdot 8 = 720$. Analogously to $n^0 = 1$ and $0! = 1$, we extend these notations with

$$n^{\underline{0}} = 1 \text{ and } n^{\overline{0}} = 1$$

for arbitrary nonnegative integers n.

Furthermore, we recall from Chapter 11 the notations $\binom{n}{m}$ (pronounced "n choose m") and $\left[\begin{smallmatrix} n \\ m \end{smallmatrix}\right]$ (pronounced "n multichoose m"): For nonnegative integers m and n,

$$\binom{n}{m} = \frac{n^{\underline{m}}}{m!} = \frac{n(n-1)\cdots(n-m+1)}{m!}$$

and

$$\left[\begin{matrix} n \\ m \end{matrix}\right] = \frac{n^{\overline{m}}}{m!} = \frac{n(n+1)\cdots(n+m-1)}{m!}.$$

According to Theorem 11.13, these quantities denote integers. The values of $\binom{n}{m}$, also known as *binomial coefficients*, are exhibited in *Pascal's Triangle* (cf. page 175); in Table 19.2 we tabulate some of these values in a table format. (Observe that, when $m > n$, the formula above yields $\binom{n}{m} = 0$; keeping the traditional shape of Pascal's Triangle, we omitted these entries from the table.)

Note that, since

Table 19.2 Some values of $\binom{n}{m}$

$\binom{n}{m}$	$m = 0$	$m = 1$	$m = 2$	$m = 3$	$m = 4$	$m = 5$	$m = 6$	$m = 7$
$n = 0$	1							
$n = 1$	1	1						
$n = 2$	1	2	1					
$n = 3$	1	3	3	1				
$n = 4$	1	4	6	4	1			
$n = 5$	1	5	10	10	5	1		
$n = 6$	1	6	15	20	15	6	1	
$n = 7$	1	7	21	35	35	21	7	1

Table 19.3 Some values of $\left[\begin{smallmatrix} n \\ m \end{smallmatrix}\right]$

$\left[\begin{smallmatrix} n \\ m \end{smallmatrix}\right]$	$m = 0$	$m = 1$	$m = 2$	$m = 3$	$m = 4$	$m = 5$	$m = 6$	$m = 7$
$n = 1$	1	1	1	1	1	1	1	1
$n = 2$	1	2	3	4	5	6	7	8
$n = 3$	1	3	6	10	15	21	28	36
$n = 4$	1	4	10	20	35	56	84	120
$n = 5$	1	5	15	35	70	126	210	330
$n = 6$	1	6	21	56	126	252	462	792
$n = 7$	1	7	28	84	210	462	924	1716

$$\frac{n(n-1)\cdots(n-m+1)}{m!} = \frac{n(n-1)\cdots(m+1)}{(n-m)!}$$

(which we can check by cross-multiplying), we have the identity

$$\binom{n}{m} = \binom{n}{n-m},$$

expressing the fact that the rows in Pascal's Triangle are "palindromic." The explanation for the term "binomial coefficient" will be clear once we discuss Theorem 19.6 below.

The first few values $\left[\begin{smallmatrix} n \\ m \end{smallmatrix}\right]$ are shown in Table 19.3.

As we can see, the two tables contain the same data—values are just shifted: the entries in column m in the first table are moved up by m rows in the second table. Indeed, since for integers n and m we clearly have

$$n^{\overline{m}} = n(n+1)\cdots(n+m-1) = (n+m-1)\cdots(n+1)n = (n+m-1)^{\underline{m}},$$

we see that values of $\left[\begin{smallmatrix} n \\ m \end{smallmatrix}\right]$ can be expressed via the more-often used binomial coefficients as

$$\left[{n \atop m} \right] = \binom{n+m-1}{m}.$$

The binomial coefficients possess many interesting properties. For example, looking at the values in Table 19.2, we may conjecture that

- the entries in each row add up to a power of 2;
- the entries in each column, above a certain row, add to an entry in the next row;
- the entries in each NE-SW diagonal add to a Fibonacci number; and
- the entries in each NW-SE diagonal, above a certain row, add to an entry in the next row.

We will prove these and some other identities in Problems 1–4.

We are now ready to "size up" our four main configurations.

Theorem 19.4 *Let A be a set of size n, and let m be a positive integer. Then*

1. *the number of m-sequences of A is n^m;*
2. *the number of m-lists of A is $n^{\underline{m}}$;*
3. *the number of m-multisubsets of A is $\left[{n \atop m} \right]$; and*
4. *the number of m-subsets of A is $\binom{n}{m}$.*

(Note that, if $n < m$, then $n^{\underline{m}} = 0$ and $\binom{n}{m} = 0$, in accordance with the fact that A has no m-lists and no m-subsets in this case.)

Proof We can easily prove claims 1 and 2 using the Multiplication Rule. Namely, for $i = 1, \ldots, m$, letting A_i and B_i denote the subsets of A from which we can select the i-th element in the m-sequence and m-list of A, respectively, we see that we have $|A_i| = n$ (any element of A can be chosen for the m-sequence), and $|B_i| = n - (i-1)$ (when choosing the i-th element for our m-list, only those elements of A can be chosen that were not chosen for the first $i - 1$ elements of the m-list). Therefore, by the Multiplication Rule, the number of m-sequences of A is

$$\prod_{i=1}^{m} |A_i| = \prod_{i=1}^{m} n = n^m,$$

and the number of m-lists of A is

$$\prod_{i=1}^{m} |B_i| = \prod_{i=1}^{m} (n - i + 1) = n^{\underline{m}}.$$

Next, we use claim 2 to prove claim 4. Let $\mathcal{L}(A, m)$ denote the collection of m-lists of A; by claim 2 we know that

$$|\mathcal{L}(A, m)| = n^{\underline{m}}.$$

We introduce a relation R on $\mathcal{L}(A, m)$ by saying that two m-lists have the relationship whenever their corresponding sets are equal; that is, for m-lists $(a_1, \ldots, a_m)$ and $(a'_1, \ldots, a'_m)$ we define

$$(a_1, \ldots, a_m) \sim_R (a'_1, \ldots, a'_m)$$

whenever

$$\{a_1, \ldots, a_m\} = \{a'_1, \ldots, a'_m\}.$$

It is easy to see that R is an equivalence relation: it is reflexive, symmetric, and transitive. Therefore, by the Fundamental Theorem of Equivalence Relations (cf. Theorem 16.11), the equivalence classes of R partition $\mathcal{L}(A, m)$.

Clearly, the number of m-subsets of A is the number of equivalence classes of R; let us denote this number by $C(A, m)$. Using claim 2 of our theorem again, we see that an m-list is equivalent to exactly

$$m^{\underline{m}} = m!$$

m-lists, so each equivalence class has $m!$ elements. Therefore, by the Addition Rule, we have

$$|\mathcal{L}(A, m)| = C(A, m) \cdot m!,$$

from which we get

$$C(A, m) = \frac{|\mathcal{L}(A, m)|}{m!} = \frac{n^{\underline{m}}}{m!} = \binom{n}{m},$$

as claimed.

Finally, we use claim 4 to prove claim 3. Let $\left[\begin{smallmatrix} A \\ m \end{smallmatrix}\right]$ denote the set of m-multisubsets of A, and let $\binom{I}{m}$ denote the set of m-subsets of the set $I = \{1, 2, \ldots, n + m - 1\}$. We will show that there is a bijection between $\left[\begin{smallmatrix} A \\ m \end{smallmatrix}\right]$ and $\binom{I}{m}$.

First we find a bijection f from $\left[\begin{smallmatrix} A \\ m \end{smallmatrix}\right]$ to the set $Q(m, n - 1)$, denoting here the set of $(m + n - 1)$-sequences of $\{0, 1\}$ that contain exactly m 0s and $n - 1$ 1s.

Let M be an m-multisubset of a set $A = \{a_1, \ldots, a_n\}$, and, for each $i = 1, \ldots, n$, let α_i denote the repetition number of a_i in M; that is, the number of times a_i appears in M. We then have

$$M = [\underbrace{a_1, \ldots, a_1}_{\alpha_1}, \underbrace{a_2, \ldots, a_2}_{\alpha_2}, \ldots, \underbrace{a_n, \ldots, a_n}_{\alpha_n}].$$

Now define $f(M)$ to be the sequence in which α_1 0s are followed by a 1, followed by α_2 0s, followed by another 1, and so on, until the $(n-1)$-st 1 is followed by α_n 0s:

$$f(M) = (\underbrace{0,\ldots,0}_{\alpha_1}, 1, \underbrace{0,\ldots,0}_{\alpha_2}, 1, \ldots, 1, \underbrace{0,\ldots,0}_{\alpha_n}).$$

Since

$$\alpha_1 + \cdots + \alpha_n = m,$$

$f(M)$ is an $(m+n-1)$-sequence containing exactly m 0s and $n-1$ 1s. Thus, we have

$$f : \begin{bmatrix} A \\ m \end{bmatrix} \to Q(m, n-1)$$

$$M \mapsto f(M);$$

it is also easy to see that f is a bijection.

Next, we find a bijection g from $Q(m, n-1)$ to $\binom{I}{m}$. Let

$$\mathbf{v} = (v_1, v_2, \ldots, v_{m+n-1})$$

be an $(m+n-1)$-sequence of $\{0, 1\}$ that contains exactly m 0s and $n-1$ 1s; we define $g(\mathbf{v})$ to be

$$g(\mathbf{v}) = \{i \in I \mid v_i = 0\}.$$

With these notations, the function

$$g : Q(m, n-1) \to \binom{I}{m}$$

$$\mathbf{v} \mapsto g(\mathbf{v})$$

is clearly a bijection.

By Proposition 8.8, the composition $g \circ f$ is a bijection from $\begin{bmatrix} A \\ m \end{bmatrix}$ to $\binom{I}{m}$, and so, by claim 4 of our theorem, the number of m-multisubsets of A equals

$$\left| \begin{bmatrix} A \\ m \end{bmatrix} \right| = \left| \binom{I}{m} \right| = \binom{m+n-1}{m} = \begin{bmatrix} n \\ m \end{bmatrix}.$$

$\square$

The counting techniques discussed in this chapter are often employed to determine the number of choices one has for selecting or arranging a given number of elements from a given set or collection of sets. For example, the Addition Rule and the Multiplication Rule can be interpreted to say that, given boxes labeled A_1, $A_2, \ldots, A_n$, if box A_i contains m_i distinct objects ($i = 1, 2, \ldots, n$), then there are

$$m_1 + m_2 + \cdots + m_n$$

ways to choose one object from one of the n boxes, and there are

$$m_1 \cdot m_2 \cdots \cdot m_n$$

ways to choose one object from each of the n boxes. In a similar manner, the four basic counting functions of Theorem 19.4 are sometimes called "choice functions"; Table 19.4 summarizes our results for the number of ways to choose m elements from a given set of n elements.

The following proposition provides a fundamental example for counting problems.

Proposition 19.5 *Let n and m be positive integers. The numbers of solutions to the equation*

$$x_1 + x_2 + \cdots + x_m = n$$

in positive integers and in nonnegative integers are $\binom{n-1}{m-1}$ and $\left[\begin{smallmatrix} n+1 \\ m-1 \end{smallmatrix}\right]$, respectively.

Note that order *does* matter when counting the number of solutions; for example, we consider $(x_1, x_2) = (3, 5)$ and $(x_1, x_2) = (5, 3)$ to be different solutions to $x_1 + x_2 = 8$. This makes our proposition more unexpected: the answers resemble the forms under "order does not matter" rather than under "order matters" in the table above.

Proof First, observe that the number of solutions to the equation

$$x_1 + x_2 + \cdots + x_m = n$$

in positive integers (the number of "positive solutions," in short) is the same as the number of positive solutions to the inequality

$$x_1 + x_2 + \cdots + x_{m-1} \leq n - 1;$$

Table 19.4 The number of ways to choose m elements from a given set of n elements

	Order matters	Order does not matter
Elements distinct	$n^{\underline{m}}$	$\binom{n}{m}$
Elements may repeat	n^m	$\left[\begin{smallmatrix} n \\ m \end{smallmatrix}\right]$

indeed, there is a one-to-one correspondence between the two sets of solutions: any positive solution $(x_1, \ldots, x_m)$ of the equation satisfies the inequality, and any positive solution $(x_1, \ldots, x_{m-1})$ to the inequality determines a unique positive solution to the equation with

$$x_m = n - (x_1 + \cdots + x_{m-1}).$$

It is easy to see that the number of positive solutions to the inequality

$$x_1 + x_2 + \cdots + x_{m-1} \leq n - 1$$

equals the number of $(m-1)$-subsets of $A = \{1, 2, \ldots, n-1\}$; indeed, any positive solution of the inequality determines a unique $(m-1)$-subset

$$\{x_1, x_1 + x_2, \ldots, x_1 + x_2 + \cdots + x_{m-1}\}$$

of A, and any $(m-1)$-subset $\{a_1, a_2, \ldots, a_{m-1}\}$ of A, where

$$1 \leq a_1 < a_2 < \cdots < a_{m-1} \leq n - 1,$$

generates a unique positive solution $(a_1, a_2 - a_1, \ldots, a_{m-1} - a_{m-2})$ of the inequality. Therefore, the number of positive solutions to the equation is $\binom{n-1}{m-1}$.

Similarly, we can see that the number of nonnegative solutions to the equation

$$x_1 + x_2 + \cdots + x_m = n$$

is the same as the number of nonnegative solutions to the inequality

$$x_1 + x_2 + \cdots + x_{m-1} \leq n,$$

which further equals the number of $(m-1)$-multisubsets of $A = \{0, 1, 2, \ldots, n\}$. Therefore, the number of nonnegative solutions of the equation is $\left[\begin{smallmatrix} n+1 \\ m-1 \end{smallmatrix}\right]$.

(For an alternative approach, note that the number of nonnegative solutions to the equation

$$x_1 + x_2 + \cdots + x_m = n$$

equals the number of positive solutions to the equation

$$y_1 + y_2 + \cdots + y_m = n + m,$$

which, by the first argument, equals $\binom{n+m-1}{m-1} = \left[\begin{smallmatrix} n+1 \\ m-1 \end{smallmatrix}\right]$. For yet another approach, we may think of the number of nonnegative solutions to the equation as the number of ways we can place n identical objects in m distinct boxes; this number is given by $\left[{m \atop n} \right]$, which also equals $\left[{n+1 \atop m-1} \right]$.)   □

We will see a variety of further counting problems in the problem set below.

The following famous theorem is a corollary to Theorem 19.4, and explains how binomial coefficients got their name.

Theorem 19.6 (The Binomial Theorem) *For all real numbers a and b and for all positive integers n we have*

$$(a+b)^n = \sum_{m=0}^{n} \binom{n}{m} a^{n-m} b^m$$

$$= a^n + na^{n-1}b + \binom{n}{2}a^{n-2}b^2 + \cdots + \binom{n}{n-2}a^2 b^{n-2} + nab^{n-1} + b^n.$$

Proof We will verify the identity above by performing some simple algebra and using appropriate binomial coefficients to combine terms. When using the distributive law to expand the expression

$$(a+b)^n = (a+b)\cdots(a+b),$$

we arrive at a sum of products of n factors, where each factor is either a or b. Using the commutative property of multiplication, each term can be arranged so that the as (if any) all come before the bs (if any). Then, using the commutative property of addition, we can collect "like" terms; that is, terms of the form $a^{n-m}b^m$ for the same $m = 0, 1, \ldots, n$. The number of such terms clearly equals the number of n-sequences of the set $\{a, b\}$ that contain exactly $n - m$ as and m bs, which, by Theorem 19.4, is exactly $\binom{n}{m}$.   □

As promised earlier, we will now return to the Addition Rule and examine what we can say in the situation when the sets are not necessarily pairwise disjoint. Recall Problem 9 of Chapter 6: there we determined that, given finite sets A and B, the size of $A \cup B$ could vary between the larger of $|A|$ and $|B|$ to $|A| + |B|$; similarly, the size of $A \cap B$ could vary between 0 and the smaller of $|A|$ and $|B|$. In short, we can write

$$0 \le |A \cap B| \le \min\{|A|, |B|\} \le \max\{|A|, |B|\} \le |A \cup B| \le |A| + |B|.$$

A bit of investigation reveals that the six quantities in this sequence satisfy a more precise—and more useful—property: The sum of the two middle quantities always equals the sum of the second and the fifth, and this further equals the sum of the first and last:

$$\min\{|A|, |B|\} + \max\{|A|, |B|\} = |A \cap B| + |A \cup B| = 0 + (|A| + |B|).$$

The fact that

$$\min\{|A|, |B|\} + \max\{|A|, |B|\} = |A| + |B|$$

is easy to see; here we prove that

$$|A \cap B| + |A \cup B| = |A| + |B|$$

always holds as well.

Proposition 19.7 *If A and B are finite sets, then we have*

$$|A \cup B| = |A| + |B| - |A \cap B|.$$

Proof Consider the sets $A \setminus (A \cap B)$, $B \setminus (A \cap B)$, and $A \cap B$. Clearly, these are pairwise disjoint sets whose union is $A \cup B$, so the Addition Rule gives us

$$|A \cup B| = |A \setminus (A \cap B)| + |B \setminus (A \cap B)| + |A \cap B|.$$

Using the Addition Rule just for the disjoint sets $A \setminus (A \cap B)$ and $A \cap B$, whose union is A, we have

$$|A| = |A \setminus (A \cap B)| + |A \cap B|;$$

similarly, we have

$$|B| = |B \setminus (A \cap B)| + |A \cap B|.$$

Our claim now follows easily from the three equations. $\square$

The situation gets more complicated as the number of sets increases. For three sets we have the following result.

Proposition 19.8 *If A, B, and C are finite sets, then we have*

$$|A \cup B \cup C| = |A| + |B| + |C| - |A \cap B| - |A \cap C| - |B \cap C| + |A \cap B \cap C|.$$

Proof Using Proposition 19.7 first for the sets $A \cup B$ and C and then again for A and B yields

$$|A \cup B \cup C| = |A \cup B| + |C| - |(A \cup B) \cap C|$$
$$= |A| + |B| - |A \cap B| + |C| - |(A \cup B) \cap C|.$$

Next, we use distributivity to write

$$(A \cup B) \cap C = (A \cap C) \cup (B \cap C);$$

Proposition 19.7 then gives

$$|(A \cup B) \cap C| = |A \cap C| + |B \cap C| - |(A \cap C) \cap (B \cap C)|.$$

But

$$(A \cap C) \cap (B \cap C) = A \cap B \cap C,$$

so we have

$$|A \cup B \cup C| = |A| + |B| - |A \cap B| + |C| - (|A \cap C| + |B \cap C| - |A \cap B \cap C|),$$

as claimed.   □

The generalized version of Propositions 19.7 and 19.8 to an arbitrary (finite) number of finite sets is called the *Inclusion–Exclusion Principle* (or sometimes the *Sieve Principle*).

Theorem 19.9 (The Inclusion–Exclusion Principle) *Suppose that* $A_1, A_2, \ldots, A_n$ *are finite sets* ($n \in \mathbb{N}$). *We then have*

$$|A_1 \cup A_2 \cup \cdots \cup A_n| = |A_1| + |A_2| + \cdots + |A_n|$$
$$-|A_1 \cap A_2| - |A_1 \cap A_3| - \cdots - |A_{n-1} \cap A_n|$$
$$+|A_1 \cap A_2 \cap A_3| + |A_1 \cap A_2 \cap A_4| + \cdots + |A_{n-2} \cap A_{n-1} \cap A_n|$$
$$- + \cdots$$
$$+(-1)^{n-1} \cdot |A_1 \cap A_2 \cap \cdots \cap A_n|.$$

The proof of the Inclusion–Exclusion Principle is left to Problem 9 (a).

As an application of Theorems 19.4 and 19.9, we prove the following generalization of Problem 6 of Chapter 8.

Theorem 19.10 *Let A be a set of size m and B be a set of size n. Then*

1. *the number of functions from A to B is* n^m;
2. *the number of injections from A to B is* $n^{\underline{m}}$; *and*
3. *the number of surjections from A to B is*

$$\binom{n}{0} n^m - \binom{n}{1} (n-1)^m + \binom{n}{2} (n-2)^m - + \cdots + (-1)^n \binom{n}{n} 0^m.$$

(There is no closed form for the number of surjections.)

Proof Claims 1 and 2 follow from parts 1 and 2 of Theorem 19.4.

To prove 3, we first enumerate functions that are not surjections. Clearly, a function from A to B is not a surjection if, and only if, at least one element of B is not in its image.

Let $B = \{b_1, b_2, \ldots, b_n\}$. For each integer $i = 1, 2, \ldots, n$, let A_i be the collection of those functions from A to B that do not contain b_i in their image. The number of functions that are not surjections is then

$$|A_1 \cup A_2 \cup \cdots \cup A_n|.$$

We will use the Inclusion–Exclusion Principle to find this quantity.

First note that $|A_i| = (n-1)^m$ for each i (each element of A can be mapped to any element of B except for b_i). For two distinct indices $i, j \in \{1, \ldots, n\}$, we have $|A_i \cap A_j| = (n-2)^m$ (each element of A can be mapped to any element of B except for b_i or b_j). Similarly, the size of the intersection of three distinct sets from $A_1, \ldots, A_n$ is $(n-3)^m$, and so on. At the end, $A_1 \cap A_2 \cap \cdots \cap A_n$ consists of $(n-n)^m = 0$ functions; in other words, there are no functions that avoid all elements of B.

Now by claim 4 of Theorem 19.4, there are $\binom{n}{1} = n$ sets of the form A_i, $\binom{n}{2}$ sets of the form $A_i \cap A_j$, and so on. We thus have

$$|A_1 \cup A_2 \cup \cdots \cup A_n| = \binom{n}{1}(n-1)^m - \binom{n}{2}(n-2)^m + - \cdots + (-1)^{n-1}\binom{n}{n}(n-n)^m$$

according to the Inclusion–Exclusion Principle.

Since there are n^m functions all together from A to B, there are

$$n^m - |A_1 \cup A_2 \cup \cdots \cup A_n|$$

surjections, which proves our claim.   □

Recall from Problem 3 of Chapter 8 that, if $m = n$, then an injection or a surjection from A to B must also be a bijection. Therefore, as a corollary to Theorem 19.10, we get the following:

Corollary 19.11 *If A and B are both sets of size n, then the number of bijections from A to B is*

$$n! = \binom{n}{0}n^n - \binom{n}{1}(n-1)^n + \binom{n}{2}(n-2)^n - + \cdots + (-1)^n\binom{n}{n}0^n.$$

Corollary 19.11 gives us a nice identity for binomial coefficients.

We close this chapter by discussing another interesting application of the Inclusion–Exclusion Principle. For a fixed positive integer n, consider a bijection f on the set $I_n = \{1, 2, \ldots, n\}$. (As we just mentioned, there are $n!$ such bijections.) We say that an element $i \in I_n$ is a *fixed point* of f, if $f(i) = i$.

Clearly, every element of I_n is a fixed point of the identity function id_{I_n}. At the other end of the spectrum, we have bijections with no fixed points at all—such bijections are called *derangements* of I_n. For example, we see that the set I_3 has two derangements: the bijections f_1 and f_2 given by Table 19.5.

Table 19.5 The two
derangements of $\{1, 2, 3\}$

i	1	2	3
$f_1(i)$	2	3	1
$f_2(i)$	3	1	2

Table 19.6 The nine derangements of $\{1, 2, 3, 4\}$

i	1	2	3	4
$f_1(i)$	2	1	4	3
$f_2(i)$	2	3	4	1
$f_3(i)$	2	4	1	3

i	1	2	3	4
$f_4(i)$	3	1	4	2
$f_5(i)$	3	4	1	2
$f_6(i)$	3	4	2	1

i	1	2	3	4
$f_7(i)$	4	1	2	3
$f_8(i)$	4	3	1	2
$f_9(i)$	4	3	2	1

Table 19.7 The proportion of derangements among bijections

n	1	2	3	4	5	6	7	8	9
d_n	0	1	2	9	44	265	1854	14833	133496
$d_n/n!$	0	0.5	0.33333	0.375	0.36667	0.36806	0.36786	0.36788	0.36788

Similarly, we find that I_4 has nine derangements, as listed in Table 19.6.
The following theorem enumerates derangements.

Theorem 19.12 *Let n be a positive integer. The number of derangements of a set
of n elements equals*

$$d_n = \frac{n!}{0!} - \frac{n!}{1!} + \frac{n!}{2!} - + \cdots + (-1)^n \frac{n!}{n!}.$$

According to this result, we have

$$d_3 = \frac{6}{1} - \frac{6}{1} + \frac{6}{2} - \frac{6}{6} = 2$$

and

$$d_4 = \frac{24}{1} - \frac{24}{1} + \frac{24}{2} - \frac{24}{6} + \frac{24}{24} = 9,$$

in agreement with our tables above. We leave the proof of Theorem 19.12 to
Problem 9 (b).

Table 19.7 exhibits the first few values of d_n and (mostly approximations for)
$d_n/n!$, the proportion of derangements among all bijections of a set of size n.

It appears that the values of $d_n/n!$ rapidly converge to a limit. Indeed, recall from
page 282 that

$$\frac{1}{e} = \frac{1}{0!} - \frac{1}{1!} + \frac{1}{2!} - \frac{1}{3!} + - \cdots,$$

where e is Euler's number. Therefore, by Theorem 19.12, we have

$$d_n \approx \frac{n!}{e}$$

(in fact, it can be shown that d_n is the integer nearest to $n!/e$). This can be strikingly illustrated by saying that, whether we have a handful of friends going out for dinner or we deal with a banquet with hundreds of participants, if everyone orders different meals and the meals are delivered to people randomly, then the chances that nobody will get their own order is about the same—just under thirty-seven percent—in either situation!

Counting questions can be quite challenging to answer. Count on countless amounts of fun while doing the problems below!

Problems

1. Let m and n be positive integers with $m < n$. Consider the following identities.

i. $\dbinom{n}{m} = \dbinom{n-1}{m} + \dbinom{n-1}{m-1}$

ii. $\dbinom{n}{m} = \dbinom{n-1}{m} + \dbinom{n-2}{m-1} + \cdots + \dbinom{n-m-1}{0}$

iii. $\dbinom{n}{m} = \dbinom{n-1}{m-1} + \dbinom{n-2}{m-1} + \cdots + \dbinom{m-1}{m-1}$

(a) Provide an illustration for $n = 7$ and $m = 3$ for each identity using Pascal's Triangle.

(b) Each of the identities could be proved using algebraic manipulations and induction. Here, however, we wish to establish these identities using counting arguments, as follows. The left-hand side of each identity equals the number of m-subsets of a given set A of n elements. For each identity, find a different way of counting the m-subsets of A that yield the right-hand side of the identity. This technique for proving identities is often called the "bijective method."

(Hint for ii: Suppose that $A = \{a_1, \ldots, a_n\}$, and consider those m-subsets of A that do not contain a_1, then those that contain a_1 but do not contain a_2, etc.)

(c) The corresponding identities for multisubsets are as follows:

i* $\begin{bmatrix} n \\ m \end{bmatrix} = \begin{bmatrix} n-1 \\ m \end{bmatrix} + \begin{bmatrix} n \\ m-1 \end{bmatrix}$

ii* $\begin{bmatrix} n \\ m \end{bmatrix} = \begin{bmatrix} n-1 \\ m \end{bmatrix} + \begin{bmatrix} n-1 \\ m-1 \end{bmatrix} + \cdots + \begin{bmatrix} n-1 \\ 0 \end{bmatrix}$

iii* $\begin{bmatrix} n \\ m \end{bmatrix} = \begin{bmatrix} n \\ m-1 \end{bmatrix} + \begin{bmatrix} n-1 \\ m-1 \end{bmatrix} + \cdots + \begin{bmatrix} 1 \\ m-1 \end{bmatrix}$

These identities could be easily reduced to the ones for subsets above. Instead, use the bijective method; that is, provide a counting argument for each identity by counting the number of m-multisubsets of an n-element set in different ways.

2. Let n be a positive integer. Explain how each of the following identities is a simple corollary to the Binomial Theorem.

(a) $\dbinom{n}{0} + \dbinom{n}{1} + \dbinom{n}{2} + \cdots + \dbinom{n}{n} = 2^n$

(b) $\dbinom{n}{0} - \dbinom{n}{1} + \dbinom{n}{2} - + \cdots + (-1)^n \dbinom{n}{n} = 0$

(c) $\dbinom{n}{0}\dbinom{n}{n} + \dbinom{n}{1}\dbinom{n}{n-1} + \dbinom{n}{2}\dbinom{n}{n-2} + \cdots + \dbinom{n}{n}\dbinom{n}{0} = \dbinom{2n}{n}$

(Hint: Consider the identity $(1+b)^n \cdot (1+b)^n = (1+b)^{2n}$.)

3. Suppose that n is a positive integer, and define

$$a_n = \dbinom{n}{0} + \dbinom{n-1}{1} + \dbinom{n-2}{2} + \cdots + \dbinom{\lceil n/2 \rceil}{\lfloor n/2 \rfloor}.$$

(a) Prove that a_n is the number of ways to cover a 2-by-n board with n 1-by-2 dominoes. (Cf. Problem 9 (b) of Chapter 2.)

(b) Prove that for all $n \in \mathbb{N}$, a_n equals the n-th Fibonacci number F_n.
(Hints: Verify that $a_1 = F_1$, $a_2 = F_2$, and that $a_{n+2} = a_n + a_{n+1}$ holds for all $n \in \mathbb{N}$. Use identity i of Problem 1.)

4. In this problem we prove the identity

$$1 \cdot \dbinom{n}{1} + 2 \cdot \dbinom{n}{2} + \cdots + n \cdot \dbinom{n}{n} = n \cdot 2^{n-1}$$

for all positive integers n using the bijective method as in Problem 1.

Let A be an arbitrary set of n elements, and consider the bijection $f : P(A) \to \{0, 1\}^n$ defined in the proof of Proposition 19.3. Construct a table with 2^n rows and n columns, in which rows are indexed by the different subsets of A, columns are indexed by the integers from 1 to n, and the entry in row X and column i equals $f_i(X)$. Count the total number of 1s in this table in two different ways (once by rows and once by columns) to arrive at the identity above.

5. The game *Poker* is played using a standard deck of fifty-two cards, each of a certain rank (2, 3, 4, 5, 6, 7, 8, 9, 10, Jack, Queen, King, or Ace) and of a certain suit (club, diamond, heart, or spade). At the beginning of the game, each player is dealt a hand consisting of five cards; the players' aim is to get the best possible hand matching one of the following distinguished hands:

- a *pair*: two cards of the same rank and three other cards of different ranks;
- a *two-pair*: two pairs of different ranks and a fifth card of a third rank;
- a *three-of-a-kind*: three cards of the same rank and two other cards of different ranks;
- a *straight*: five cards of consecutive ranks, not all the same suit;
- a *flush*: all five cards of the same suit, but not a straight;
- a *full house*: a pair and a three-of-a-kind;
- a *four-of-a-kind*: four cards of the same rank and one other card;
- a *straight flush*: all five cards of the same suit and of consecutive ranks, but not a Royal flush;
- a *Royal flush*: a 10, a Jack, a Queen, a King, and an Ace of the same suit.

For the purposes of a straight or straight flush, an Ace can be placed before a 2 (so that it plays the role of a 1).

The winner of the game is the player who has the least likely type of distinguished hand from the list above. (For the situation in which more than one player achieves the same type, additional rules are in place. For exact rules of *Poker*, see, for example, www.poker.com.)

For each of the nine distinguished types of poker hands above, find the number of possible hands of that type. What is more likely: that a hand is a distinguished hand or that it is not?

6. The College Bookstore is selling n different books (each in "unlimited" supply; two copies of the same book are indistinguishable). Alvin decided to purchase m books at the Bookstore, and wants to place them in his brand new cabinet that contains k different shelves. (Assume that there is no limitation on how many books the shelves can house.) In how many ways can he purchase and arrange the books given the following conditions? Give each of your answers either as one of the following quantities or as the product of two of these quantities:

    n^m, $n^{\underline{m}}$, $n^{\overline{m}}$, k^m, $k^{\underline{m}}$, $k^{\overline{m}}$, $\binom{n}{m}$, $\left[\begin{smallmatrix}n\\m\end{smallmatrix}\right]$, $\binom{k}{m}$, $\left[\begin{smallmatrix}k\\m\end{smallmatrix}\right]$, $\binom{nk}{m}$, $\left[\begin{smallmatrix}nk\\m\end{smallmatrix}\right]$.

    (a) The m books are all different, and the order of the books on each shelf matters.

    (b) The m books are all different, and the order of the books on the shelves does not matter.

    (c) The m books are not necessarily different, and the order of the books on each shelf matters.

    (d The m books are not necessarily different, and the order of the books on the shelves does not matter.

    (Hint: Suppose that the Bookstore makes each of the n books available in k different varieties, labeled "shelf 1 only," "shelf 2 only," etc.)

Remark This problem provides a generalization to Theorem 19.4; namely, when $k = 1$, the answers to the questions above should agree with the corresponding values in Table 19.4.

7. In a certain city, streets run in the East–West direction and avenues run in the North–South direction; avenues are numbered starting with 1st Avenue on the East side of the city and streets are numbered starting with 1st Street at the Southern border. An exception to this perfect grid is a grassy and wooded city park with a rectangular shape between 5th and 8th Avenues and 59th and 110th Streets.

In how many ways can one walk from the corner of 5th Avenue and 34th Street to the corner of 10th Avenue and 116th Street avoiding the park and walking always either directly North or directly West? (Walking alongside the park is fine.)

Remarks This problem is loosely based on a possible walk in Manhattan, NY, from the CUNY Graduate Center at 34th Street and Fifth Avenue, to Columbia University, at 116th Street and Tenth Avenue. (The Empire State building is also at 34th Street at Fifth Avenue.) The park mentioned above is Central Park.

8. Recall from Problem 10 of Chapter 4 that in the Plutonian alphabet there are only four letters: A, B, C, and D, and that every finite string containing these letters is a Plutonian word.

    (a) Prove that there are

$$\frac{(a+b+c+d)!}{a! \cdot b! \cdot c! \cdot d!}$$

    Plutonian words with exactly a As, b Bs, c Cs, and d Ds. (Here and below, a, b, c, d are nonnegative integers.)
    (b) Prove that there are

$$\frac{(a+b+c+d+1)!}{(a+b+c+1) \cdot a! \cdot b! \cdot c! \cdot d!}$$

    Plutonian words with exactly a As, b Bs, c Cs, and at most d Ds.
    (c) Recall from Problem 4 (e) of Chapter 10 that a Plutonian word is a "D-lite" if at most half of its letters are Ds. How many D-lites are there with exactly a As, b Bs, and c Cs?

9. (a) Prove Theorem 19.9, the Inclusion–Exclusion Principle.
    (b) Prove Theorem 19.12 on the number of derangements.

10. Let n be an integer that is greater than 1, and suppose that the prime factorization of n is

$$n = \prod_{i=1}^{r} p_i^{\alpha_i}.$$

(Here $r \geq 1$ is the number of distinct positive prime divisors of n; $p_1, \ldots, p_r$ are these prime divisors; and for an $i \in \{1, \ldots, r\}$, α_i is the "multiplicity" of p_i in the prime factorization of n.)

(a) Prove that the number of positive divisors of n is

$$d(n) = \prod_{i=1}^{r} (\alpha_i + 1).$$

Use this to verify that n is prime if, and only if, $d(n) = 2$ (cf. Definition 2.1).

(b) Prove that the sum of the positive divisors of n is

$$\sigma(n) = \prod_{i=1}^{r} \frac{p_i^{\alpha_i + 1} - 1}{p_i - 1}.$$

Use this to verify that if n is of the form $n = 2^{k-1}(2^k - 1)$ where k is a positive integer for which $2^k - 1$ is a prime number, then n is a perfect number. (See Theorem 4.1.)

(c) Prove that the number of integers between 1 and n, inclusive, that are relatively prime to n is

$$\phi(n) = \prod_{i=1}^{r} p_i^{\alpha_i - 1}(p_i - 1).$$

(Hint: Use the Inclusion–Exclusion Principle.)

Remark $\phi(n)$ is the famous *Euler ϕ-function*; it appears in various places in mathematics. For example, the group U_n (cf. Problem 5 in Chapter 15) has order $\phi(n)$.

11. In this problem we briefly discuss "discrete" analogues of some familiar concepts of "continuous" calculus.

Definition 19.13 *Let* $\mathbf{a} = (a_n)_{n=1}^{\infty} = (a_1, a_2, a_3, \ldots)$ *be an infinite sequence of real numbers.*

*   *The* difference sequence *of* $\mathbf{a}$ *is the sequence*

$$\Delta \mathbf{a} = (a_{n+1} - a_n)_{n=1}^{\infty} = (a_2 - a_1, a_3 - a_2, \ldots).$$

*   *If* $\mathbf{a}$ *is the difference sequence of a sequence* $\mathbf{b}$, *then* $\mathbf{b}$ *is called an* antidifference sequence *of* $\mathbf{a}$. *The set of all antidifference sequences of* $\mathbf{a}$ *is denoted by* $\Delta^{-1}\mathbf{a}$.

Remarks The difference $a_{n+1} - a_n$ equals the familiar difference quotient

$$\frac{a_{n+1} - a_n}{(n+1) - n}.$$

Thus we may think of $\Delta \mathbf{a}$ and $\Delta^{-1} \mathbf{a}$ as the discrete analogues of the continuous derivative and antiderivative, respectively.

(a) Prove the following proposition.

> **Proposition 19.14** *Let $c \in \mathbb{R}$. If $\mathbf{b} = (b_n)_{n=1}^{\infty}$ is an antidifference sequence of the sequence $\mathbf{a}$, then the sequence $\mathbf{b}' = (b_n + c)_{n=1}^{\infty}$ is also an antidifference sequence of $\mathbf{a}$.*
>
> *Conversely, if $\mathbf{b} = (b_n)_{n=1}^{\infty}$ and $\mathbf{b}' = (b_n')_{n=1}^{\infty}$ are both antidifference sequences of a sequence $\mathbf{a}$, then there is a constant $c \in \mathbb{R}$ for which $b_n - b_n' = c$ holds for all $n \in \mathbb{N}$.*

Remark The antidifference sequence $\mathbf{b}$ of $\mathbf{a}$ that has first element $b_1 = c$ is denoted by $\Delta_c^{-1} \mathbf{a}$. According to Proposition 19.14, $\Delta_c^{-1} \mathbf{a}$ exists and is unique for all sequences $\mathbf{a}$ and constants c.

(b) Prove the following proposition.

> **Proposition 19.15** *Let $\mathbf{a} = (a_1, a_2, a_3, \dots)$ be a sequence and $c \in \mathbb{R}$ be a constant. Prove that*
>
> $$\Delta \Delta_c^{-1} \mathbf{a} = \mathbf{a}$$
>
> *and*
>
> $$\Delta_{a_1}^{-1} \Delta \mathbf{a} = \mathbf{a}.$$

Remark Proposition 19.15 is the discrete version of the Fundamental Theorem of Calculus.

(c) Let m be a fixed positive integer and $c \in \mathbb{R}$ be a constant. Prove that

$$\Delta \left(n^{\underline{m}} \right)_{n=1}^{\infty} = \left(m \cdot n^{\underline{m-1}} \right)_{n=1}^{\infty}$$

and that

$$\Delta_c^{-1} \left(n^{\underline{m}} \right)_{n=1}^{\infty} = \left(c + \frac{n^{\underline{m+1}}}{m+1} \right)_{n=1}^{\infty}.$$

(d) Let $\mathbf{a} = (1)_{n=1}^{\infty} = (1, 1, 1, 1, \dots)$. Prove that

$$\Delta_1^{-1}\Delta_1^{-1}\Delta_1^{-1}\Delta_1^{-1}\mathbf{a} = \left(\binom{n}{4} + \binom{n}{2} + 1 \right)_{n=1}^{\infty}$$

$$= (1, 2, 4, 8, 16, 31, 57, \ldots).$$

(Cf. Problem 4 (d) of Chapter 3 and Problem 12 below.)

12. In this problem we analyze Problem 4 (d) of Chapter 3. (This problem uses graph theory; cf. Appendix E).

   (a) Suppose that n points are given on a circle in general position (no three chords meet at the same point inside the circle). Prove that the number of regions that the chords determine inside the circle is

   $$\binom{n}{4} + \binom{n}{2} + 1.$$

   (Hints: Let G be the planar graph created by the circle and the chords; let V, E, and F be the number of vertices, edges, and faces of G, respectively. Prove that

   $$V = \binom{n}{4} + n \text{ and } E = \frac{4\binom{n}{4} + (n+1)n}{2},$$

   then use Euler's Formula.)
   (b) Prove that

   $$\binom{n}{4} + \binom{n}{2} + 1 = 2^{n-1} - \sum_{i=5}^{n-1} \binom{n-1}{i}.$$

   Remark This shows that the claim of Problem 4 (d) of Chapter 3 is false if $n \geq 6$.

13. Given a nonnegative integer k, let S_k denote the diagram that results from taking the first 2^k rows of Pascal's Triangle (that is, rows n with $n = 0, 1, \ldots, 2^k - 1$) and replacing each odd integer by a "full" circle and each even integer with an "open" circle. For example, S_3 is shown in Figure 19.1.

   Remarks Our diagram is essentially the *Sierpiński Triangle* (also known as the *Sierpiński Sieve* or *Sierpiński Gasket*), named after the Polish mathematician who introduced it first in 1915.

   The infinite version of the Sierpiński Triangle is a well-known example of a *fractal* or self-similar set (cf. page 291). In our diagram of S_3, above, we may note that the full circles form (the boundaries of) three smaller triangles (corresponding to three images of S_2), and each of these triangles can be divided

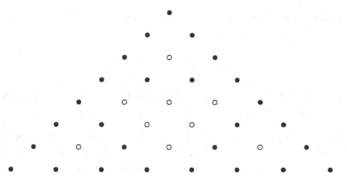

Fig. 19.1 Part of the Sierpiński Triangle

into three smaller triangles (each one corresponding to S_1). Beautiful images of the Sierpiński Triangle can be found, for example, at http://mathworld.wolfram.com/SierpinskiSieve.html, and mesmerizing animated images may be viewed on YouTube.

In this problem we investigate some interesting properties of S_k.

(a) Let n and m be nonnegative integers with $n \le 2^k - 1$ and $m \le n$. Prove that the exponent of 2 in the prime factorization of $\binom{n}{m}$ equals

$$\sum_{i=0}^{k-1} \left(\left\lfloor \frac{n}{2^i} \right\rfloor - \left\lfloor \frac{m}{2^i} \right\rfloor - \left\lfloor \frac{n-m}{2^i} \right\rfloor \right).$$

(b) Let n and m be nonnegative integers with $n \le 2^k - 1$ and $m \le n$. Let

$$n = n_{k-1} \ldots n_1 n_0$$

and

$$m = m_{k-1} \ldots m_1 m_0$$

be the binary representations of n and m, respectively (cf. Theorem 12.10; here we insert initial bits of 0s if the numbers have fewer than k bits). Prove that $\binom{n}{m}$ is odd if, and only if, $m_i \le n_i$ holds for all $i = 0, 1, \ldots, k - 1$.

(c) Find all values of n between 0 and $2^k - 1$ for which the n-th row of S_k contains only full circles, and those values of n for which the n-th row contains only open circles other than the two full circles at each end.

(d) Prove that S_k contains exactly 3^k full circles.

Remark For a nonnegative integer n, let f_n denote the number of odd values in the first n rows of Pascal's Triangle. The first few terms of the sequence $\mathbf{f} = (f_0, f_1, f_2, \ldots)$ are 0, 1, 3, 5, 9, 11, 15, 19, 27. According to our claim, we have $f_{2^k} = 3^k$. It can be shown that the terms of $\mathbf{f}$ satisfy the recursion $f_0 = 0$, $f_1 = 1$,

$$f_n = 2f(\lfloor n/2 \rfloor) + f(\lceil n/2 \rceil)$$

for $n \geq 2$. This more general fact clearly implies our claim for $n = 2^k$. (However, your proof should not rely on this fact.)

(e) Define $A(S_k)$ (the "density" of full circles in the diagram) to be the number of full circles in S_k divided by the total number of circles in S_k. For example, as the diagram above demonstrates, we have

$$A(S_3) = \frac{27}{36} = 0.75.$$

Prove that

$$\lim_{k \to \infty} A(S_k) = 0.$$

Chapter 20
Infinite Delights

In Chapter 18 we defined what it means for a sequence to have an infinite limit. In this chapter we discuss a different aspect of the intriguing concept of infinity; namely, as promised in Chapter 19, we study sets of infinite size. As we will soon see, not all infinite sets are created equal: some are "larger" ("much larger") than others. On the other hand, as it was already realized by the Italian mathematician and physicist Galileo Galilei (1564–1642), even when an infinite set is a proper subset of another infinite set, we may have to accept the notion that they have the "same size." Among other things, in this chapter we will make precise the following statements about the sizes of our five basic number sets:

- $\mathbb{N}$, $\mathbb{Z}$, and $\mathbb{Q}$ are the "same size";
- $\mathbb{R}$ and $\mathbb{C}$ are the "same size"; but
- $\mathbb{R}$ and $\mathbb{C}$ are "larger than" $\mathbb{N}$, $\mathbb{Z}$, and $\mathbb{Q}$.

As we pursue our development, we will be confronted with some of the most fundamental, yet least understood, questions in abstract mathematics today.

Let us start by first pointing out what is arguably the main characteristic of finite sets and which distinguishes them from infinite sets: the elements of a finite set—except for the empty-set that we also consider finite—can be listed from first to last. (Of course, we are talking here about a theoretical possibility only: a list description may not be very practical if the set is large or if the elements are not given explicitly.)

Thus, if a set A has n elements for some $n \in \mathbb{N}$, then it is possible to find $a_1, a_2, \ldots, a_n \in A$ so that $\{a_1, a_2, \ldots, a_n\} = A$. More precisely, we can say that, for every nonempty finite set A, there is a positive integer n and a surjection f from $\{1, 2, \ldots, n\}$ to A. (Actually, we can require that f be a bijection—and that n be as small as possible—but here we want to emphasize the fact that the set $\{a_1, a_2, \ldots, a_n\}$ "covers" all of A and not that the elements are distinct.)

© Springer Nature Switzerland AG 2020
B. Bajnok, *An Invitation to Abstract Mathematics*, Undergraduate Texts in Mathematics, https://doi.org/10.1007/978-3-030-56174-1_20

For an infinite set A, however, there is no positive integer n and elements $a_1, a_2, a_3, \ldots, a_n$ for which $\{a_1, a_2, a_3, \ldots, a_n\} = A$; that is, a function from $\{1, 2, \ldots, n\}$ to A is never a surjection. It is possible to restate this description of infinite sets differently; rather than saying what one cannot do, we can say what we can do: For an infinite set A, we can create an infinite list $(a_1, a_2, a_3, \ldots)$ with pairwise distinct elements. Here a_1 is the first element in the list, a_2 is the second (different from the first), a_3 is the third (different from the first two), and so on, and the list does not contain a last element. For a set A to possess an infinite list consisting of pairwise distinct elements can be concisely stated as being able to find an injection $g : \mathbb{N} \to A$. (As we will see later, requiring that g be a bijection is *not* equivalent to this!)

We can summarize our observations as follows:

Proposition 20.1 *For any set A, exactly one of the following three possibilities holds:*

- *A is the empty-set; or*
- *there is a positive integer n for which there is a surjection from $\{1, \ldots, n\}$ to A; or*
- *there is an injection from $\mathbb{N}$ to A.*

In the first two cases A is finite; if the third possibility holds, then A is infinite. While we hope that our explanations for Proposition 20.1 above were convincing, a precise proof based on the axioms of set theory (including the Axiom of Choice) would take considerable effort.

Next, and for most of the rest of this chapter, we discuss how large an infinite set may be. At the most elementary level, we may just say that an infinite set A has size $|A| = \infty$. It turns out, however, that it is possible—indeed, desirable—to distinguish among different "kinds of infinities."

Let us return to the issue of trying to arrange the elements of a set in a list. As we discussed above, every nonempty finite set can be listed (in theory). It turns out that some—but not all!—infinite sets can also be arranged in list notation—the list would have to be infinitely long, but every element of the set must appear in the list in a "finite position." That is, an infinite set A can be listed, if there is a sequence $(a_1, a_2, a_3, \ldots)$ so that $\{a_1, a_2, a_3, \ldots\} = A$; more concisely, if there is a surjection from $\mathbb{N}$ to A. In this sense infinite sets that can be listed are not all that different from finite sets.

So we find that it is possible to distinguish between two types of sets: those that can be arranged as a list and those that cannot be. We make this precise with the following definition.

Definition 20.2 *We say that a set A is* countable *if it is the empty-set or if there is a surjection from $\mathbb{N}$ to A. If A is not countable, we say that it is* uncountable.

As the most obvious example, we have:

Proposition 20.3 *The set $\mathbb{N}$ of natural numbers is countable.*

Proof The identity map $id_{\mathbb{N}}$ is clearly a surjection from $\mathbb{N}$ to $\mathbb{N}$.   □

It is a bit less obvious that the set $\mathbb{Z}$ of all integers and the set $\mathbb{Q}$ of all rational numbers are also countable; we will establish these results below (cf. also page 98 and Problem 4 in Chapter 6, respectively). In Chapter 6 we also mentioned that the set $\mathbb{R}$ of real numbers is uncountable, and we will prove this surprising fact below as well.

It is important to note that the surjection in Definition 20.2 need not be a bijection and, therefore, a countable set may be either finite or infinite. In fact, every nonempty finite set is countable: if $A = \{a_1, \ldots, a_n\}$ is a finite set, then the function

$$f : \mathbb{N} \to \quad A$$
$$i \mapsto a_{\min\{i,n\}}$$

—corresponding to the sequence $(a_1, a_2, \ldots, a_n, a_n, a_n, \ldots)$—is a surjection. Therefore, we now have four types of sets:

* the empty set;
* nonempty finite sets (which are all countable);
* sets that are infinite and countable (in short, *countably infinite* sets); and
* uncountable sets (which are all infinite).

Heuristically, we want to say that our four types of sets are listed here in increasing order by size: in particular, an infinite set (whether countable or uncountable) is larger than a finite set, and an uncountable set is larger than a countable set (whether finite or countably infinite). This is indeed feasible, according to the following proposition.

Proposition 20.4 *Let A and B be sets for which $A \subseteq B$. Then*

1. if A is infinite, then B is infinite, and
2. if A is uncountable, then B is uncountable.

Proof To prove our first claim, we use Proposition 20.1. Assume that $f : \mathbb{N} \to A$ is an injection. Define the function g as

$$g : A \to B$$
$$a \mapsto a;$$

clearly, g is an injection. Therefore, by Proposition 8.8, $g \circ f : \mathbb{N} \to B$ is an injection, and thus B is infinite.

We prove our second claim by showing that if B is countable, then A must be countable as well. Since the empty-set is countable, we may assume that $A \neq \emptyset$; therefore, there exists a surjection $f : \mathbb{N} \to B$. Since $A \neq \emptyset$ we can choose an $a_0 \in A$, with which we define the function

Table 20.1 Is $A \subseteq B$ possible?

	B finite	B countably infinite	B uncountable
A finite	Yes	Yes	Yes
A countably infinite	No	Yes	Yes
A uncountable	No	No	Yes

$$g : \mathbb{N} \to A$$

$$n \mapsto \begin{cases} f(n) \text{ if } f(n) \in A \\ \\ a_0 \quad \text{ if } f(n) \notin A \end{cases}$$

Then, since $A \subseteq B$ and f is a surjection, g is also a surjection, and thus A is countable, as claimed.   □

Table 20.1 summarizes the possibilities for A and B allowed by Proposition 20.4.

Later in this chapter we discuss a much more refined method of ranking sets by how large they are, but for now we just focus on the two sizes: countable and uncountable.

We turn to the infinite versions of the Addition Rule and Multiplication Rule (cf. Lemmas 19.1 and 19.2).

Lemma 20.5 (The Addition Rule for Infinite Sets) *If A and B are countable sets, then $A \cup B$ is also countable.*

More generally, if $A_1, A_2, \ldots, A_n$ are countable sets $(n \in \mathbb{N})$, then

$$A_1 \cup A_2 \cup \cdots \cup A_n$$

is also countable.

If at least one of $A_1, A_2, \ldots, A_n$ is uncountable, then their union is also uncountable.

Note that, unlike we did for the finite case (cf. Lemma 19.1), here we do not need to assume that the sets are pairwise disjoint.

Proof If A and B are countable sets, then we have surjections $f : \mathbb{N} \to A$ and $g : \mathbb{N} \to B$. It is not difficult to see that

$$\{f(1), g(1), f(2), g(2), \ldots\} = A \cup B.$$

More precisely, we can prove that the function

$$h : \mathbb{N} \to A \cup B$$

$$i \mapsto \begin{cases} f\left(\frac{i+1}{2}\right) \text{ if } i \text{ is odd} \\ \\ g\left(\frac{i}{2}\right) \quad \text{ if } i \text{ is even} \end{cases}$$

is a surjection, as follows. Let $x \in A \cup B$; we need to show that x is in $\mathrm{Im}(h)$. Let us suppose that $x \in A$—the case when $x \in B$ is quite similar. Since f is a surjection, we have some $j \in \mathbb{N}$ for which $f(j) = x$. But then $h(2j - 1) = f(j) = x$.

The claim that, for any positive integer n, the union of n countable sets is countable can be easily established by induction from the case of $n = 2$. We omit the details.

To prove the last statement, assume wlog that A_1 is uncountable. Since

$$A_1 \subseteq A_1 \cup A_2 \cup \cdots \cup A_n,$$

our claim follows directly from Proposition 20.4.   □

As a corollary, we get that $\mathbb{Z}$ is countable.

Proposition 20.6 *The set $\mathbb{Z}$ of all integers is countable.*

Proof Note that

$$\mathbb{Z} = \mathbb{N} \cup \{0\} \cup (-\mathbb{N})$$

where $-\mathbb{N}$ denotes the set of negative integers. Our claim follows from the Addition Rule above, since all three components are countable.   □

Lemma 20.7 (The Multiplication Rule for Infinite Sets) *If A and B are countable sets, then $A \times B$ is also countable.*

More generally, if $A_1, A_2, \ldots, A_n$ are countable sets ($n \in \mathbb{N}$), then

$$A_1 \times A_2 \times \cdots \times A_n$$

is also countable.

If none of the sets $A_1, A_2, \ldots, A_n$ is empty and at least one of them is uncountable, then their Cartesian product is uncountable.

Proof Let us start with the first claim. Since A and B are countable sets, we have surjections $g_A : \mathbb{N} \to A$ and $g_B : \mathbb{N} \to B$. It is then easy to see that the function

$$g : \mathbb{N} \times \mathbb{N} \to A \times B$$

$$(i, j) \mapsto ((g_A(i), g_B(j))$$

is a surjection from $\mathbb{N} \times \mathbb{N}$ to $A \times B$. Indeed, for an arbitrary element (a, b) of $A \times B$, we have some $i \in \mathbb{N}$ for which $g_A(i) = a$ and some $j \in \mathbb{N}$ for which $g_B(j) = b$. This yields $g(i, j) = (a, b)$.

In order to find a surjection from $\mathbb{N}$ to $A \times B$, we will use several results from Chapter 8. First, by Problem 8 of Chapter 8, we have a bijection $f : \mathbb{N} \times \mathbb{N} \to \mathbb{N}$. Therefore, by Proposition 8.9, f has an inverse $f^{-1} : \mathbb{N} \to \mathbb{N} \times \mathbb{N}$; and, by

Proposition 8.7, f^{-1} is also invertible, and thus is also a bijection. In particular, f^{-1} is a surjection from $\mathbb{N}$ to $\mathbb{N} \times \mathbb{N}$.

But, if f^{-1} is a surjection from $\mathbb{N}$ to $\mathbb{N} \times \mathbb{N}$ and g is a surjection from $\mathbb{N} \times \mathbb{N}$ to $A \times B$, then, by Proposition 8.8, $g \circ f^{-1}$ is a surjection from $\mathbb{N}$ to $A \times B$, and so $A \times B$ is countable.

Analogously to the Addition Rule for Infinite Sets, the second claim follows from the first by induction.

To prove our last claim, suppose wlog that A_1 is uncountable. If $n = 1$, there is nothing to prove, so assume that $n \geq 2$. Since the set A_i is not empty for any $i \in \{2, \ldots, n\}$, we can choose elements $a_i \in A_i$. Consider the set

$$A_1^* = A_1 \times \{a_2\} \times \cdots \times \{a_n\}.$$

It is then quite clear that there is a bijection $f : A_1^* \to A_1$, so, since A_1 is uncountable, A_1^* must be uncountable as well. Indeed, if $g : \mathbb{N} \to A_1^*$ were to be a surjection, then, by Proposition 8.8, $f \circ g : \mathbb{N} \to A_1$ would also be a surjection, which is impossible since A_1 is uncountable.

Since

$$A_1^* \subseteq A_1 \times A_2 \times \cdots \times A_n,$$

our claim follows via Proposition 20.4.   □

As a corollary, we get that $\mathbb{Q}$ is countable.

Proposition 20.8 *The set $\mathbb{Q}$ of all rational numbers is countable.*

Proof Let us define the function

$$f : \mathbb{Z} \times \mathbb{N} \to \mathbb{Q}$$

$$(a, b) \mapsto \frac{a}{b}$$

Clearly, f is a surjection (though not a bijection!). By Propositions 20.6 and 20.3, $\mathbb{Z}$ and $\mathbb{N}$ are countable, so by Lemma 20.7, there is some surjection g from $\mathbb{N}$ to $\mathbb{Z} \times \mathbb{N}$. But then, by Proposition 8.8, $f \circ g$ is a surjection from $\mathbb{N}$ to $\mathbb{Q}$, and so $\mathbb{Q}$ is countable.   □

It is quite useful to also consider the versions of the Addition and Multiplication Rules for countably many sets.

Lemma 20.9 (The Addition Rule for Countably Many Sets) *Suppose that for each $n \in \mathbb{N}$, A_n is a countable set. Then their union*

$$A = \bigcup_{n=1}^{\infty} A_n = A_1 \cup A_2 \cup \cdots$$

is countable.

If at least one of the sets $A_1, A_2, \ldots$ is uncountable, then their union is also uncountable.

Note that, by the axioms of set theory, we know that the infinite union above is indeed a set (cf. Appendix C).

Proof Suppose that f_n is a surjection from $\mathbb{N}$ to A_n. Define the function

$$f : \mathbb{N} \times \mathbb{N} \to A$$

$$(n, m) \mapsto f_n(m)$$

We can easily show that f is a surjection. Indeed, if $a \in A$, then, by definition, there is an $n \in \mathbb{N}$ for which $a \in A_n$. Since f_n is a surjection from $\mathbb{N}$ to A_n, there is an $m \in \mathbb{N}$ for which $f_n(m) = a$, and, therefore, we have $f(n, m) = a$.

By Lemma 20.7, $\mathbb{N} \times \mathbb{N}$ is countable, so there is a surjection g from $\mathbb{N}$ to $\mathbb{N} \times \mathbb{N}$. Then, by Proposition 8.8, $f \circ g$ is a surjection from $\mathbb{N}$ to A, and so A is countable.

The proof that the union will be uncountable if at least one of the sets $A_1, A_2, \ldots$ is uncountable follows immediately from Proposition 20.4.   □

So far, none of our "rules" allows us to create an uncountable set from a countable (finite or countably infinite) set of countable sets. The following rule does just that.

Lemma 20.10 (The Multiplication Rule for Countably Many Sets) *Suppose that for each $n \in \mathbb{N}$, A_n is a set with at least two (distinct) elements. Then their Cartesian product*

$$A = \prod_{n=1}^{\infty} A_n$$

is uncountable.

Recall that we defined infinite direct products in Definition 16.9; in particular, the set A above denotes the set of all sequences $(a_1, a_2, \ldots)$ with $a_i \in A_i$ for $i = 1, 2, \ldots$. We should note again that the axioms of set theory guarantee that this product is indeed a set (cf. Appendix C).

Proof Let us assume indirectly that A is countable; we then have a surjection

$$f : \mathbb{N} \to A = \prod_{n=1}^{\infty} A_n$$

$$k \mapsto f(k) = (f(k)_1, f(k)_2, \ldots)$$

where, for each $k \in \mathbb{N}$ and $n \in \mathbb{N}$, $f(k)_n \in A_n$. In particular, for each $n \in \mathbb{N}$, $f(n)_n \in A_n$.

Since A_n has at least two elements for each n, we can find an element $b_n \in A_n \setminus \{f(n)_n\}$. Now let $\mathbf{b} = (b_1, b_2, \dots)$. Clearly, $\mathbf{b} \in A$; therefore, since f is a surjection, we must have a $k \in \mathbb{N}$ for which $f(k) = \mathbf{b}$. Then, $f(k)_k = b_k$, but this is a contradiction since $f(k)_k \in A_k$ but $b_k \in A_k \setminus \{f(k)_k\}$.   □

We are now ready to prove that $\mathbb{R}$ is uncountable.

Theorem 20.11 *The set $\mathbb{R}$ of all real numbers is uncountable.*

We provide two proofs as they are both beautiful.

Proof I Let us consider the set A of real numbers between 0 and 1 that have only decimal digits 3 or 8; that is,

$$A = \{0.d_1 d_2 \cdots \mid \forall n \in \mathbb{N}, d_n \in D\}$$

where $D = \{3, 8\}$. Since $A \subset \mathbb{R}$, by Proposition 20.4, it is enough to prove that A is uncountable.

It is easy to see that there is a bijection f from A to the set $D \times D \times \cdots$ mapping the real number $0.d_1 d_2 \dots$ to the sequence $(d_1, d_2, \dots)$. So if there were to be a surjection g from $\mathbb{N}$ to A, then $f \circ g$ would be a surjection from $\mathbb{N}$ to $D \times D \times \cdots$, implying that the latter set is countable. This is a contradiction with Lemma 20.10.   □

Proof II Let $f : \mathbb{N} \to \mathbb{R}$ be any function. Then $f(n)$ is a real number for each $n \in \mathbb{N}$, so we can choose an interval I_n of length $\frac{1}{2^n}$ so that $f(n) \in I_n$, thus

$$\mathrm{Im}(f) \subseteq \bigcup_{n=1}^{\infty} I_n.$$

Recall that, by Problem 11 (a) of Chapter 18,

$$\sum_{n=1}^{\infty} \frac{1}{2^n} = 1,$$

so we cannot have

$$\bigcup_{n=1}^{\infty} I_n = \mathbb{R},$$

and, therefore, f cannot be a surjection.   □

Since the set of real numbers is a subset of the set of complex numbers, Theorem 20.11 immediately implies the following:

Corollary 20.12 *The set $\mathbb{C}$ of all complex numbers is uncountable.*

So far, we have classified infinite sets into two categories: those that are countable—for example, $\mathbb{N}$, $\mathbb{Z}$, and $\mathbb{Q}$—and those that are uncountable—such as $\mathbb{R}$ and $\mathbb{C}$. Our stated goal at the beginning of the chapter, however, involved making comparisons by "size": we talked about some sets having the "same size" and some being "larger" than others. In order to make these notions precise, we need to carefully define what we mean by these comparisons.

The approach we follow is due to the German mathematician Georg Cantor (1845–1918) who developed much of the foundations for modern set theory. (As we will soon see, not all of Cantor's efforts were fruitful.)

We already have efficient methods for comparing finite sets. Namely, by Problem 3 of Chapter 8, for finite sets A and B, the inequality $|A| \leq |B|$ is equivalent to both of the following statements:

- there is an injection from A to B;
- there is a surjection from B to A.

Furthermore, the equality $|A| = |B|$ is equivalent to both of the following:

- there is an injection from A to B and an injection from B to A;
- there is a bijection from A to B.

Our goal is to verify that the two pairs of predicates are equivalent for infinite sets A and B as well; by doing so, we can safely use these conditions to *define* when two infinite sets have the "same size" or when one is "larger" than the other.

Our first claim can be established easily.

Theorem 20.13 *Suppose that A and B are sets. Then there is an injection from A to B if, and only if, there is a surjection from B to A.*

Proof Suppose first that there is an injection f from A to B. By Proposition 8.9, f then has a left inverse; that is, there is a function $g : B \to A$ for which $g \circ f = id_A$. But then g has a right inverse, thus g is a surjection using Proposition 8.9 again.

The other direction can be proved similarly. □

Now we turn to the proof of our second claim, which is considerably more challenging. The first proofs for the fact that, given any two sets, having injections from each set to the other set is equivalent to having a bijection between the two sets were given independently by German mathematicians Felix Bernstein (1878–1956) and Ernst Schröder (1841–1902) during the last decade of the nineteenth century. (Bernstein was only 21 years old when his proof appeared.) Since the claim was first conjectured by Cantor, the statement is now usually known as the Cantor–Schröder–Bernstein Theorem.

Theorem 20.14 (The Cantor–Schröder–Bernstein Theorem) *Suppose that A and B are sets. Then there is a bijection from A to B if, and only if, there is both an injection from A to B and an injection from B to A.*

Theorem 20.14 is considered a major milestone in set theory, and there is a variety of different proofs known for it; our proof here is based on Ignace Kolodner's 1967 article in the *American Mathematical Monthly*. We will need the following lemma.

Lemma 20.15 *Let X be a complete lattice for the partial order $\preceq$. Suppose that f is an order-preserving function on X; that is, whenever we have $x_1 \preceq x_2$ for elements $x_1, x_2 \in X$, we also have $f(x_1) \preceq f(x_2)$. Then f has a fixed point; that is, there is an element $s \in X$ for which $f(s) = s$.*

Proof Consider the set

$$S = \{x \in X \mid x \preceq f(x)\}.$$

Note that S is nonempty; the minimum element m of X (which exists by Proposition 17.5) is an element of S since obviously $m \preceq f(m)$.

Since X is a complete lattice, every subset of X must have a supremum. In particular, S must have a supremum s. We will prove that $f(s) = s$.

First we show that $s \preceq f(s)$. Let $x \in S$ be arbitrary; then $x \preceq f(x)$. Since $s = \sup S$, it is also an upper bound for S, and thus we have $x \preceq s$. Since f is order-preserving, $f(x) \preceq f(s)$, and, therefore, by transitivity, we get $x \preceq f(s)$. But x was an arbitrary element of S, so this means that $f(s)$ is an upper bound for S. By the definition of supremum, we get $s \preceq f(s)$ (and, consequently, that $s \in S$).

Next, we show that $f(s) \preceq s$. We just saw that we have $s \preceq f(s)$; therefore, since f is order-preserving, we get $f(s) \preceq f(f(s))$, so $f(s) \in S$. Since $s = \sup S$, we get $f(s) \preceq s$.

Since $\preceq$ is a partial order on X, it must be antisymmetric, so $s \preceq f(s)$ and $f(s) \preceq s$ imply that $f(s) = s$ as claimed. $\square$

Proof of the Cantor–Schröder–Bernstein Theorem One direction is easy: if f is a bijection from A to B, then obviously f is also an injection from A to B; since f^{-1} is a bijection from B to A, it is an injection from B to A.

For the other direction, let $f : A \to B$ and $g : B \to A$ be injections. By Proposition 8.9, g has a left inverse that we denote by g^*. Recall that for a function $\alpha : C \to D$ and a subset X of C, the image $\alpha(X)$ of X is defined as $\alpha(X) = \{\alpha(x) \mid x \in X\}$.

We use the Boolean lattice $(P(A), \subseteq)$ that, as we discussed in Chapter 17, is a complete lattice (cf. page 263). Consider the function

$$\phi : P(A) \to P(A)$$

$$X \mapsto A \setminus g(B \setminus f(X))$$

First we show that ϕ is order-preserving; that is, if X_1 and X_2 are subsets of A with $X_1 \subseteq X_2$, then $\phi(X_1) \subseteq \phi(X_2)$. Indeed, we have

$$X_1 \subseteq X_2$$

$$\Downarrow$$

$$f(X_1) \subseteq f(X_2)$$

$$\Downarrow$$

$$B \setminus f(X_2) \subseteq B \setminus f(X_1)$$

$$\Downarrow$$

$$g(B \setminus f(X_2)) \subseteq g(B \setminus f(X_1))$$

$$\Downarrow$$

$$A \setminus g(B \setminus f(X_1)) \subseteq A \setminus g(B \setminus f(X_2))$$

$$\Downarrow$$

$$\phi(X_1) \subseteq \phi(X_2)$$

Therefore, by Lemma 20.15, we have a subset S of A for which $\phi(S) = S$. (As the proof of Lemma 20.15 shows,

$$S = \bigcup \{X \in P(A) \mid X \subseteq \phi(X)\}$$

is one such set.)

We can now define a bijection from A to B as follows:

$$h : A \to B$$

$$a \mapsto \begin{cases} f(a) & \text{if } a \in S \\ g^*(a) & \text{if } a \notin S \end{cases}$$

The rest of the proof will be the verification that h is a bijection.

Before doing so, we note that for an element a of A, $a \notin S$ holds if, and only if, there is some $b \in B \setminus f(S)$ for which $a = g(b)$. Indeed,

$$S = \phi(S) = A \setminus g(B \setminus f(S)),$$

so for an element $a \in A$, $a \notin S$ is equivalent to $a \in g(B \setminus f(S))$.

To prove that h is injective, let us assume that we have some $a_1, a_2 \in A$ for which $h(a_1) = h(a_2)$. We need to address three possibilities:

i. $a_1 \in S$ and $a_2 \in S$;
ii. $a_1 \notin S$ and $a_2 \in S$; and
iii. $a_1 \notin S$ and $a_2 \notin S$.

In the first case, $h(a_1) = f(a_1)$ and $h(a_2) = f(a_2)$. Since f is an injection, $h(a_1) = h(a_2)$ implies that $a_1 = a_2$.

In the second case, by our note above, we have an element $b \in B \setminus f(S)$ for which $a_1 = g(b)$. Then

$$h(a_1) = g^*(a_1) = g^*(g(b)) = (g^* \circ g)(b) = id_B(b) = b,$$

so $h(a_1) \notin f(S)$. But $a_2 \in S$, so $h(a_2) = f(a_2) \in f(S)$, and thus we cannot have $h(a_1) = h(a_2)$; this case cannot occur.

Finally, in the third case, we have some $b_1, b_2 \in B \setminus f(S)$ for which $a_1 = g(b_1)$ and $a_2 = g(b_2)$. Therefore,

$$h(a_1) = g^*(a_1) = g^*(g(b_1)) = (g^* \circ g)(b_1) = id_B(b_1) = b_1$$

and

$$h(a_2) = g^*(a_2) = g^*(g(b_2)) = (g^* \circ g)(b_2) = id_B(b_2) = b_2,$$

so $h(a_1) = h(a_2)$ implies $b_1 = b_2$, from which we get

$$a_1 = g(b_1) = g(b_2) = a_2.$$

This completes the proof that h is an injection.

It remains to be shown that h is a surjection. Let $b \in B$; we need to find an $a \in A$ for which $h(a) = b$. We have two possibilities: $b \in f(S)$ or $b \notin f(S)$.

If $b \in f(S)$, then we have an $a \in S$ for which $b = f(a)$; in this case we have $h(a) = b$.

If $b \notin f(S)$, then $b \in B \setminus f(S)$, so $g(b) \in g(B \setminus f(S))$, and thus

$$g(b) \notin A \setminus g(B \setminus f(S)).$$

But

$$A \setminus g(B \setminus f(S)) = \phi(S) = S,$$

so $g(b) \notin S$. Then, by the definition of h, we have

$$h(g(b)) = g^*(g(b)) = (g^* \circ g)(b) = id_B(b) = b,$$

so, with $a = g(b)$, we have $h(a) = b$. This completes the proof that h is a surjection.

Since h is both an injection and a surjection, it is a bijection, and we are done.
$\square$

We are now ready to address the issue of comparing sets by their "sizes." We start by introducing the following relation on sets.

Definition 20.16 *Suppose that A and B are sets. We say that A and B are* equinumerous *if there is a bijection from A to B; this will be denoted by A ≈ B. If A and B are not equinumerous, we write A ≉ B.*

As our term suggests, we have the following:

Proposition 20.17 *The relation of equinumerosity is an equivalence relation on* $P(U)$, *the set of all subsets of a (universal) set U. In particular, the following properties hold.*

1. *For any subset A of U, we have A ≈ A.*
2. *If A and B are two subsets of U for which A ≈ B, then we also have B ≈ A.*
3. *If A, B, and C are three subsets of U for which A ≈ B and B ≈ C, then we also have A ≈ C.*

Note that the collection of all sets is not a set, so we need a universal set U in Proposition 20.17 (see the Axiom of Separation in Chapter 6). To make our discussion as meaningful as possible, we assume that U is "very large"; namely, that it contains all sets that we discuss. We leave the proof of Proposition 20.17 to Problem 10.

As a consequence of Proposition 20.17 and the Fundamental Theorem of Equivalence Relations, we can partition the power set $P(U)$ of U into equivalence classes: two sets will belong to the same equivalence class if, and only if, they are equinumerous. The equivalence classes determined this way are called *cardinality classes*; we denote the cardinality class of a set A by $[A]$.

The situation for finite subsets of U is easy to understand. Clearly, the only element of $[\emptyset]$ is $\emptyset$. Recall that there is a bijection between two finite sets if, and only if, they have the same size (cf. Problem 3 of Chapter 8). Therefore, for each positive integer n, the n-element subsets of U will form a single cardinality class (a distinct class for each n).

Cardinality classes of infinite sets are considerably more complicated. Let us now examine $[\mathbb{N}]$, the cardinality class of the set of positive integers.

Proposition 20.18 *A set A is equinumerous with* $\mathbb{N}$ *if, and only if, it is countably infinite.*

Proof If $A \approx \mathbb{N}$, then there is a bijection f from A to $\mathbb{N}$. In particular, f is a surjection from A to $\mathbb{N}$, which, according to Theorem 20.13, implies that there is an injection from $\mathbb{N}$ to A, so A is infinite by Proposition 20.1. Furthermore, f^{-1} is a surjection from $\mathbb{N}$ to A, thus A is countable by definition. Therefore, A is countably infinite.

In the other direction, assume that A is countably infinite. Therefore, we have an injection f from $\mathbb{N}$ to A and a surjection g from $\mathbb{N}$ to A. By Theorem 20.13, the latter fact implies that there is also an injection h from A to $\mathbb{N}$. The Cantor–Schröder–Bernstein Theorem then guarantees that we also have a bijection f from A to $\mathbb{N}$, and thus $A \approx \mathbb{N}$. □

According to Proposition 20.18, all countably infinite sets are in the same cardinality class. As the next theorem shows, this is definitely not the case for uncountable sets. The proof is not very difficult, but the claim is both historic and far reaching, so it deserves the "theorem" designation.

Theorem 20.19 *For any set A, there is no surjection from A to $P(A)$. In particular, we have $A \not\approx P(A)$.*

Proof Suppose, indirectly, that there is a surjection f from A to $P(A)$. Let

$$Z = \{a \in A \mid a \notin f(a)\}.$$

Since $Z \in P(A)$ and f is a surjection, we must have an element $z \in A$ for which $f(z) = Z$.

There are now two possibilities: either $z \in Z$ or $z \notin Z$; we can easily see that both of these possibilities lead to a contradiction. Indeed, in the first case, by the definition of Z, we get $z \notin f(z)$, which contradicts $f(z) = Z$; the other case is analogous.   □

As a consequence of Theorem 20.19, we immediately see that there are infinitely many different infinite cardinality classes: for example,

$$[\mathbb{N}], \ [P(\mathbb{N})], \ [P(P(\mathbb{N}))], \ [P(P(P(\mathbb{N})))],$$

and so on, are all distinct. It is interesting to see where the sets $\mathbb{R}$ and $\mathbb{C}$ fall into this sequence or whether, perhaps, they represent other cardinality classes. Our next proposition addresses the case of the reals, leaving the set $\mathbb{C}$ to Problem 12.

Theorem 20.20 *The set of real numbers is equinumerous with the power set of the set of positive integers; that is, $\mathbb{R} \approx P(\mathbb{N})$.*

Proof By Problem 10 of Chapter 8, the interval $[0, 1)$ is equinumerous with $\mathbb{R}$; therefore, since $\approx$ is a transitive relation, it suffices to show that $[0, 1) \approx P(\mathbb{N})$. By the Cantor–Schröder–Bernstein Theorem, this can be reduced to finding injections from $[0, 1)$ to $P(\mathbb{N})$ and from $P(\mathbb{N})$ to $[0, 1)$.

We start by finding an injection from $P(\mathbb{N})$ to $[0, 1)$. As in the proof of Theorem 20.11, we consider the set A of real numbers between 0 and 1 that have only decimal digits 3 or 8; that is,

$$A = \{0.d_1 d_2 \cdots \mid \forall n \in \mathbb{N}, d_n \in \{3, 8\}\}.$$

It is easy to see that there is a bijection from $P(\mathbb{N})$ to A: to each subset S of $\mathbb{N}$, we can assign the real number $0.d_1 d_2 \cdots \in A$ by setting $d_n = 3$ if $n \in S$ and $d_n = 8$ if $n \notin S$. Since $A \subset [0, 1)$, this determines an injection from $P(\mathbb{N})$ to $[0, 1)$.

For the other direction, rather than using decimal representations for real numbers, it is more convenient to use binary representations. According to Theorem 18.24, every real number $a \in [0, 1)$ can be written in the form

$$a = 0.d_1 d_2 \ldots$$

where $d_n \in \{0, 1\}$ for all $n \in \mathbb{N}$; if we exclude the possibility of having an index $n \in \mathbb{N}$ for which $d_k = 1$ for each $k \geq n$, then the representation is unique. In other words, there is a bijection between the interval $[0, 1)$ and the set

$$\Delta = \{(d_1, d_2, \ldots) \mid \forall n \in \mathbb{N}, d_n \in \{0, 1\} \text{ and } \nexists n \in \mathbb{N} \text{ for which } \forall k \geq n, d_k = 1\}.$$

Therefore, it suffices to find an injection f from Δ to $P(\mathbb{N})$. This is simple: for any given $\mathbf{a} = (d_1, d_2, \ldots) \in \Delta$, let

$$f(\mathbf{a}) = \{n \in \mathbb{N} \mid d_n = 1\}.$$

Then f is an injection (but not a bijection!) from Δ to $P(\mathbb{N})$, and thus we are done.   □

Our final topic of this chapter is the introduction of *cardinalities* (or *cardinal numbers*)—one of Cantor's ingenious notions. Here we will not provide a formal definition for cardinalities; instead, we treat it as a fundamental concept and just say that we can assign to each cardinality class its own cardinality. Analogously to the sizes of finite sets, cardinalities will indicate how "large" the sets in the various cardinality classes are. The cardinality of a set A (which is then shared by any set in the cardinality class of A) will be denoted by $\|A\|$.

In order to be consistent with the notion of size for finite sets, in the case when A is finite, we let $\|A\|$ be simply the size $|A|$ of A. In particular, we write $\|\emptyset\| = 0$ and, if A has n elements ($n \in \mathbb{N}$), then $\|A\| = n$.

For infinite sets, we start by introducing the following notations:

$$\|\mathbb{N}\| = \beth_0,$$

$$\|P(\mathbb{N})\| = \beth_1,$$

$$\|P(P(\mathbb{N}))\| = \beth_2,$$

$$\|P(P(P(\mathbb{N})))\| = \beth_3,$$

and so on; the cardinality $\beth_n$ can be defined recursively for all $n \in \mathbb{N}$. (Here $\beth$, pronounced "beth," is the second letter of the Hebrew alphabet.) As we have seen, these cardinalities are all distinct.

We should note that there are (many) cardinalities besides the ones we listed here. For example, we can set T to be

$$T = \mathbb{N} \cup P(\mathbb{N}) \cup P(P(\mathbb{N})) \cup \cdots ;$$

it is now easy to see that $||T||$, denoted usually by $\beth_\omega$, is different from all previous cardinalities. And, we don't need to stop with $\beth_\omega$: we can consider $||P(T)|| = \beth_{\omega+1}$, $||P(P(T))|| = \beth_{\omega+2}$, and so on. (Here the indices ω, $\omega + 1$, etc. are *ordinal numbers*—see Problem 14.)

By Propositions 20.6, 20.8, and 20.18, we have $||\mathbb{Z}|| = ||\mathbb{Q}|| = \beth_0$. The traditional notation for the cardinality of the set of real numbers is $||\mathbb{R}|| = \mathbf{c}$, abbreviating the word "continuum" that is sometimes used for the points on the real line. According to Theorem 20.20, we have $\mathbf{c} = \beth_1$. As we mentioned before, the set of real numbers is larger than the set of rationals; we now define precisely what we mean by this.

Definition 20.21 *Let A and B be sets. We say that the cardinality of A is* less than or equal to *the cardinality of B, if there is an injection from A to B; this will be denoted by $||A|| \leq ||B||$. Equivalently, we say that the cardinality of B is* greater than or equal to *the cardinality of A, and write $||B|| \geq ||A||$.*

Furthermore, we say that the cardinality of A is less than *the cardinality of B, if $||A|| \leq ||B||$ but $||A|| \neq ||B||$; in other words, if there is an injection from A to B but there is no bijection from A to B. This will be denoted by $||A|| < ||B||$. Equivalently, we say that the cardinality of B is* greater than *the cardinality of A, and write $||B|| > ||A||$.*

Before anything else, it is important to note that Definition 20.21 is logically correct: since we are comparing cardinalities and not sets, we need to verify that the existence of the required injection (or the non-existence of the relevant bijection) does not depend on which representatives of the cardinality classes we picked. The easy proof is left for Problem 10.

We should also verify that our usage of the terms "less than or equal to" and "less than" are in line with our expectations for such terminology. Namely, we want to prove that if Ψ is the set of cardinalities of the sets in a (universal) set of sets U, then the $\leq$ relation in Definition 20.21 defines a partial order on Ψ. The proof of this is left for Problem 10.

In Theorem 20.19, we proved that there is never a bijection from a set to its power set. At the same time, an injection clearly exists: for example, we can map each element a of the set A to $\{a\} \in P(A)$. Therefore, for any set A we have $||A|| < ||P(A)||$. Thus, among the cardinalities we have mentioned, we have the total order

$$0 < 1 < 2 < 3 < \cdots < \beth_0 < \beth_1 < \beth_2 < \beth_3 < \cdots < \beth_\omega < \beth_{\omega+1} < \beth_{\omega+2} < \cdots .$$

According to Proposition 20.1, for any infinite set A, there is an injection from $\mathbb{N}$ to A. Furthermore, by Proposition 20.18, for any countably infinite set A, there is a bijection from A to $\mathbb{N}$, but if A is uncountable, no such bijection exists. Using our new terminology, we can state these as follows:

Proposition 20.22 *For any infinite cardinality α, we have $\beth_0 \leq \alpha$, and for any uncountable cardinality α, we have $\beth_0 < \alpha$.*

Therefore, it makes sense to define $\aleph_0$ to be the smallest infinite cardinality; by Proposition 20.22, we have $\aleph_0 = \beth_0$. (Here $\aleph$, pronounced "aleph," is the first letter of the Hebrew alphabet.)

It is also natural to wonder if there is a smallest uncountable cardinality. This is indeed the case, but its proof is beyond our scope here. We state the following results without proof.

Theorem 20.23 *The partial order relation defined on a set Ψ of cardinalities is a total order; that is, for any two cardinalities $\alpha, \beta \in \Psi$, we have exactly one of the following:*

$$\alpha < \beta, \quad \alpha = \beta, \quad \alpha > \beta.$$

By Theorem 20.23, any cardinality α can be inserted into our chain of cardinalities above. The following theorem states an even stronger result.

Theorem 20.24 *The partial order relation defined on a set of cardinalities Ψ is a well-order; that is, for any nonempty $\Xi \subseteq \Psi$, we have a cardinality $\alpha \in \Xi$ that is less than or equal to any cardinality in Ξ.*

Therefore, among all cardinalities that are larger than $\aleph_0$, there is a smallest cardinality; this cardinality is denoted by $\aleph_1$. By Proposition 20.22, we can say that $\aleph_1$ is the smallest uncountable cardinality. At this point, one may conjecture the following:

Axiom 20.25 (The Continuum Hypothesis) $\aleph_1 = \beth_1$; *that is, there is no set A for which $||\mathbb{N}|| < ||A|| < ||\mathbb{R}||$.*

Georg Cantor dedicated enormous efforts to trying to prove the Continuum Hypothesis—with no success. (Sadly, this contributed to a debilitating mental illness during much of the last part of his life.) In 1900, the German mathematician David Hilbert stated a famous list of twenty-three open questions ranging over most branches of mathematics. The first problem on his list was the Continuum Hypothesis (cf. page 56).

By now we know that the task of proving the Continuum Hypothesis was to be hopeless. In 1938, Kurt Gödel proved that the Continuum Hypothesis is consistent with the usual axioms of set theory; that is, it can be added without a contradiction. In 1963 Paul Cohen showed that the negation of the Continuum Hypothesis is also consistent with the axioms of set theory; that is:

Theorem 20.26 *The Continuum Hypothesis is independent from the usual axioms of set theory.*

By Theorem 20.24 we can also define $\aleph_2$ to be the smallest cardinality that is greater than $\aleph_1$, $\aleph_3$ to be the smallest cardinality that is greater than $\aleph_2$, and so on. We can then extend the Continuum Hypothesis as follows:

Axiom 20.27 (The Generalized Continuum Hypothesis) $\aleph_n = \beth_n$ *for all* $n \in \mathbb{N}$.

(Even more generally, the axiom is usually stated to say that the above equation holds for all *ordinals* n; for example, $\aleph_\omega = \beth_\omega$, etc.; cf. Problem 14.)

At the present time, it is unclear how the Continuum Hypothesis and the Generalized Continuum Hypothesis are to be viewed by mathematicians. While many accept them just like any of the other axioms of set theory, others believe that it's better to assume their negations. And, of course, there are also those who argue for a middle ground: let's develop two theories, one based on the acceptance and one on the denial of the hypotheses. (This two-fold approach is usually followed when it comes to the Parallel Postulate.)

In closing, we point out what has evidently become an unfortunate irony in the hope of laying a solid foundation of abstract mathematics. Toward the end of the nineteenth century, many mathematicians looked to the newly developing branches of logic and set theory to cement the foundations of much of, even all of, mathematics. For example, David Hilbert, the prolific and highly influential giant of the era, was hoping for and advocated such efforts. But today, at the beginning of the twenty-first century, after numerous "disappointing" results by Gödel, Cohen, and others, it seems that not only would we have to give up on the effectiveness of such axiomatic systems, but we also have to—at least for a while—live with the fact that there may be more than one legitimate foundation. Thus, set theory not only failed at uniting and synchronizing all of mathematics, it actually caused vast divisions. We will see what the future holds for us in this regard—so much for the "absolute truth" in mathematics!

Problems

1. Let $P(X, Y)$ denote the predicate that there exists an injection from the set X to the set Y, and let $Q(X, Y)$ denote the predicate that there exists a surjection from the set X to the set Y. Furthermore, for a positive integer n, let I_n denote the set $\{1, \ldots, n\}$. Using these and other standard notations, below we form a variety of statements about a certain non-empty set A. For each such statement, write another statement that is equivalent to it and that is in the format $||A|| * \alpha$, where α is a specific cardinality and $*$ is one of the following relation signs: $=$, $\leq$, $\geq$, $<$, or $>$. (For example, if the statement holds for all sets A, you may answer $||A|| \geq 1$; if it holds for no set A, you may write $||A|| < 1$.)

   (a) $\exists n \in \mathbb{N},\ P(I_n, A)$
   (b) $\exists n \in \mathbb{N},\ Q(I_n, A)$
   (c) $\exists n \in \mathbb{N},\ P(A, I_n)$
   (d) $\exists n \in \mathbb{N},\ Q(A, I_n)$
   (e) $\forall n \in \mathbb{N},\ P(I_n, A)$
   (f) $\forall n \in \mathbb{N},\ Q(I_n, A)$
   (g) $\forall n \in \mathbb{N},\ P(A, I_n)$

(h) $\forall n \in \mathbb{N}, Q(A, I_n)$

(i) $P(\mathbb{N}, A)$

(j) $Q(\mathbb{N}, A)$

(k) $P(A, \mathbb{N})$

(l) $Q(A, \mathbb{N})$

2. In Lemma 20.9, we proved that if sets A and B can be listed, then so can $A \times B$. Our proof took advantage of the fact that the function $f : \mathbb{N} \times \mathbb{N} \to \mathbb{N}$, defined by

$$f(m, n) = 2^{m-1}(2n - 1),$$

is a bijection. (Here, as customary, we simply write $f(m, n)$ instead of $f((m, n))$.) Given the lists $(a_1, a_2, \ldots)$ and $(b_1, b_2, \ldots)$ for (infinite) sets A and B, respectively, our method results in the list

$$((a_1, b_1), (a_2, b_1), (a_1, b_2), (a_3, b_1), (a_1, b_3), (a_2, b_2), \ldots)$$

for $A \times B$. We can visualize this list with the help of Table 20.2.

The fact that f above is a bijection can be rephrased to say that each positive integer appears exactly once in the table.

Of course, there are other ways to list $A \times B$—in fact, it can be shown that the set of all lists is uncountable. In this problem, we examine two other lists, given by their schematic tables.

(a) Find the bijection corresponding to Table 20.3 explicitly.
    (Hint: Use Problem 4 (a) of Chapter 12.)

Table 20.2 The bijection from $\mathbb{N} \times \mathbb{N}$ to $\mathbb{N}$, given by $(m, n) \mapsto 2^{m-1}(2n - 1)$

	b_1	b_2	b_3	b_4	b_5	...
a_1	1	3	5	7	9	...
a_2	2	6	10	14	18	...
a_3	4	12	20	28	36	...
a_4	8	24	40	56	72	...
a_5	16	48	80	112	144	...
⋮	⋮	⋮	⋮	⋮	⋮	⋱

Table 20.3 Another bijection from $\mathbb{N} \times \mathbb{N}$ to $\mathbb{N}$

	b_1	b_2	b_3	b_4	b_5	...
a_1	1	2	5	10	17	...
a_2	4	3	6	11	18	...
a_3	9	8	7	12	19	...
a_4	16	15	14	13	20	...
a_5	25	24	23	22	21	...
⋮	⋮	⋮	⋮	⋮	⋮	⋱

Table 20.4 Yet another
bijection from $\mathbb{N} \times \mathbb{N}$ to $\mathbb{N}$

	b_1	b_2	b_3	b_4	b_5	...
a_1	1	2	4	7	11	...
a_2	3	5	8	12	17	...
a_3	6	9	13	18	24	...
a_4	10	14	19	25	32	...
a_5	15	20	26	33	41	...
⋮	⋮	⋮	⋮	⋮	⋮	⋱

    (b) Find the bijection corresponding to Table 20.4 explicitly.
        (Hint: Use Problem 4 (b) of Chapter 12.)

3. Let $A = \{0, 1, 2\}$, and let

$$\mathcal{A} = \prod_{n=1}^{\infty} A = A \times A \times \cdots .$$

Then, by Lemma 20.10, $\mathcal{A}$ is an uncountable set.
Let $\mathcal{B}$ be the set of those infinite sequences in $\mathcal{A}$ in which consecutive terms are
always distinct. Decide whether $\mathcal{B}$ is countable or uncountable.
4. Prove that the Cantor set is uncountable.
   (Hint: Use Problem 13 (b) of Chapter 18.)
5. Let X be an infinite set of nonzero intervals (that is, intervals of positive
lengths).

    (a) Prove that if the intervals in X are pairwise disjoint, then X is countable.
        (Hint: Use the fact that $\mathbb{Q}$ is dense in $\mathbb{R}$; see Problem 9 (a) of Chapter 18.)
    (b) Suppose that no pair of intervals in X is disjoint. Is X necessarily
        countable?
    (c) Suppose that no interval in X contains another. Is X necessarily countable?

6. Recall from Appendix B that we say that a real number a is *algebraic* if there is
a not-identically-zero polynomial f with integer coefficients for which $f(a) =$
0; if a is not algebraic, we say that it is *transcendental*.
Let $\mathcal{A}$ be the set of algebraic numbers and let $\mathcal{T}$ be the set of transcendental
numbers. Use the following lemma to prove Cantor's Theorem B.5 that $\mathcal{A}$ is
countable but $\mathcal{T}$ is uncountable.

Lemma 20.28 *A polynomial with real number coefficients and degree $n \in \mathbb{N}$*
has at most n roots in $\mathbb{R}$.

Remarks It is easy to verify that certain numbers, such as 0, -7, 3.14, $\sqrt{11}$,
and $(\sqrt[3]{11} + \sqrt{7})/5$ are all algebraic. Even though, as this problem asks you
to verify, "almost all" numbers are transcendental, there are very few specific
numbers that are known to be transcendental. We have known for over a century
that e and π are transcendental (the proofs are not easy); more recently, it was
proven that $2^{\sqrt{2}}$, e^{π}, and even

$$e^{\pi\sqrt{163}}$$

are also transcendental. (More on this last number, called Ramanujan's constant, in Chapter 21.) But, at the present time, we do not know whether $e + \pi$, $e \cdot \pi$, and π^e are algebraic or transcendental (most likely they are transcendental).

7. The claims below are all false. Find the mistakes—shall we say, the cardinal mistakes—in their arguments. Be as specific as possible.

(a) **Claim.** Every nonempty set is countable.

   Argument. Since every nonempty finite set is clearly countable, it suffices to prove our claim for infinite sets.

   Let A be an infinite set. Then, by definition, there is an injection $f : \mathbb{N} \to A$. Consider the set

   $$\text{Im}(f) = \{f(1), f(2), \dots\}.$$

   If $\text{Im}(f) = A$, then f is a surjection, and A is countable. Assume then that $\text{Im}(f)$ is a proper subset of A, and let $a \in A \setminus \text{Im}(f)$. Define a function g as follows:

   $$g : \mathbb{N} \to A$$

   $$n \mapsto \begin{cases} a & \text{if } n = 1 \\ f(n-1) & \text{if } n \geq 2 \end{cases}$$

   Note that

   $$\text{Im}(g) = \{g(1), g(2), \dots\} = \{a, f(1), f(2), \dots\}.$$

   If $\text{Im}(g) = A$, then g is a surjection, so A is countable. If $\text{Im}(g)$ is a proper subset of A, then we proceed as above; continuing the process until we eliminate all of A, we finally arrive at a function h from $\mathbb{N}$ to A whose image is all of A, and thus A is countable.

(b) **Claim.** The set $P(\mathbb{N})$ is countable.

   Argument. Let A be the set of all functions from $\mathbb{N}$ to the two-element set $\{0, 1\}$. We first show that $A \approx P(\mathbb{N})$. To see this, consider the function

   $$f : A \to P(\mathbb{N})$$

   $$g \mapsto \{n \in \mathbb{N} \mid g(n) = 1\}$$

   Then it is not hard to see that f is a bijection.

Note that any function g in A can be uniquely identified with its graph

$$G(g) = \{(n, g(n)) \mid n \in \mathbb{N}\}.$$

Now for each $g \in A$, the graph $G(g)$ is a subset of $\mathbb{N} \times \{0, 1\}$. However, by Lemma 20.7, the entire set $\mathbb{N} \times \{0, 1\}$ is countable, so certainly any subset is. Putting it together: we have shown that $P(\mathbb{N}) \approx A$ and that A is countable; therefore, $P(\mathbb{N})$ is also countable.

(c) **Claim.** The set $P(\mathbb{N})$ is countable.

Argument. Let $P = \{2, 3, 5, 7, 11, \dots\}$ be the set of positive primes. Note that, by the Fundamental Theorem of Arithmetic, every positive integer n can be written in the form

$$n = 2^{\alpha_1} \cdot 3^{\alpha_2} \cdot 5^{\alpha_3} \cdots$$

with nonnegative integers $\alpha_1, \alpha_2, \alpha_3, \dots$, and the expression of n in this form is unique.

Using the expression of n above, we can define the function

$$f : \mathbb{N} \to P(\mathbb{N})$$

$$n \mapsto \{\alpha_i \mid \alpha_i \geq 1\}$$

(For example, we have $f(1) = \emptyset$, $f(2) = f(6) = \{1\}$, $f(63) = \{1, 2\}$, and $f(63, 000, 000) = \{1, 2, 6\}$.)

It is easy to see that f is a surjection (though clearly not an injection), thus $P(\mathbb{N})$ is countable.

(d) Recall from Problem 10 of Chapter 4 that in the Plutonian alphabet there are only four letters: A, B, C, and D, and that every finite string containing these letters is a Plutonian word.

Claim. The set of Plutonian words is uncountable.

Argument. Let $S = \{A, B, C, D, \smile\}$ be the set consisting of the four letters in the Plutonian alphabet and the symbol $\smile$. Each Plutonian word can be thought of as an infinite sequence of elements of S where a finite string of the four letters is followed by infinitely many $\smile$s. For example, the word $AABCDA$ can be identified with the infinite sequence $AABCDA \smile\smile\smile$ $\cdots$. Therefore, the set of Plutonian words is essentially the same as the Cartesian product $S \times S \times \cdots$, which, by Lemma 20.10, is uncountable.

(e) **Claim.** The set of real numbers in the interval $[0, 1)$ is countable.

Argument. Write each real number $x \in [0, 1)$ in its binary representation (cf. Theorem 18.24):

$$x = 0.d_1 d_2 \dots$$

where the binary digits (bits) $d_1, d_2, \ldots$ all equal 0 or 1. Note that certain numbers have two such representations; namely, if in the representation of x above, there is a $k \in \mathbb{N}$ for which the k-th bit is 0 and it is followed by infinitely many 1 bits, then x is unchanged if we replace the k-th bit by a 1 and each successive bit by 0. Therefore, we may assume that each real number between 0 and 1 has a binary representation with only finitely many 1 bits.

We can now create a list of all real numbers between 0 and 1, as follows. The list will start with the only real number in $[0, 1)$ with no 1 bit: $0 = 0.00000\ldots$. It is followed by the other number that has no 1 bits beyond the first bit, $\frac{1}{2} = 0.100000\ldots$. Then, we list the two numbers in $[0, 1)$ that have no 1 bits beyond the second bit (and that have not been listed before): $\frac{1}{4} = 0.010000\ldots$ and $\frac{3}{4} = 0.110000\ldots$, and so on. Note that, for each positive integer n, there is only a finite number of binary representations that have no 1 bits beyond the n-th bit; these can obviously be arranged in a finite list. Therefore, proceeding like this for successive values of n creates a list that contains all real numbers with only finitely many 1 bits and, therefore, all real numbers in the interval $[0, 1)$.

8. The aim of this problem is to facilitate a thorough understanding of the proof of the Cantor–Schröder–Bernstein Theorem. In each part below, a pair of sets A and B is given, together with functions $f : A \to B$ and $g : B \to A$. Verify that f and g are injections (but not bijections), then find an explicit description for the key elements of the proof of the theorem: the set S and the function $h : A \to B$. In each case, verify that $\phi(S) = S$ and that h is a bijection.

(a) $A = \mathbb{N}$, $B = \mathbb{N}$; $f : n \mapsto n + 1$, and $g : n \mapsto n + 1$.
   (Hint: S is the set of all odd positive integers.)
(b) $A = (0, 1)$, $B = (0, 1]$; $f : x \mapsto x$, and $g : x \mapsto x/2$.
   (Hint: S is the set of all real numbers between 0 and 1, except for those that have a binary representation with exactly one 1 bit.)

9. Use results obtained in this chapter to prove that the collection of all sets is not a set.

10. In this problem we fulfill promises made earlier in this chapter.

(a) Prove Proposition 20.17.
(b) Prove that Definition 20.21 is logically correct. In particular, prove that if sets A_1, A_2, B_1, and B_2 satisfy $A_1 \approx A_2$ and $B_1 \approx B_2$, then there is an injection from A_1 to B_1 if, and only if, there is an injection from A_2 to B_2.
(a) Suppose that U is a (universal) set of sets, and let Ψ be the set of cardinalities of the sets in U:

$$\Psi = \{\|A\| \mid A \in U\}.$$

Prove that the $\leq$ relation in Definition 20.21 defines a partial order on Ψ; that is, prove that the $\leq$ relation is reflexive, antisymmetric, and transitive.

Fig. 20.1 Nine points that
form many equilateral
triangles

11. For a given cardinality α, we say that a partition Π of the Euclidean plane $\mathbb{R}^2$ is an α-coloring of the plane if the cardinality of Π is $||\Pi|| = \alpha$. For example, a partition of the plane into four quadrants (with each boundary point in a unique part) is a 4-coloring, and a partition into "horizontal" lines is a c-coloring as there are uncountably many such lines.

Furthermore, for a fixed cardinality α, let $P(\alpha)$ be the predicate that every α-coloring of the plane contains a monochromatic triangle (that is, three points, not all on the same line, that are in the same part of the partition); similarly, let $Q(\alpha)$ be the predicate that every α-coloring of the plane contains a monochromatic equilateral triangle, and let $R(\alpha)$ be the predicate that every α-coloring of the plane contains a monochromatic right triangle.

(a) Prove that $P(n)$ is true for every $n \in \mathbb{N}$.

(b) Prove that $Q(2)$ is true.
     (Hint: Consider Figure 20.1.)

     Remark It is an open question whether there are any integers $n \geq 3$ for which $Q(n)$ holds.

(c) Prove that $R(n)$ is true for every $n \in \mathbb{N}$.
     (Hints: $R(1)$ is trivially true. For $n \geq 2$, consider a $(2n-1)$-by-$(2n-1)$ grid, and use Problem 9 of Chapter 11.)

(d) Prove that $P(\aleph_0)$ is true.

     Remarks By a result of Jack Ceder, we know that $Q(\aleph_0)$ is false; that is, it is possible to color each point of the plane with one of $\aleph_0$-many colors without creating a monochromatic equilateral triangle.

     A deep result of Paul Erdős and Péter Komjáth says that $R(\aleph_0)$ holds if, and only if, the Continuum Hypothesis is false.

12. Prove that each of the following sets is equinumerous with $\mathbb{R}$ and, therefore, has the cardinality $c = \beth_1$. Do not use the Continuum Hypothesis. (For parts (f) and (g) recall that X^∞ stands for the set of infinite sequences of elements from X.)

(a) $\mathbb{R} \setminus \mathbb{N}$

(b) $\mathbb{R} \setminus \mathbb{Q}$

(c) $[0, 1] \times [0, 1]$

(d) $\mathbb{C}$

(e) $\mathbb{R}^3$

(f) $\mathbb{N}^\infty$

(g) $\mathbb{R}^\infty$

Remarks According to part (c), one can find a surjection from $\mathbb{R}$ (or from the interval $[0, 1]$) to the square region $[0, 1] \times [0, 1]$. It is very surprising that, in fact, a *continuous* surjection can also be found; that is, it is possible to draw a curve that "fills" the entire two-dimensional region. (By this we mean that there exists a continuous function with domain $[0, 1]$ and image $[0, 1] \times [0, 1]$.) Such functions (which, according to a result of Eugen Netto, cannot be bijections) are called *space filling curves*. The first such curve was discovered by Giuseppe Peano in 1890; a year later, a simpler example was given by David Hilbert. The interactive web page http://www.cs.utexas.edu/users/vbb/misc/sfc/0index.html allows one to generate approximations for Peano's and Hilbert's curves (as well as ones given by others).

13. In this problem we develop the basics of what is known as *cardinal arithmetic*. We make the following definitions.

Definition 20.29 *Let α and β be cardinalities, and suppose that A and B are disjoint sets with $||A|| = \alpha$ and $||B|| = \beta$. Then $\alpha + \beta$ is defined as the cardinality of $A \cup B$.*

We should note that, for any cardinalities α and β, one can always find disjoint sets A and B with $||A|| = \alpha$ and $||B|| = \beta$. Without changing their cardinalities, we can always modify A and B and set $A^* = \{(1, a) \mid a \in A\}$ and $B^* = \{(2, b) \mid b \in B\}$; these sets are then disjoint.

Definition 20.30 *Let α and β be cardinalities, and suppose that A and B are sets with $||A|| = \alpha$ and $||B|| = \beta$. Then $\alpha \cdot \beta$ is defined as the cardinality of $A \times B$.*

Definition 20.31 *Let α and β be cardinalities, and suppose that A and B are sets with $||A|| = \alpha$ and $||B|| = \beta$. Then α^β is defined as the cardinality of the set $\text{Fun}(B \to A)$ of functions from B to A.*

As before, we can easily verify that the three operations are well defined; that is, they do not depend on which sets A and B one uses.

(a) Show that these definitions extend the corresponding notions for finite sets; that is, verify that for finite—in the case of addition, disjoint—sets A and B we have

$$||A|| + ||B|| = |A| + |B|,$$

$$||A|| \cdot ||B|| = |A| \cdot |B|,$$

and

$$||A||^{||B||} = |A|^{|B|}.$$

(b) Explain how the validity of the entries in the following tables follows from previous results.

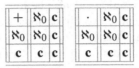

(c) Explain how the identities

$$2^{\beth_m} = \beth_{m+1},$$

$$\aleph_0^{\aleph_0} = \mathbf{c},$$

and

$$\mathbf{c}^{\aleph_0} = \mathbf{c}$$

follow from previous results.

(d) Prove that the addition and multiplication of cardinalities are commutative and associative operations and that multiplication is distributive with respect to addition.

(e) Prove that for arbitrary cardinalities α, β, and γ, we have

   i.

$$\alpha^\beta \cdot \alpha^\gamma = \alpha^{\beta+\gamma};$$

   ii.

$$\alpha^\gamma \cdot \beta^\gamma = (\alpha \cdot \beta)^\gamma;$$

   iii.

$$\left(\alpha^\beta\right)^\gamma = \alpha^{\beta\cdot\gamma}.$$

(f) Prove that, for all nonnegative integers n and m, we have

   i.

$$\beth_n + \beth_m = \beth_{\max\{n,m\}};$$

ii.

$$\beth_n \cdot \beth_m = \beth_{\max\{n,m\}};$$

iii.

$$(\beth_n)^{\beth_m} = \beth_{\max\{n,m+1\}}.$$

Remark More generally, one can prove the following:

Theorem 20.32 *For any two infinite cardinalities α and β, we have*

$$\alpha + \beta = \alpha \cdot \beta = \max\{\alpha, \beta\}.$$

According to Theorem 20.32, addition and multiplication of cardinalities are rather trivial operations; exponentiation, however, without assuming the Generalized Continuum Hypothesis, is considerably more complicated and is still the subject of much study.

14. In this problem we briefly investigate another—as we will see, much more refined—method of measuring the magnitude of sets; namely, we introduce Cantor's concept of *ordinals* or *ordinal numbers*. At a very intuitive level, an ordinal measures how "long" a set is (while a cardinal measures how "large" it is).

Consider, for example, the following well-orders of $\mathbb{N}$:

$$1 < 2 < 3 < 4 < 5 < 6 < \cdots,$$

$$4 \prec 5 \prec 6 \prec 7 \prec \cdots \prec 1 \prec 2 \prec 3,$$

$$1 \prec 3 \prec 5 \prec 7 \prec \cdots \prec 2 \prec 4 \prec 6 \prec 8 \prec \cdots,$$

and

$$1 \prec 2 \prec 4 \prec 8 \prec \cdots \prec 3 \prec 6 \prec 12 \prec 24 \prec \cdots \prec 5 \prec 10 \prec 20 \prec 40 \prec \cdots.$$

As we discussed in Problem 14 of Chapter 8, each of these well-orders of $\mathbb{N}$ is of a different order type—this distinction allows us to differentiate among sets based on their "lengths." Rather than providing a formal definition for ordinals, we say that the order type of a well-ordered set (cf. Definition 17.20) is called an *ordinal number* or *ordinal*, for short.

The ordinal number of a finite set of size n—which, according to Problem 14 of Chapter 8, is unique—is denoted simply by n; the ordinal numbers of the four distinct well-orders for $\mathbb{N}$ above are denoted, in order, by

$$\omega,$$

$$\omega + 3,$$

$$\omega + \omega \text{ (or } \omega \cdot 2),$$

and

$$\omega \cdot \omega \text{ (or } \omega^2),$$

respectively. We explain these notations by introducing arithmetic operations on ordinals, as follows.

Recall from Problem 13 above that for a given pair of sets A and B we introduced $A^* = \{(1, a) \mid a \in A\}$ and $B^* = \{(2, b) \mid b \in B\}$. (Here A^* and B^* are disjoint, and we have $||A^*|| = ||A||$ and $||B^*|| = ||B||$.)

Definition 20.33 *Let κ and λ be ordinals, and suppose that wosets $(A, \preceq_A)$ and $(B, \preceq_B)$ have order types κ and λ, respectively. The* sum $\kappa + \lambda$ *is then defined as the order type of the lexicographic order on $A^* \cup B^*$.*

Note that the lexicographic order on $A^* \cup B^*$ puts all elements of A^* before any of the elements of B^* but keeps the order of the elements in both A^* and B^* unchanged (see Problem 11 of Chapter 6).

Definition 20.34 *Let κ and λ be ordinals, and suppose that wosets $(A, \preceq_A)$ and $(B, \preceq_B)$ have order types κ and λ, respectively. The* product $\kappa \cdot \lambda$ *is then defined as the order type of the co-lexicographic order on $A \times B$.*

Recall that, in co-lexicographic order, the element $(a, b) \in A \times B$ comes before an element $(a', b') \in A \times B$ if, and only if, $b \prec_B b'$ or $b = b'$ and $a \preceq_A a'$. We then see that the co-lexicographic order on $A \times B$ starts with the set of elements

$$\{(a, b_1) \mid a \in A\}$$

arranged in order according to $\preceq_A$ (here $b_1 = \min B$); this is followed by the elements of

$$\{(a, b_2) \mid a \in A\}$$

in similar order (here $b_2 = \min B \setminus \{b_1\}$), etc.

Next, we define exponentiation, but only in the case in which the exponent is a finite ordinal.

Definition 20.35 *Let n be a positive integer and κ be any ordinal; we define κ^n recursively by $\kappa^1 = \kappa$ and $\kappa^n = \kappa^{n-1} \cdot \kappa$ for $n \geq 2$.*

The definition of ordinal exponentiation for non-finite exponents is considerably more complicated and will not be given here.

As usual, we need to verify that the operations do not depend on which sets A and B one uses—this can, indeed, be accomplished easily; we skip the details.

(a) Verify that, given Cantor's notation ω for the order type of $\mathbb{N}$ with the usual $\leq$ order, the other well-orders on $\mathbb{N}$ above, denoted by $\omega+3, \omega+\omega, \omega\cdot2, \omega\cdot\omega$, and ω^2, are consistent with the operation of addition and multiplication.

(b) Prove that for all $n \in \mathbb{N}$ we have

$$n + \omega = \omega \neq \omega + n$$

and

$$n \cdot \omega = \omega \neq \omega \cdot n.$$

(c) As we have just seen, addition and multiplication are not commutative. (However, they are associative.) Does any form of distributivity hold?

(d) Given an infinite sequence of ordinal numbers $(\kappa_1, \kappa_2, \dots)$, provide a definition for the sum

$$\sum_{n=1}^{\infty} \kappa_n = \kappa_1 + \kappa_2 + \cdots .$$

Verify that your definition gives

$$\omega + \omega + \omega + \cdots = \omega^2.$$

(e) Provide an explicit well order on $\mathbb{N}$ that has order type

$$\omega + \omega^2 + \omega^3 + \cdots .$$

Remark The ordinal number corresponding to this order type is denoted by ω^ω.

Chapter 21
Number Systems Systematically

Throughout this book, we frequently considered our familiar sets of numbers:

- $\mathbb{N}$: the set of natural numbers;
- $\mathbb{Z}$: the set of integers;
- $\mathbb{Q}$: the set of rational numbers;
- $\mathbb{R}$: the set of real numbers; and
- $\mathbb{C}$: the set of complex numbers.

We have already seen many of their distinguishing attributes, including some elaborate and far-reaching properties. But we have never, actually, defined these number sets; up till now, we treated them as primitives (cf. page 15). In this chapter we discuss how one can "build up" these number sets in the order above; we will also explain why, in a certain sense, $\mathbb{C}$ cannot be—and needn't be—enlarged further.

In Chapter 20, we studied these sets from the viewpoint of set theory: we focused only on their cardinalities and not on their algebraic or analytical properties. In this chapter, we look deeper into the structure of these number sets and discuss such features as their distinguished elements (e.g., 0 or 1), specific operations (e.g., addition, multiplication), and relations (e.g., order). A number set together with some specific additional structure is referred to as a *number system*.

Before we proceed, we should explain why we would want to "build" these—already very familiar—number systems and, furthermore, why we waited until almost the end of this book to do so.

There are at least two reasons for studying the development of number systems. Firstly, numbers, of course, are at the heart of mathematics, and thus it is interesting to see how one can "create" them precisely using as few primitives and axioms as possible. Secondly, with a constructive approach, we verify that a certain model for these number systems indeed exists. For example, without providing a legitimate model for the set of real numbers (using the previously constructed model of the rationals), a statement such as "the square root of two is irrational" would need to be rephrased to say that, "if the square root of two exists at all, then it cannot be a

© Springer Nature Switzerland AG 2020

B. Bajnok, *An Invitation to Abstract Mathematics*, Undergraduate Texts in Mathematics, https://doi.org/10.1007/978-3-030-56174-1_21

rational number." Of course, $\sqrt{2}$ can be uniquely characterized as the positive real number whose square equals 2, but wishing for such a positive number to exist is not the same as creating one using previously constructed numbers (in this case, the rational numbers).

As for having to wait until now to study number systems, we will soon see that a precise treatment is quite involved, using many of the previously studied concepts. Therefore, in a way, this chapter can be thought of as a unifying capstone for much of the material discussed previously in this book. Some of the difficulty in this chapter comes from the fact that, when constructing a number system, we need to avoid the mistake of assuming, without proof, that the numbers in question exist or that they have the familiar properties.

We should emphasize that our constructions will be highly abstract; while it is certainly pleasing to see that it is possible to construct each system from, literally, nothing more than the empty-set (and properties of logic and set theory), a drawback is that the constructions yield virtually unrecognizable systems. Thus, we need to admit that this development, although of high theoretical importance, does not produce practical applications.

So, if our constructions are so abstract, how can we claim that the number systems that we create are the ones we want? Even if we check that they satisfy some of the properties that we expect of the systems, can we be sure that they are *the* number systems we have always been working with?

For each system, we address this issue with the following two-fold approach. First, we assemble a collection of axiomatic properties that are *categorical* of the number system in question; that is, we exhibit a list of characteristic properties that, collectively, identify the system "up to isomorphism" (we will describe this precisely below). Second, we construct an explicit model for each number system; this then also proves that the axiomatic properties are *consistent*.

For example, in the case of the real number system, our axiomatic properties can be summarized by saying that the system is a *complete ordered field*; in other words, any complete ordered field is *isomorphic* to (essentially the same as) the real number system. Then, we construct an explicit model for a complete ordered field; the resulting model of this construction can then legitimately be called *the* real number system.

We carry out our development following the order of containment:

$$\mathbb{N} \subset \mathbb{Z} \subset \mathbb{Q} \subset \mathbb{R} \subset \mathbb{C}.$$

The construction of the natural number system will use only the concepts and axioms of set theory (cf. Appendix C); the construction of each subsequent system will be based on the one preceding it in the list above. Since we insist on a constructive approach, extending a number system to a larger one cannot use "imaginary" elements. For example, moving from $\mathbb{Z}$ to $\mathbb{Q}$, we cannot just "add" the fractions to the already constructed set of integers, since, at that point, we do not know what "fractions" are. We circumvent this difficulty by thinking of rational numbers as equivalence classes of ordered pairs of integers. Therefore, technically, $\mathbb{Q}$ will not contain $\mathbb{Z}$; instead, we construct $\mathbb{Q}$ in such a way that it contains a set $\mathbb{Z}'$ that is isomorphic to $\mathbb{Z}$. Similarly, for each of our number systems, except for the natural numbers, we identify a *subsystem* contained in it that is isomorphic to the previously constructed system.

We will clarify all this below; however, since a precise and thorough development would be rather long and tedious, we will sometimes omit the details. (Some of the notable steps will be left to the problems at the end of the chapter.) Our construction of the natural numbers will be particularly delicate and laborious. Paradoxically—but perhaps not surprisingly—the most fundamental system requires the most advanced techniques.

So let us get started with our first number system: the natural numbers. Our first task is to identify the axioms that categorically determine the set of natural numbers.

At the most fundamental level, our notion of the natural numbers is simply a list of distinct elements, with a first element, a second element, a third element, and so on, indefinitely. It turns out that, once we make this concept more precise, we arrive at a categorical set of axioms for the natural number system.

Let N be an arbitrary set. (To make it clear that we don't yet want to rely on the set of natural numbers $\mathbb{N}$, we chose a slightly different notation.) The fact that the elements of N form a list that continues "indefinitely" can be expressed by saying that each element will have a *successor* in the list; we denote the successor of an element $n \in N$ by $S(n)$. (It is important to note that, in spite of the suggestive term, we do not attach additional meaning to the successor function other than the fact that it is a function on N; later, we will *define* $n + 1$ to be $S(n)$.) Furthermore, our list of natural numbers has an initial element, and this element is denoted by 1. Thus, our natural number system involves a set N, a function $S : N \to N$, and a special element $1 \in N$; we now attempt to collect the properties of the system $(N, 1, S)$ that categorically identify the natural number system.

The first such property is that 1 is not the successor of any other element in the list.

Axiom 21.1 (P1) *There is no $n \in N$ for which $S(n) = 1$.*

This property alone is not yet sufficient for fully characterizing the kind of list we have in mind. For example, it does not exclude the possibility that our list is "periodic" and contains only finitely many distinct elements such as the list $1, 2, 3, 2, 3, 2, 3, \ldots$ does. It is not even necessarily the case that we arrive at a list: for example, we may have two or more lists merging into one. The following property aims to forbid such situations.

Axiom 21.2 (P2) *The successor function is injective; that is, for any pair of distinct elements $m \in N$ and $n \in N$, we have $S(m) \neq S(n)$.*

These two properties go a long way toward describing our notion of N, but they allow for sets much larger than what we need. The next property expresses the fact that N has no additional elements besides the ones in a single list. To make our statement more succinct, let us introduce the term *1-inductive subset of N* for a subset that contains 1 and has the property that, whenever it contains an element $n \in N$, it also contains its successor $S(n)$. We can now state our final property as follows:

Axiom 21.3 (P3) *The only 1-inductive subset of N is N itself.*

We have already seen axiom (P3) in Chapter 10 as the Induction Axiom.

These three axioms collectively are called the *Peano axioms*, named after the Italian mathematician Giuseppe Peano (1858–1932) who, at the beginning of the twentieth century, developed and studied them. All further concepts and properties of the natural numbers, including their addition and multiplication, can be built up from the Peano axioms. We will see some of this development shortly.

The Peano axioms uniquely characterize N (and thus $\mathbb{N}$). What we mean by this is the following. For any set N, particular element $1 \in N$, and function $S : N \to N$ for which the Peano axioms above hold, the triple $(N, 1, S)$ is called a *Peano system*.

We say that the Peano systems $(N, 1, S)$ and $(N', 1', S')$ are *isomorphic* if there is a bijection $f : N \to N'$ with the following two properties:

- $f(1) = 1'$, and
- $f(S(n)) = S'(f(n))$ holds for all $n \in N$.

In other words, $(N, 1, S)$ and $(N', 1', S')$ are isomorphic if there is a one-to-one correspondence between N and N' where 1 and $1'$ correspond to each other, as do $S(1)$ and $S'(1')$, $S(2)$ and $S'(2')$, and so on. Given these definitions, one can prove that any two Peano systems are isomorphic; that is, the Peano axioms are categorical. (We omit the details.)

Our next task is to construct a Peano system. Since the Peano axioms are categorical, all Peano systems are essentially the same. Therefore, it is legitimate to identify $\mathbb{N}$ with the set we are about to construct.

Our construction will yield a list of sets, all of which are sets of sets. Our initial set is the set containing the empty-set, $\{\emptyset\}$. The next set on our list will be the union of this set and the set containing it, that is,

$$\{\emptyset\} \cup \{\{\emptyset\}\} = \{\emptyset, \{\emptyset\}\}.$$

Then, the next set on our list will be the union of this set and the set containing it:

$$\{\emptyset, \{\emptyset\}\} \cup \{\{\emptyset, \{\emptyset\}\}\} = \{\emptyset, \{\emptyset\}, \{\emptyset, \{\emptyset\}\}\}.$$

We continue to form our list of sets in the same fashion: the set following a set A in our sequence will be the set $A \cup \{A\}$. For example, the fourth and fifth sets on our list will be

$$\{\emptyset, \{\emptyset\}, \{\emptyset, \{\emptyset\}\}, \{\emptyset, \{\emptyset\}, \{\emptyset, \{\emptyset\}\}\}\}$$

and

$$\{\emptyset, \{\emptyset\}, \{\emptyset, \{\emptyset\}\}, \{\emptyset, \{\emptyset\}, \{\emptyset, \{\emptyset\}\}\}, \{\emptyset, \{\emptyset\}, \{\emptyset, \{\emptyset\}\}, \{\emptyset, \{\emptyset\}, \{\emptyset, \{\emptyset\}\}\}\}\}.$$

We can introduce notations for the sets we just constructed as follows:

$1_N = \{\emptyset\}$,

$2_N = \{\emptyset, \{\emptyset\}\}$,

$3_N = \{\emptyset, \{\emptyset\}, \{\emptyset, \{\emptyset\}\}\}$,

$4_N = \{\emptyset, \{\emptyset\}, \{\emptyset, \{\emptyset\}\}, \{\emptyset, \{\emptyset\}, \{\emptyset, \{\emptyset\}\}\}\}$,

$5_N = \{\emptyset, \{\emptyset\}, \{\emptyset, \{\emptyset\}\}, \{\emptyset, \{\emptyset\}, \{\emptyset, \{\emptyset\}\}\}, \{\emptyset, \{\emptyset\}, \{\emptyset, \{\emptyset\}\}, \{\emptyset, \{\emptyset\}, \{\emptyset, \{\emptyset\}\}\}\}\}$,

and so on; more concisely, we may write

$$1_N = \{\emptyset\},$$
$$2_N = \{\emptyset, 1_N\},$$
$$3_N = \{\emptyset, 1_N, 2_N\},$$
$$4_N = \{\emptyset, 1_N, 2_N, 3_N\},$$
$$5_N = \{\emptyset, 1_N, 2_N, 3_N, 4_N\}.$$

We note in passing that, following Cantor, we can differentiate between ordinal numbers (denoting positions in a list) and cardinal numbers (denoting sizes of sets). Our construction for the natural numbers has the convenient property that the ordinal numbers correspond to the appropriate cardinal numbers: for example, the fifth natural number, 5_N, is a set of size five.

Now we want to say that the set of natural numbers is

$$\mathbb{N} = \{1_N, 2_N, 3_N, \dots\}$$

but, unfortunately, the meaning of the "..." cannot be made precise without relying on some form of induction and, at this point, we do not have that tool yet. We can instead proceed as follows.

We say that a set of sets $\mathcal{A}$ is $\{\emptyset\}$-*inductive* if $\{\emptyset\} \in \mathcal{A}$ and it has the property that whenever $A \in \mathcal{A}$ for some set A, we also have $A \cup \{A\} \in \mathcal{A}$. Heuristically, we see that, by definition, every $\{\emptyset\}$-inductive set must contain $\{\emptyset\}$, $\{\emptyset, \{\emptyset\}\}$, $\{\emptyset, \{\emptyset\}, \{\emptyset, \{\emptyset\}\}\}$, and so on (while, perhaps, containing other elements not at all in this form). This leads us to the following definition.

Definition 21.4 A natural number *is a set that is an element of every $\{\emptyset\}$-inductive set. The set of natural numbers is denoted by* $\mathbb{N}$.

Admittedly, Definition 21.4 is considerably more complicated and less explicit than one would have hoped. We cannot even make the somewhat simpler declaration that the set of natural numbers is formed by the intersection of the collection of all $\{\emptyset\}$-inductive sets: the collection of all $\{\emptyset\}$-inductive sets is not a set (there are too many $\{\emptyset\}$-inductive sets), and, therefore, we are unable to rely on their intersection set!

Since the definition of a set $\mathcal{A}$ being $\{\emptyset\}$-inductive explicitly requires that $\{\emptyset\}$ be an element of $\mathcal{A}$, $\{\emptyset\}$ must be in every $\{\emptyset\}$-inductive set; so, by Definition 21.4, $\{\emptyset\}$ is a natural number. Thus $1_{\mathbb{N}} \in \mathbb{N}$. Furthermore, if a natural number $n_{\mathbb{N}}$ is in every $\{\emptyset\}$-inductive set, then so is $n_{\mathbb{N}} \cup \{n_{\mathbb{N}}\}$. Therefore, the function $S_{\mathbb{N}}$, which assigns $n_{\mathbb{N}} \cup \{n_{\mathbb{N}}\}$ to $n_{\mathbb{N}}$, is a map from $\mathbb{N}$ to $\mathbb{N}$.

Next, we prove that $(\mathbb{N}, 1_{\mathbb{N}}, S_{\mathbb{N}})$ is a Peano system. It will be helpful to introduce a term for the inverse of the successor function: we say that $m_{\mathbb{N}} \in \mathbb{N}$ is a *predecessor* of $n_{\mathbb{N}}$ if $S_{\mathbb{N}}(m_{\mathbb{N}}) = n_{\mathbb{N}}$. As we are about to prove, every natural number, other than $1_{\mathbb{N}}$, has a unique predecessor, and the predecessor function is indeed the inverse of the successor function.

Theorem 21.5 *With our definitions and notations as above, $(\mathbb{N}, 1_{\mathbb{N}}, S_{\mathbb{N}})$ is a Peano system. Furthermore, if $n_{\mathbb{N}} \in \mathbb{N}$ and $n_{\mathbb{N}} \neq 1_{\mathbb{N}}$, then $n_{\mathbb{N}}$ has a unique predecessor.*

Proof We start by proving axiom (P2). Suppose that we have distinct natural numbers $m_{\mathbb{N}}$ and $n_{\mathbb{N}}$ for which $S_{\mathbb{N}}(m_{\mathbb{N}}) = S_{\mathbb{N}}(n_{\mathbb{N}})$; that is,

$$m_{\mathbb{N}} \cup \{m_{\mathbb{N}}\} = n_{\mathbb{N}} \cup \{n_{\mathbb{N}}\}.$$

Clearly, $m_{\mathbb{N}} \in \{m_{\mathbb{N}}\}$, thus $m_{\mathbb{N}}$ is an element of the left-hand side. Therefore, it must be an element of the right-hand side as well; since $m_{\mathbb{N}} \notin \{n_{\mathbb{N}}\}$, we must have $m_{\mathbb{N}} \in n_{\mathbb{N}}$. Similarly, we get $n_{\mathbb{N}} \in m_{\mathbb{N}}$. Now consider the set $\{m_{\mathbb{N}}, n_{\mathbb{N}}\}$. Since $m_{\mathbb{N}} \in n_{\mathbb{N}}$ and $n_{\mathbb{N}} \in m_{\mathbb{N}}$, neither $m_{\mathbb{N}}$ nor $n_{\mathbb{N}}$ is disjoint from $\{m_{\mathbb{N}}, n_{\mathbb{N}}\}$, contradicting the Axiom of Regularity (cf. page 138).

Next, we prove axiom (P3). Let M be a subset of $\mathbb{N}$ with the property that $1_{\mathbb{N}} \in M$ and that $n_{\mathbb{N}} \in M$ implies $S_{\mathbb{N}}(n_{\mathbb{N}}) \in M$; we need to prove that $M = \mathbb{N}$. But, since $1_{\mathbb{N}} = \{\emptyset\}$ and $S_{\mathbb{N}}(n_{\mathbb{N}}) = n_{\mathbb{N}} \cup \{n_{\mathbb{N}}\}$, this means that M is $\{\emptyset\}$-inductive. Let $n_{\mathbb{N}} \in \mathbb{N}$ be arbitrary. By definition, $n_{\mathbb{N}}$ is an element of every $\{\emptyset\}$-inductive set; in particular, $n_{\mathbb{N}} \in M$. Therefore, $M = \mathbb{N}$.

We now turn to the proof of our second statement. First we prove the existence. Let P be a set of all elements of $\mathbb{N}$ that have a predecessor, and let $M = \{1_{\mathbb{N}}\} \cup P$. Our claim is equivalent to proving that $M = \mathbb{N}$.

We show that M is a $1_{\mathbb{N}}$-inductive subset of $\mathbb{N}$. Clearly, $1_{\mathbb{N}} \in M$. Let $n_{\mathbb{N}} \in M$. Since $S_{\mathbb{N}}(n_{\mathbb{N}})$ has a predecessor (namely, $n_{\mathbb{N}}$), we have $S_{\mathbb{N}}(n_{\mathbb{N}}) \in P$ and thus $S_{\mathbb{N}}(n_{\mathbb{N}}) \in M$. Therefore, M is a $1_{\mathbb{N}}$-inductive subset of $\mathbb{N}$, so, by (P3), $M = \mathbb{N}$. The uniqueness of the predecessor follows from (P2).

At last, we prove (P1). Assume indirectly that $n_{\mathbb{N}} \cup \{n_{\mathbb{N}}\} = 1_{\mathbb{N}}$ holds for some $n_{\mathbb{N}} \in \mathbb{N}$. But $1_{\mathbb{N}} = \{\emptyset\}$ contains only one element, so we must have $n_{\mathbb{N}} = \emptyset$. But then $n_{\mathbb{N}} \neq 1_{\mathbb{N}}$, so by the statement we just proved, $n_{\mathbb{N}}$ has a (unique) predecessor $m_{\mathbb{N}}$. Therefore, $S_{\mathbb{N}}(m_{\mathbb{N}}) = n_{\mathbb{N}}$; that is, $m_{\mathbb{N}} \cup \{m_{\mathbb{N}}\} = \emptyset$, which is a contradiction.

$\square$

We have thus achieved both of our goals about the natural number system that we stated at the beginning of this chapter. We identified its categorical set of axioms as those of a Peano system, and we proved that the system $(\mathbb{N}, 1_{\mathbb{N}}, S_{\mathbb{N}})$ we constructed is a Peano system. Therefore, it is legitimate to think of the natural number system

as a "generic" Peano system; since this is what we do from now on, we will omit
the subscripts and just use $(\mathbb{N}, 1, S)$; we will also let $2 = S(1)$, $3 = S(2)$, etc.

According to Theorem 21.5, we can see that we have bijections $S : \mathbb{N} \to \mathbb{N} \setminus \{1\}$
and $P : \mathbb{N} \setminus \{1\} \to \mathbb{N}$ so that $P \circ S = id_{\mathbb{N}}$ and $S \circ P = id_{\mathbb{N} \setminus \{1\}}$; that is, P and S are
inverses of each other.

We use these important facts to define addition and multiplication in $\mathbb{N}$. We start
with addition.

Definition 21.6 *We define the sum $n_1 + n_2$ of natural numbers n_1 and n_2 as follows:*

$$n_1 + n_2 = \begin{cases} S(n_1) & \text{if } n_2 = 1 \\ S(n_1 + P(n_2)) & \text{if } n_2 \neq 1 \end{cases}$$

We then use addition to define multiplication.

Definition 21.7 *We define the product $n_1 \cdot n_2$ of natural numbers n_1 and n_2 as
follows:*

$$n_1 \cdot n_2 = \begin{cases} n_1 & \text{if } n_2 = 1 \\ (n_1 \cdot P(n_2)) + n_1 & \text{if } n_2 \neq 1 \end{cases}$$

Our definitions are recursive; to be thorough, we would need to prove that they
indeed define binary operations on $\mathbb{N}$ (functions from $\mathbb{N} \times \mathbb{N}$ to $\mathbb{N}$). We omit these
rather technical proofs; instead, we provide an example for how these definitions
yield the expected answer.

Proposition 21.8 *We have $2 \cdot 2 = 4$.*

Proof Using Definition 21.7, we have

$$2 \cdot 2 = (2 \cdot P(2)) + 2;$$

since $P(2) = 1$, this says that

$$2 \cdot 2 = (2 \cdot 1) + 2.$$

Using Definition 21.7 again for $2 \cdot 1$, we get

$$2 \cdot 2 = 2 + 2.$$

Next, we use Definition 21.6 on the right-hand side, which gives us

$$2 \cdot 2 = S(2 + P(2));$$

which simplifies to

$$2 \cdot 2 = S(2 + 1).$$

We use Definition 21.6 once more, to get

$$2 \cdot 2 = S(S(2)).$$

But $S(2) = 3$ and $S(3) = 4$, which proves our claim.   □

It is important to point out that in the proof of Proposition 21.8 we did not use associativity, commutativity, or distributivity. These properties, however, can be proven:

Proposition 21.9 *Both the addition and multiplication operations of natural numbers, defined above, are associative and commutative, and multiplication is distributive with respect to addition. Furthermore,* 1 *is a multiplicative identity in* $\mathbb{N}$.

We assign the proof in Problem 2.

Let us now move from the natural numbers to the integers. Before we proceed, it is worth mentioning the historical fact that the integers appeared relatively late in the development of mathematics and civilization at large. Long after the acceptance of fractions and even irrational numbers such as $\sqrt{2}$, e, and π, people still had difficulty accepting the notion of zero and of a negative number. One can, indeed, easily circumvent expressions such as "minus 10 degrees Celsius" by saying instead "10 degrees below freezing." The main advantage of the introduction of negative integers is that, this way, the operation of subtraction is closed: for any $a \in \mathbb{Z}$ and $b \in \mathbb{Z}$, we have $a - b \in \mathbb{Z}$. (Thus, $\mathbb{Z}$ is a group for addition.) Of course, the integer system has many other convenient properties; our first goal is to determine those that constitute its categorical axioms.

The most important property of the integers, as we discussed in Chapter 13, is that they form an integral domain for their usual operations of addition and multiplication; that is, the system $(\mathbb{Z}, +, \cdot)$ is a commutative ring with identity that satisfies the nonzero product property (cf. Definition 13.1). Furthermore, recall that the integers also form an ordered integral domain; that is, there is a subset $P \subset \mathbb{Z}$ that is closed for addition and multiplication, and for which for every integer a, exactly one of $a = 0$, $a \in P$, or $-a \in P$ holds (cf. Definition 13.12). For integers a and b, we then defined $a > b$ to mean that $a - b \in P$.

Being an ordered integral domain does not identify the integers: after all, $\mathbb{Q}$ and $\mathbb{R}$ are also ordered integral domains. The latter two systems are, of course, ordered fields—but, saying that we want $\mathbb{Z}$ to be an ordered integral domain that is not a field is not satisfactory either. To see this, consider the set of finite decimals, $\mathbb{Z}[\frac{1}{10}]$; cf. Problem 4 in Chapter 13. This set is also an ordered integral domain that is not a field, but it is "too big."

What we want to say is that the integers form the "smallest" system with these properties. Indeed, if a subset S of $\mathbb{Z}$ were to contain 1 and were to be an integral domain for the same operations, then, in order for addition and subtraction to be closed in S, it would also need to contain $2, 3, 4, \ldots$ and $0, -1, -2, -3, \ldots$; therefore, we would need to have $S = \mathbb{Z}$.

We will make this all precise as follows. Given an integral domain Z, we say that a subset S is a *subdomain* of Z if S is an integral domain for the same operations. For example, $\mathbb{Z}$ is a subdomain of $\mathbb{Q}$, and $\mathbb{Q}$ is a subdomain of $\mathbb{R}$; $\mathbb{Z}$ is also a subdomain of the set $\mathbb{Z}[\frac{1}{10}]$ defined above. Trivially, any integral domain is a subdomain of itself.

We need to emphasize that, in order to be a subdomain, S has to be an integral domain for the same operations. For example, the set $\{0, 1\}$, with addition and multiplication defined "mod 2," is not considered to be a subdomain of $\mathbb{Z}$ even though $\{0, 1\}$ is a subset of $\mathbb{Z}$, and "mod 2" addition and multiplication make $\{0, 1\}$ into an integral domain (usually denoted by $\mathbb{Z}_2$; cf. Chapter 13).

We can now define the *integral system* $(Z, +, \cdot, 1, P)$ as a system consisting of an ordered integral domain Z for binary operations $+$ and $\cdot$, with multiplicative identity element 1 and set of positives P, for which it is true that every subdomain of Z that contains 1 is Z itself. Note that the condition that the only subdomain of an integral system that contains 1 is the system itself is the analogue of property (P3) of a Peano system: both these properties serve to assure that our system is the smallest one possible possessing the other characteristics.

As our terminology suggests, there is essentially only one integral system: the integer system. More precisely, any two integral systems $(Z, +, \cdot, 1, P)$ and $(Z', +', \cdot', 1', P')$ are *isomorphic*, meaning that there is a bijection $f : Z \to Z'$ with the following properties:

- $f(z_1 + z_2) = f(z_1) +' f(z_2)$ holds for all $z_1, z_2 \in Z$,
- $f(z_1 \cdot z_2) = f(z_1) \cdot' f(z_2)$ holds for all $z_1, z_2 \in Z$,
- $f(1) = 1'$, and
- $p \in P$ if, and only if, $f(p) \in P'$.

We omit the rather straightforward (but tedious) proof.

Let us now turn to the construction of an integral system. It helps to recall that our goal for enlarging the set of natural numbers to the set of integers is that for any pair of integers (and, thus, natural numbers) a and b, we wish to have an integer c for which $a = b + c$; that is, we want to define the difference $a - b$ for any pair of integers a and b. Of course, the difference $2 - 5$, for example, should be defined to be the same integer as $3 - 6$ or $8 - 11$, etc.; in general, we want to consider $a - b$ and $c - d$ to be the same as long as—and we say this now without using subtraction at all—we have $a + d = b + c$. This brings to mind Problem 7 (a) of Chapter 16, which we here restate and reprove.

Proposition 21.10 *The relation defined by*

$$R = \{((a, b), (c, d)) \in (\mathbb{N}^2)^2 \mid a + d = b + c\}$$

is an equivalence relation.

In the proof below, we want to be careful to not use subtraction (which was permissible in Chapter 16 but not here since subtraction has not yet been defined).

Proof We need to prove that R is reflexive, symmetric, and transitive.

By Proposition 21.9, addition of natural numbers is commutative, so for each $a, b \in \mathbb{N}$, we have $a + b = b + a$. But this means that $((a, b), (a, b)) \in R$ holds for all $(a, b) \in \mathbb{N}^2$, and thus R is reflexive. Similarly, if $((a, b), (c, d)) \in R$, then $a + d = b + c$, so $c + b = d + a$, thus $((c, d), (a, b)) \in R$, and R is symmetric.

To prove transitivity, assume that $((a, b), (c, d)) \in R$ and $((c, d), (e, f)) \in R$. Therefore, $a + d = b + c$ and $c + f = d + e$; adding these equations, we get

$$(a + d) + (c + f) = (b + c) + (d + e).$$

Using the associative and commutative properties of natural number addition, this equation yields

$$(a + f) + (c + d) = (b + e) + (c + d),$$

and this, using Proposition 21.32 (cf. Problem 3), implies that $a + f = b + e$, or $((a, b), (e, f)) \in R$, as needed.   □

According to the Fundamental Theorem of Equivalence Relations, the equivalence classes formed by the relation in Proposition 21.10 create a partition of $\mathbb{N}^2$.

Definition 21.11 *The equivalence classes of the relation in Proposition 21.10 are called* integers. *The set of integers is denoted by* $\mathbb{Z}$.

So, according to Definition 21.11, an integer is a set consisting of ordered pairs of certain natural numbers. For example, we can verify that the set

$$\{(1, 4), (2, 5), (3, 6), (4, 7), (5, 8), (6, 9), \ldots\}$$

is one such equivalence class, to be denoted as $-3_{\mathbb{Z}}$. Indeed, we introduce the following notations:

$$
\begin{aligned}
0_{\mathbb{Z}} &= [(1, 1)] = \{(n, n) \mid n \in \mathbb{N}\}, \\
1_{\mathbb{Z}} &= [(2, 1)] = \{(n + 1, n) \mid n \in \mathbb{N}\}, \\
-1_{\mathbb{Z}} &= [(1, 2)] = \{(n, n + 1) \mid n \in \mathbb{N}\}, \\
2_{\mathbb{Z}} &= [(3, 1)] = \{(n + 2, n) \mid n \in \mathbb{N}\}, \\
-2_{\mathbb{Z}} &= [(1, 3)] = \{(n, n + 2) \mid n \in \mathbb{N}\}, \\
3_{\mathbb{Z}} &= [(4, 1)] = \{(n + 3, n) \mid n \in \mathbb{N}\}, \\
-3_{\mathbb{Z}} &= [(1, 4)] = \{(n, n + 3) \mid n \in \mathbb{N}\},
\end{aligned}
$$

and so on. We should note that, at this point, the negative signs here are just notations; soon, however, we shall see that they are well chosen.

Next, we define the binary operations of addition and multiplication in $\mathbb{Z}$.

Definition 21.12 *Let $a_{\mathbb{Z}}$ and $b_{\mathbb{Z}}$ be two integers, and suppose that $(a_1, a_2) \in a_{\mathbb{Z}}$ and $(b_1, b_2) \in b_{\mathbb{Z}}$. We define their* sum *as*

$$a_{\mathbb{Z}} + b_{\mathbb{Z}} = [(a_1 + b_1, a_2 + b_2)]_R$$

and their product *as*

$$a_{\mathbb{Z}} \cdot b_{\mathbb{Z}} = [(a_1 \cdot b_1 + a_2 \cdot b_2, a_1 \cdot b_2 + a_2 \cdot b_1)]_R.$$

Since we defined these binary operations on integers by taking (arbitrary) elements from the equivalence classes represented by the integers, we need to verify that the operations are *well-defined*; that is, the results do not depend on which particular elements we chose. This is left for Problem 4.

Our final definition regarding the integers is for the set of positive elements.

Definition 21.13 *Let $a_{\mathbb{Z}}$ be an integer, and suppose that $(a_1, a_2) \in a_{\mathbb{Z}}$. We say that $a_{\mathbb{Z}}$ is* positive *if there is a natural number k for which $a_1 = a_2 + k$. The set of positive integers is denoted by $P_{\mathbb{Z}}$.*

As with Definition 21.12, we need to make sure that this is not element-dependent; see Problem 4. We see that, for example, the integers $1_{\mathbb{Z}}$, $2_{\mathbb{Z}}$, and $3_{\mathbb{Z}}$ are positive, while $0_{\mathbb{Z}}$, $-1_{\mathbb{Z}}$, and $-2_{\mathbb{Z}}$ are not. To see, for example, that $-1_{\mathbb{Z}} = [(1, 2)] \notin P_{\mathbb{Z}}$, we can verify that there is no natural number k for which $2 + k = 1$. Indeed, $2 + 1 = 3$, so $2 + k = 1$ cannot occur with $k = 1$, and for all other $k \in \mathbb{N}$ we have $2 + k = S(2 + P(k))$, which cannot equal 1 by Axiom (P1).

And now the expected result:

Theorem 21.14 *With our definitions and notations above, $(\mathbb{Z}, +, \cdot, 1_{\mathbb{Z}}, P_{\mathbb{Z}})$ is an integral system.*

Proof Commutativity, associativity, and distributivity follow easily from the definitions and the corresponding properties of the natural numbers. The additive and multiplicative identities in $\mathbb{Z}$ are easily seen to be $0_{\mathbb{Z}}$ and $1_{\mathbb{Z}}$, respectively: for the integer $a_{\mathbb{Z}} = [(a_1, a_2)]$ we have

$$a_{\mathbb{Z}} + 0_{\mathbb{Z}} = [(a_1, a_2)] + [(1, 1)] = [(a_1 + 1, a_2 + 1)] = [(a_1, a_2)] = a_{\mathbb{Z}}$$

and

$$a_{\mathbb{Z}} \cdot 1_{\mathbb{Z}} = [(a_1, a_2)] \cdot [(2, 1)] = [(2a_1 + a_2, a_1 + 2a_2)] = [(a_1, a_2)] = a_{\mathbb{Z}}.$$

The negative of an integer $a_{\mathbb{Z}} = [(a_1, a_2)]$ is $-a_{\mathbb{Z}} = [(a_2, a_1)]$; we then get

$$a_{\mathbb{Z}} + (-a_{\mathbb{Z}}) = [(a_1 + a_2, a_2 + a_1)] = 0_{\mathbb{Z}}.$$

Therefore, $\mathbb{Z}$ is a commutative ring with identity.

We can prove the nonzero product property as follows. Assume that $a_{\mathbb{Z}} \in \mathbb{Z}$, $b_{\mathbb{Z}} \in \mathbb{Z}$, $(a_1, a_2) \in a_{\mathbb{Z}}$, $(b_1, b_2) \in b_{\mathbb{Z}}$, and $a_{\mathbb{Z}} \cdot b_{\mathbb{Z}} = 0_{\mathbb{Z}}$. Let us assume that $a_{\mathbb{Z}} \neq 0_{\mathbb{Z}}$; we then need to prove that $b_{\mathbb{Z}} = 0_{\mathbb{Z}}$. Since $a_{\mathbb{Z}} \neq 0_{\mathbb{Z}}$, we have $a_1 \neq a_2$; by Proposition 21.33, we know that there is a natural number k for which either $a_1 = a_2 + k$ or $a_2 = a_1 + k$. We only consider the first case; the second case is similar.

By the definition of multiplication, $a_{\mathbb{Z}} \cdot b_{\mathbb{Z}} = 0_{\mathbb{Z}}$ means that

$$a_1 \cdot b_1 + a_2 \cdot b_2 = a_1 \cdot b_2 + a_2 \cdot b_1;$$

substituting $a_1 = a_2 + k$ yields

$$(a_2 + k) \cdot b_1 + a_2 \cdot b_2 = (a_2 + k) \cdot b_2 + a_2 \cdot b_1$$

or

$$a_2 \cdot b_1 + k \cdot b_1 + a_2 \cdot b_2 = a_2 \cdot b_2 + k \cdot b_2 + a_2 \cdot b_1.$$

By Proposition 21.32, we can cancel the terms $a_2 \cdot b_1$ and $a_2 \cdot b_2$, yielding

$$k \cdot b_1 = k \cdot b_2.$$

Using Proposition 21.34 this time, we get $b_1 = b_2$, from which $b_{\mathbb{Z}} = 0_{\mathbb{Z}}$, as claimed. Therefore, $(\mathbb{Z}, +, \cdot, 1_{\mathbb{Z}})$ is an integral domain.

Next, we verify the order properties. The proof of the fact that the sets $\{0_{\mathbb{Z}}\}$, $P_{\mathbb{Z}}$, and $-P_{\mathbb{Z}} = \{-p_{\mathbb{Z}} \mid p_{\mathbb{Z}} \in P_{\mathbb{Z}}\}$ form a partition of $\mathbb{N}^2$ follows directly from Proposition 21.33.

To prove that $P_{\mathbb{Z}}$ is closed for addition and multiplication, let $a_{\mathbb{Z}} \in P_{\mathbb{Z}}, b_{\mathbb{Z}} \in P_{\mathbb{Z}}$, $(a_1, a_2) \in a_{\mathbb{Z}}$, and $(b_1, b_2) \in b_{\mathbb{Z}}$. By definition, we have natural numbers k and l such that $a_1 = a_2 + k$ and $b_1 = b_2 + l$.

We then have

$$a_1 + b_1 = (a_2 + b_2) + (k + l),$$

which means that $a_{\mathbb{Z}} + b_{\mathbb{Z}} \in P_{\mathbb{Z}}$. Furthermore,

$$\begin{aligned}
a_1 \cdot b_1 + a_2 \cdot b_2 &= (a_2 + k) \cdot (b_2 + l) + a_2 \cdot b_2 \\
&= a_2 \cdot b_2 + a_2 \cdot l + k \cdot b_2 + k \cdot l + a_2 \cdot b_2 \\
&= (a_2 + k) \cdot b_2 + a_2 \cdot (b_2 + l) + k \cdot l \\
&= a_1 \cdot b_2 + a_2 \cdot b_1 + k \cdot l,
\end{aligned}$$

so $a_{\mathbb{Z}} \cdot b_{\mathbb{Z}} \in P_{\mathbb{Z}}$ as well.

Finally, to see that every subdomain that contains $1_\mathbb{Z}$ is the entire set, note that the subdomain is an Abelian group and thus must contain $n_\mathbb{Z}$ and $-n_\mathbb{Z}$ for each natural number n, and, therefore, each integer.   □

Before leaving the integer system, let us point out that the set of positive integers $P_\mathbb{Z}$ can be identified with the set of natural numbers $\mathbb{N}$; cf. Problem 5.

With that, let us now turn to $\mathbb{Q}$. The development of the rational number system is quite similar to the development of the integer system. The main difference, of course, is that the rational numbers form a field and not just an integral domain: every nonzero rational number has a multiplicative inverse (cf. Definition 13.1). The rational number system, formed by the set of rational numbers, the operations of addition and multiplication, and the set of positive elements, is an ordered field (cf. Definition 13.12) that has no ordered subfields other than itself; furthermore, it is the only such system up to isomorphism. We can construct the rational numbers from the integers using the relation given in Problem 7 (b) of Chapter 16. We leave the details to Problem 6.

Next, we move to the discussion of the real numbers—a substantially more challenging task. We have already seen that the real numbers form an ordered field; furthermore, in Chapter 17, we stated that this ordered field satisfies the Completeness Axiom: every nonempty bounded set of real numbers has a supremum and an infimum in $\mathbb{R}$. (As we pointed out in Chapter 17, this axiom fails to hold in $\mathbb{Q}$.) It turns out that we do not need anything else to characterize the reals; it can be proven that any two ordered fields that satisfy the Completeness Axiom are Isomorphic. Thus, the categorical set of axioms for the real numbers is quite evident. The construction of the real number system, however, takes more effort.

Before defining the real numbers, let us note that, to define integers, we partitioned the set of ordered pairs of natural numbers, $\mathbb{N}^2$, into equivalence classes via a certain equivalence relation R, then identified the set of integers $\mathbb{Z}$ with the set of the equivalence classes generated by R. Our construction of the rational numbers was similar. However, we cannot expect the same kind of construction for the real numbers, because, as we learned in Chapter 20, while the cardinality of $\mathbb{Z}$ is the same as the cardinality of $\mathbb{N}^2$, and the cardinality of $\mathbb{Q}$ is the same as the cardinality of $\mathbb{Z}^2$, the cardinality of $\mathbb{R}$ is greater than the cardinality of $\mathbb{Q}^2$ (and thus also greater than the cardinality of the equivalence classes that any relation on $\mathbb{Q}$ creates). We know, though, that the cardinality of $\mathbb{R}$ equals the cardinality of the power set $P(\mathbb{Q})$ (see Theorem 20.20), so we can hope to identify the real numbers with the subsets of rational numbers. It turns out that we do not need all subsets of $\mathbb{Q}$—only the so-called Dedekind cuts.

Definition 21.15 *A nonempty proper subset* C *of rational numbers is called a real number if for every* $q \in C$,

• *C contains every rational number that is less than q, and*
• *C contains some rational number that is greater than q.*

The set of real numbers is denoted by $\mathbb{R}$.

Thus we define real numbers as "initial segments" of rational numbers, also known as *Dedekind cuts*, named after Gauss's last doctoral student, Julius Dedekind (1831–1916). As an example, we see that the set of negative rational numbers, that is, the set

$$\{q \in \mathbb{Q} \mid q < 0\}$$

is a real number; we denote this real number by $0_\mathbb{R}$.

Since $\mathbb{Q}$ does not satisfy the Completeness Axiom, a given subset of rational numbers—such as a real number—may not have a supremum. We make the following definition.

Definition 21.16 *A real number C is called a* rational real number *if it has a supremum in* $\mathbb{Q}$; *otherwise it is called an* irrational real number.

It is easy to see that $0_\mathbb{R}$ is a rational real number. (Below we will see that rational real numbers correspond nicely to rational numbers, but the distinction here is important: here a rational real number, like any real number, is defined as an infinite set of rational numbers!) The following proposition provides a generalization of this example.

Proposition 21.17 *For a given rational number r, define the set*

$$\{r\}^{\Downarrow} = \{q \in \mathbb{Q} \mid q < r\}.$$

Then $\{r\}^{\Downarrow}$ *is a real number for every rational number r; furthermore, a real number is a rational real number if, and only if, it is of the form* $\{r\}^{\Downarrow}$ *for some rational number r.*

Note that we defined $\{r\}^{\Downarrow}$ as the set of *strict* lower bounds of $\{r\}$; that is, we have $r \notin \{r\}^{\Downarrow}$.

Proof We see that $\emptyset \neq \{r\}^{\Downarrow} \neq \mathbb{Q}$: for example, $r - 1 \in \{r\}^{\Downarrow}$ but $r + 1 \notin \{r\}^{\Downarrow}$. Note that if $a \in \{r\}^{\Downarrow}$, $b \in \mathbb{Q}$, and $b < a$, then $b \in \{r\}^{\Downarrow}$ by transitivity. Furthermore, if $a \in \{r\}^{\Downarrow}$, then a quick verification shows that $b = (a + r)/2$ (for example) is an element of $\{r\}^{\Downarrow}$ and $a < b$. Indeed, $b \in \mathbb{Q}$; and $a < r$ implies that

$$a < \frac{a + r}{2} < r.$$

This proves our first claim.

We now turn to our second claim. Clearly, $\sup \{r\}^{\Downarrow} = r$ in $\mathbb{Q}$, so $\{r\}^{\Downarrow}$ is a rational real number for every rational number r. Conversely, suppose that C is a rational real number, so there exists an $r \in \mathbb{Q}$ for which $\sup C = r$. We will prove that $C = \{r\}^{\Downarrow}$.

To see that $\{r\}^{\Downarrow} \subseteq C$, let $q \in \{r\}^{\Downarrow}$. Then $q < r$, so q is not an upper bound of C, and, therefore, we can find an $s \in C$ for which $q < s$. But, by definition, we then have $q \in C$; this proves $\{r\}^{\Downarrow} \subseteq C$.

Now let $c \in C$. Since C is a real number, we must have some $c' \in C$ with $c < c'$. Since r is an upper bound of C, we must have $c' \leq r$. But then $c < r$, and, therefore, $c \in \{r\}^\Downarrow$, proving that $C \subseteq \{r\}^\Downarrow$ as well.   □

We can see that the set

$$\mathbb{Q}' = \{\{r\}^\Downarrow \mid r \in \mathbb{Q}\}$$

is essentially the same as the set of rational numbers; more precisely—once we define addition, multiplication, and order in $\mathbb{R}$—one can prove that $\mathbb{Q}'$ is isomorphic to $\mathbb{Q}$. By Proposition 21.17, we may safely use the notation $r_\mathbb{R}$ for the real number $\{r\}^\Downarrow$. In fact, although real numbers are defined as (infinite) sets, keeping with tradition, rather than capital letters, from now on we will use lower-case letters (usually, from the end of the alphabet) and the subscript $\mathbb{R}$ to denote real numbers. Thus, when writing, for example, $x_\mathbb{R}$, we imply that it is a real number; we do not, however, assume that $x_\mathbb{R} = \{x\}^\Downarrow$ as $x_\mathbb{R}$ is not necessarily a rational real number. (Of course, as we learned in Chapter 20, there are not enough rational numbers to index all real numbers.)

By Proposition 21.17, we have a good understanding of which real numbers are rational. We now prove that a certain familiar real number is irrational.

Proposition 21.18 *Let us define the set*

$$\sqrt{2}_\mathbb{R} = \{q \in \mathbb{Q} \mid q \leq 0 \text{ or } q^2 < 2\}.$$

Then $\sqrt{2}_\mathbb{R}$ is an irrational real number.

At this point, $\sqrt{2}_\mathbb{R}$ is just a notation for this particular set; in Proposition 21.24 below, we prove that the notation is well chosen.

Proof Note that $\mathbb{Q}$ is an ordered field, so we may use the statements proved in Problem 11 of Chapter 14.

First we show that $\sqrt{2}_\mathbb{R} \in \mathbb{R}$. Since $1 \in \sqrt{2}_\mathbb{R}$ and $2 \notin \sqrt{2}_\mathbb{R}$, we have $\emptyset \neq \sqrt{2}_\mathbb{R} \neq \mathbb{Q}$.

Suppose now that $a \in \sqrt{2}_\mathbb{R}, b \in \mathbb{Q}$, and $b < a$. If $b \leq 0$, then $b \in \sqrt{2}_\mathbb{R}$. If $b > 0$, then $b^2 < a^2 < 2$, so again $b \in \sqrt{2}_\mathbb{R}$.

Suppose again that $a \in \sqrt{2}_\mathbb{R}$; we must find an element $b \in \sqrt{2}_\mathbb{R}$ so that $a < b$. If $a \leq 0$, we may choose $b = 1$. Suppose then that $a > 0$. Note that, since $a \in \sqrt{2}_\mathbb{R}$, we must also have $a^2 < 2$ and thus $a < 2$.

Choose a natural number n so that

$$\frac{5}{2 - a^2} < n.$$

(Note that we do not need to use the Archimedean property here—a property that depends on the Completeness Axiom of $\mathbb{R}$ and which we may not yet use. Since $5/(2 - a^2)$ is a positive rational number, it can be written as a quotient of two natural numbers; we can choose, for example, any natural number n that is greater than the numerator.)

Let $b = a + \frac{1}{n}$; clearly $b \in \mathbb{Q}$. We then have

$$b^2 = \left(a + \frac{1}{n}\right)^2 = a^2 + 2 \cdot \frac{a}{n} + \frac{1}{n^2} < a^2 + 2 \cdot \frac{2}{n} + \frac{1}{n} = a^2 + \frac{5}{n} < 2,$$

so $b \in \sqrt{2}_{\mathbb{R}}$.

Next, we need to prove that $\sqrt{2}_{\mathbb{R}} \notin \mathbb{Q}'$. Assume, indirectly, that there is a rational number r for which

$$\sqrt{2}_{\mathbb{R}} = \{r\}^{\Downarrow} = \{q \in \mathbb{Q} \mid q < r\}.$$

Since $r \notin \{r\}^{\Downarrow}$, we must have $r > 0$ and $r^2 \geq 2$. But we know that there is no rational number whose square is 2, so $r > 0$ and $r^2 > 2$.

Similarly to the method above, we can choose a natural number m so that

$$\frac{2r}{r^2 - 2} < m.$$

This gives

$$\left(r - \frac{1}{m}\right)^2 = r^2 - 2 \cdot \frac{r}{m} + \frac{1}{m^2} > r^2 - 2 \cdot \frac{r}{m} > 2.$$

Therefore, $r - \frac{1}{m} \notin \sqrt{2}_{\mathbb{R}}$, contradicting our assumption that every rational number less than r is in $\sqrt{2}_{\mathbb{R}}$.   □

We now turn to the addition and multiplication operations and order relation of real numbers. Defining addition is quite simple.

Definition 21.19 *The sum of real numbers $x_{\mathbb{R}}$ and $y_{\mathbb{R}}$ is defined as*

$$x_{\mathbb{R}} + y_{\mathbb{R}} = \{q_1 + q_2 \mid q_1 \in x_{\mathbb{R}} \text{ and } q_2 \in y_{\mathbb{R}}\}.$$

It is not hard to prove that addition of real numbers yields a real number—see Problem 7. Commutativity and associativity of addition follow immediately from the same properties of rational numbers. It is also quite straightforward to verify that the real number $0_{\mathbb{R}}$ serves as the additive identity. The proof that every real number has an additive inverse, however, needs some preparation.

It may be helpful to examine some examples. After a bit of contemplation, we see that the additive inverse of

$$3_{\mathbb{R}} = \{q \in \mathbb{Q} \mid q < 3\}$$

should be

$$-3_{\mathbb{R}} = \{q \in \mathbb{Q} \mid q < -3\},$$

while the additive inverse of

$$\sqrt{2}_{\mathbb{R}} = \{q \in \mathbb{Q} \mid q \leq 0 \text{ or } q^2 < 2\}$$

has to be

$$-\sqrt{2}_{\mathbb{R}} = \{q \in \mathbb{Q} \mid q < 0 \text{ and } q^2 > 2\}.$$

(With a bit of routine work, one can verify that these are real numbers and that they serve as additive inverses for the relevant numbers.) It is not immediately clear how one can define additive inverses in general.

It turns out that the following approach works for all real numbers. First some useful notation. For a real number $x_{\mathbb{R}}$, we let $x_{\mathbb{R}}^{\Uparrow}$ denote the set of strict upper bounds for $x_{\mathbb{R}}$ in $\mathbb{Q}$; that is,

$$x_{\mathbb{R}}^{\Uparrow} = \{r \in \mathbb{Q} \mid \forall q \in x_{\mathbb{R}}, q < r\}.$$

We then define

$$\widehat{x_{\mathbb{R}}} = \begin{cases} x_{\mathbb{R}}^{\Uparrow} \setminus \{\sup x_{\mathbb{R}}\} & \text{if } x_{\mathbb{R}} \text{ is rational;} \\ \\ x_{\mathbb{R}}^{\Uparrow} & \text{if } x_{\mathbb{R}} \text{ is irrational.} \end{cases}$$

In other words, $\widehat{x_{\mathbb{R}}}$ consists of all upper bounds of $x_{\mathbb{R}}$ in $\mathbb{Q}$, except for the supremum of $x_{\mathbb{R}}$ in $\mathbb{Q}$, if it exists. For example, for the rational real number $3_{\mathbb{R}}$ we have

$$\widehat{3_{\mathbb{R}}} = \{q \in \mathbb{Q} \mid q > 3\},$$

and for the irrational real number $\sqrt{2}_{\mathbb{R}}$ we get

$$\widehat{\sqrt{2}_{\mathbb{R}}} = \{q \in \mathbb{Q} \mid q > 0 \text{ and } q^2 > 2\}.$$

We are now ready to define additive inverses of real numbers.

Definition 21.20 *We define the additive inverse of the real number $x_{\mathbb{R}}$ as*

$$-x_{\mathbb{R}} = \{-q \mid q \in \widehat{x_{\mathbb{R}}}\}.$$

We can prove that the additive inverse of a real number is a real number and that the sum of a real number and its additive inverse is $0_{\mathbb{R}}$—see Problem 7.

For our previous examples, we can verify that Definition 21.20 yields the right answer:

$$-3_{\mathbb{R}} = \{-q \mid q \in \widehat{3_{\mathbb{R}}}\}$$
$$= \{-q \mid q \in \mathbb{Q} \text{ and } q > 3\}$$
$$= \{q \in \mathbb{Q} \mid q < -3\},$$

and

$$-\sqrt{2}_{\mathbb{R}} = \{-q \mid q \in \widehat{\sqrt{2}_{\mathbb{R}}}\}$$
$$= \{-q \mid q \in \mathbb{Q}, \text{ and } q > 0, \text{ and } q^2 > 2\}$$
$$= \{q \in \mathbb{Q} \mid q < 0 \text{ and } q^2 > 2\}.$$

We still need to define multiplication and order in $\mathbb{R}$. Since multiplication will be defined first for positive real numbers and then extended to the cases in which one or both terms are nonpositive, we first define order (as usual, via describing the set of positive numbers).

Definition 21.21 *We say that a real number $x_{\mathbb{R}}$ is positive whenever $0_{\mathbb{R}} \subset x_{\mathbb{R}}$. The set of positive real numbers is denoted by $P_{\mathbb{R}}$.*

Note that for $x_{\mathbb{R}}$ to be positive, we require $0_{\mathbb{R}}$ to be a proper subset of $x_{\mathbb{R}}$; therefore, $0_{\mathbb{R}}$ itself is not positive. It is not hard to see that our definition is equivalent to saying that $x_{\mathbb{R}}$ contains some positive rational number (and, therefore, infinitely many).

Given Definition 21.21, we can define negative real numbers as those whose additive inverse is positive; furthermore, we can define the order relations "less than" and "greater than" in the usual way: $x_{\mathbb{R}} < y_{\mathbb{R}}$ when $x_{\mathbb{R}} - y_{\mathbb{R}}$ is negative, and $x_{\mathbb{R}} > y_{\mathbb{R}}$ when $x_{\mathbb{R}} - y_{\mathbb{R}}$ is positive. It is a fairly routine exercise to verify that the usual properties hold—cf. Problem 7.

Let us now examine how we could define the product of positive real numbers $x_{\mathbb{R}}$ and $y_{\mathbb{R}}$. First we note that the set

$$\{q_1 \cdot q_2 \mid q_1 \in x_{\mathbb{R}} \text{ and } q_2 \in y_{\mathbb{R}}\},$$

which would be the direct analogue of how we defined sums, would not work: this set actually equals $\mathbb{Q}$, which is not a real number! For example, with

$$2_{\mathbb{R}} = \{q_1 \in \mathbb{Q} \mid q_1 < 2\}$$

and

$$3_{\mathbb{R}} = \{q_2 \in \mathbb{Q} \mid q_2 < 3\},$$

the products $q_1 \cdot 1$ with $q_1 < 0$ yield all negative rationals, and $q_1 \cdot (-1)$ with $q_1 \leq 0$ yield all nonnegative rationals.

We would, of course, like to have $2_{\mathbb{R}} \cdot 3_{\mathbb{R}} = 6_{\mathbb{R}}$, or

$$\{q_1 \in \mathbb{Q} \mid q_1 < 2\} \cdot \{q_2 \in \mathbb{Q} \mid q_2 < 3\} = \{q \in \mathbb{Q} \mid q < 6\}.$$

Note that

$$\{q \in \mathbb{Q} \mid q < 6\} = \{q \in \mathbb{Q} \mid q \leq 0\} \cup \{q_1 \cdot q_2 \mid q_1 \in 2_{\mathbb{R}}, q_1 \in 3_{\mathbb{R}}, q_1 > 0, q_2 > 0\}.$$

This prompts us to introduce the notation

$$x_{\mathbb{R}}^+ = \{q \in x_{\mathbb{R}} \mid q > 0\},$$

which enables us to define multiplication as follows:

Definition 21.22 *The product of real numbers $x_{\mathbb{R}}$ and $y_{\mathbb{R}}$ is defined as*

$$x_{\mathbb{R}} \cdot y_{\mathbb{R}} = \begin{cases} \{q \in \mathbb{Q} \mid q \leq 0\} \cup \{q_1 \cdot q_2 \mid q_1 \in x_{\mathbb{R}}^+, q_2 \in y_{\mathbb{R}}^+\} & \textit{if } x_{\mathbb{R}} > 0_{\mathbb{R}}, y_{\mathbb{R}} > 0_{\mathbb{R}}; \\[2mm] 0_{\mathbb{R}} & \textit{if } x_{\mathbb{R}} = 0_{\mathbb{R}} \textit{ or } y_{\mathbb{R}} = 0_{\mathbb{R}}; \\[2mm] -((-x_{\mathbb{R}}) \cdot y_{\mathbb{R}}) & \textit{if } x_{\mathbb{R}} < 0_{\mathbb{R}}, y_{\mathbb{R}} > 0_{\mathbb{R}}; \\[2mm] -(x_{\mathbb{R}} \cdot (-y_{\mathbb{R}})) & \textit{if } x_{\mathbb{R}} > 0_{\mathbb{R}}, y_{\mathbb{R}} < 0_{\mathbb{R}}; \\[2mm] (-x_{\mathbb{R}}) \cdot (-y_{\mathbb{R}}) & \textit{if } x_{\mathbb{R}} < 0_{\mathbb{R}}, y_{\mathbb{R}} < 0_{\mathbb{R}}. \end{cases}$$

Definition 21.22 satisfies all of our expectations; in particular, we can prove that the product of two real numbers is a real number, $1_{\mathbb{R}}$ is the multiplicative identity, and every nonzero real number has a multiplicative inverse — see Problem 7.

Let us now return to our set

$$\sqrt{2}_{\mathbb{R}} = \{q \in \mathbb{Q} \mid q \leq 0 \text{ or } q^2 < 2\}.$$

As we promised, we now use Definition 21.22 to prove that the notation is well chosen and we have

$$\sqrt{2}_{\mathbb{R}} \cdot \sqrt{2}_{\mathbb{R}} = 2_{\mathbb{R}}.$$

We will be careful to avoid using that the real number $2_{\mathbb{R}}$ has a square root. Our proof below will rely solely on the properties of $\mathbb{Q}$ without even mentioning real numbers.

We need the following result that is of interest in its own right.

Proposition 21.23 *Let a and b be positive rational numbers with a < b. Then there exists a positive rational number r such that*

$$a < r^2 < b.$$

We should emphasize again that, though Proposition 21.23 could be established by methods similar to that of Proposition 12.1 (where we assumed the existence of square roots) or Problem 9 of Chapter 18 (which relied on the Archimedean Property, a consequence of the Completeness Axiom of $\mathbb{R}$), here we provide a proof that is entirely within the confines of the set of rational numbers.

Proof We first assume that $1 \le a < b$. Let n be any natural number for which

$$n > \frac{3a}{b - a}.$$

(Note that we do not need to use the Archimedean property here; see the relevant comment in the proof of Proposition 21.18.) Furthermore, let m be the smallest natural number for which $m^2 > an^2$. (The existence of such an m follows from Theorem 10.6.)

We will show that $r = m/n$ satisfies our requirements. By our choice of m, we have $m^2 > an^2$, so $a < r^2$ holds; the rest of the argument below will establish that $r^2 < b$ holds as well.

We first observe that the assumption $a \ge 1$ and our choices for m and n imply that

$$9(m - 1)^2 \le 9an^2 \le 9a^2n^2 = (b - a)^2n^2 \left(\frac{3a}{b - a} \right)^2 < (b - a)^2n^4,$$

and therefore

$$3(m - 1) < (b - a)n^2.$$

Since $m^2 > an^2 \ge n^2$, we must have $m \ge 2$ and thus

$$2m - 1 = 3(m - 1) - (m - 2) \le 3(m - 1) < (b - a)n^2.$$

Using this, we get

$$m^2 = an^2 + m^2 - an^2 \le an^2 + m^2 - (m-1)^2 = an^2 + 2m - 1 < an^2 + (b-a)n^2 = bn^2,$$

which yields $r^2 < b$ as claimed.

The case $0 < a < 1 < b$ is trivial: $r = 1$ works.

Finally, assume that $0 < a < b \le 1$. In this case we have $1 \le 1/b < 1/a$, so by the argument above, we can find a rational number $\hat{r}$ with

$$\frac{1}{b} < \hat{r}^2 < \frac{1}{a};$$

setting $r = 1/\hat{r}$ will then satisfy our claim.   □

Proposition 21.24 *For the real number*

$$\sqrt{2}_\mathbb{R} = \{q \in \mathbb{Q} \mid q \leq 0 \text{ or } q^2 < 2\}$$

we have $\sqrt{2}_\mathbb{R} \cdot \sqrt{2}_\mathbb{R} = 2_\mathbb{R}$.

Proof Recall that

$$2_\mathbb{R} = \{q \in \mathbb{Q} \mid q < 2\}$$

and, by Definition 21.22,

$$\sqrt{2}_\mathbb{R} \cdot \sqrt{2}_\mathbb{R} = \{q \in \mathbb{Q} \mid q \leq 0\} \cup \{q_1 \cdot q_2 \mid q_1 \in \mathbb{Q}, q_2 \in \mathbb{Q}, q_1 > 0, q_2 > 0, q_1^2 < 2, q_2^2 < 2\}.$$

Clearly, for any $q \in \mathbb{Q}$ with $q \leq 0$, we have $q \in \sqrt{2}_\mathbb{R} \cdot \sqrt{2}_\mathbb{R}$ and $q \in 2_\mathbb{R}$, so it suffices to prove that any positive rational number that is an element of one of these sets is an element of both.

A positive rational number in $\sqrt{2}_\mathbb{R} \cdot \sqrt{2}_\mathbb{R}$ is of the form $q_1 \cdot q_2$ where q_1 and q_2 are positive rational numbers and $q_1^2 < 2$ and $q_2^2 < 2$. Therefore, we have $q_1^2 \cdot q_2^2 < 4$, so $0 < q_1 \cdot q_2 < 2$.

For the other direction, let $q \in \mathbb{Q}$ with $0 < q < 2$. By Proposition 21.23, we have a positive rational number r for which $q < r^2 < 2$. Setting $q_1 = r$ and $q_2 = q/r$, we have $q = q_1 \cdot q_2$, $q_1^2 < 2$, and

$$q_2^2 = \frac{q^2}{r^2} = q \cdot \frac{q}{r^2} < q < 2,$$

and therefore $q \in \sqrt{2}_\mathbb{R} \cdot \sqrt{2}_\mathbb{R}$.   □

We complete our discussion of real numbers by stating the following theorem.

Theorem 21.25 *With our definitions as above, $(\mathbb{R}, +, \cdot, P_\mathbb{R})$ is an ordered field that satisfies the Completeness Axiom.*

The proof of Theorem 21.25 entails verifying the required properties of addition, multiplication, and order—see Problem 7.

Before moving on to the set of complex numbers, we should point out the main benefits of extending the set of natural numbers to the integers, the rational numbers, and then to the real numbers. Going from $\mathbb{N}$ to $\mathbb{Z}$ we gained the important property of being able to subtract any two elements; similarly, enlarging $\mathbb{Z}$ to $\mathbb{Q}$ enabled us to divide any two elements (except for dividing by zero, of course). Thus, in $\mathbb{Q}$, any linear polynomial $f(x) = ax + b$ has a root. The main advantage of extending $\mathbb{Q}$

to $\mathbb{R}$ is that the order relation became "complete": any bounded subset in $\mathbb{R}$ has an infimum and a supremum, and, therefore, the real numbers "fill in" an entire line without "gaps." The set $\mathbb{R}$ also has the advantage that many polynomials that have no root in $\mathbb{Q}$ have a root in $\mathbb{R}$; for example, in $\mathbb{R}$ we can take the square root (and the cube root, the fourth root, etc.) of any nonnegative real number. However, even in $\mathbb{R}$, some simple polynomials, such as $f(x) = x^2 + 1$, have no roots. The beauty of extending $\mathbb{R}$ to $\mathbb{C}$ is that, in $\mathbb{C}$, every non-constant polynomial has a root.

In Problem 6 of Chapter 20 we introduced the notion of algebraic numbers as those numbers that are roots of some polynomials whose coefficients are integers (and not all zero) and the term transcendental number for those numbers for which no such polynomial exists. There we listed several examples for algebraic and transcendental numbers; for example, 3.14 (the root of $f(x) = 100x - 314$) and $\sqrt{11}$ (the root of $f(x) = x^2 - 11$) are algebraic, while π is transcendental (the proof of this is not easy). Note that whether we require that the coefficients be integers or rational numbers makes no difference: if f is a polynomial whose coefficients are rational numbers, then its roots are also roots of the polynomial $d \cdot f$ where d is any nonzero common multiple of the denominators of the coefficients of f (note that $d \cdot f$ has the same degree as f). Since here we talk about fields, we prefer to use the condition that the coefficients be rational numbers.

We introduce the following general concepts.

Definition 21.26 *Suppose that F is a field and K is a subfield of F.*
We say that an element $u \in F$ is

- algebraic *over K, if u is a root of some nonzero polynomial f with coefficients in K;*
- transcendental *over K, if no such polynomial exists.*

Furthermore, we say that F is

- *an* algebraic extension *of K if all elements of F are algebraic over K;*
- algebraically closed *if its only algebraic extension is itself; and*
- *an* algebraic closure *of K if it is an algebraic extension of K that is algebraically closed.*

Note that every field K is an algebraic extension of itself; for example, $\mathbb{Q}$ is an algebraic extension of itself as every rational number u is the root of the polynomial $f(x) = x - u$ (among others). It turns out, however, that $\mathbb{Q}$ is not algebraically closed; in fact, it has infinitely many algebraic extensions (cf. Problem 10). One can prove that the set of algebraic numbers $\mathbb{A}$ forms an algebraic extension of $\mathbb{Q}$ as well; furthermore, $\mathbb{A}$ is algebraically closed and thus it is an algebraic closure of $\mathbb{Q}$. In 1910, the German mathematician Ernst Steinitz proved that every field has an algebraic closure and that this algebraic closure is (up to isomorphism) unique.

So we may wonder what the algebraic closure of the field of real numbers is. As the following proposition shows, $\mathbb{R}$ is not algebraically closed.

Proposition 21.27 *The field of complex numbers is an algebraic extension of the field of real numbers.*

Proof It is easy to see that the complex number $a + bi$ (with $a, b \in \mathbb{R}$ and i the imaginary unit with the property that $i^2 = -1$) is the root of the quadratic polynomial $f(x) = x^2 - (2a)x + (a^2 + b^2)$:

$$
\begin{aligned}
f(a + bi) &= (a + bi)^2 - 2a(a + bi) + a^2 + b^2 \\
&= a^2 + 2abi - b^2 - 2a^2 - 2abi + a^2 + b^2 \\
&= 0.
\end{aligned}
$$

$\square$

Therefore, $\mathbb{C}$ is an algebraic extension of $\mathbb{R}$. But is $\mathbb{C}$ also the algebraic closure of $\mathbb{R}$? The affirmative answer is given by the following important theorem.

Theorem 21.28 (The Fundamental Theorem of Algebra) *The complex number field is algebraically closed.*

There is a variety of beautiful proofs for the Fundamental Theorem of Algebra, some analytic, some algebraic, and some topological. All proofs, however, rely at least in part on some tools from analysis.

We have thus arrived at the characterization of $\mathbb{C}$.

Theorem 21.29 *The set of complex numbers, $\mathbb{C}$, has the categorical property that it is the algebraic closure of $\mathbb{R}$.*

It is worth pointing out that the Fundamental Theorem of Algebra implies not only that the algebraic closure of $\mathbb{R}$ is $\mathbb{C}$, but that the algebraic closure of $\mathbb{C}$ is $\mathbb{C}$ itself. Namely, adding i to the set of real numbers (as well as all linear combinations $a + bi$ with $a, b \in \mathbb{R}$ to ensure that the resulting set $\mathbb{C}$ is a field) assures that

- every non-constant polynomial with real number coefficients has all of its roots in $\mathbb{C}$, and
- every non-constant polynomial with complex number coefficients has all of its roots in $\mathbb{C}$.

This is what we meant when we said that there is no need to enlarge $\mathbb{C}$ further. (Nevertheless, we investigate the possibility of extending $\mathbb{C}$ below.)

Of course, the *imaginary number* i is purely a symbol at this point; we still need to construct the complex numbers. As is well known, we identify the set of complex numbers with the set of points in the Euclidean plane. (Thus, the construction of our final number system is the easiest.)

Definition 21.30 *The elements of $\mathbb{R}^2$ are called complex numbers; $\mathbb{C}$ is just another notation for $\mathbb{R}^2$.*

The sum *of complex numbers (a_1, b_1) and (a_2, b_2) is defined as*

$$
(a_1 + a_2, b_1 + b_2).
$$

The product *of complex numbers* (a_1, b_1) *and* (a_2, b_2) *is defined as*

$$(a_1 \cdot a_2 - b_1 \cdot b_2, a_1 \cdot b_2 + b_1 \cdot a_2).$$

We also introduce the notations $i_\mathbb{C} = (0, 1)$ and $(a + bi)_\mathbb{C} = (a, b)$. Furthermore, to emphasize that $\mathbb{C}$ contains the subfield $\mathbb{R} \times \{0\}$ that is isomorphic to $\mathbb{R}$, for a real number a we let $a_\mathbb{C}$ denote the complex number $(a, 0)$.

While the definition of multiplication seems strange, using it we get

$$\begin{aligned}
i_\mathbb{C} \cdot i_\mathbb{C} &= (0, 1) \cdot (0, 1) \\
&= (0 \cdot 0 - 1 \cdot 1, 0 \cdot 1 + 1 \cdot 0) \\
&= (-1, 0) = (-1)_\mathbb{C},
\end{aligned}$$

as desired.

Naturally, Definition 21.30 is made to ensure that the system $(\mathbb{C}, +, \cdot)$ is the algebraic closure of $\mathbb{R}$. We omit the proof.

Note that we did not mention an order relation in $\mathbb{C}$; as we pointed out in Problem 2 of Chapter 17, it is not possible to define order in $\mathbb{C}$ that would make it into an ordered field. So, extending $\mathbb{R}$ to $\mathbb{C}$, we gained the convenient properties of $\mathbb{C}$ being an algebraically closed field but we had to give up on the notion of order.

Now that we have extended $\mathbb{R}$ to $\mathbb{R}^2$, we may wonder if we can go further. As the previous paragraph suggests, further extensions may come with additional sacrifices. It is, in fact, possible to define addition and multiplication in $\mathbb{R}^4$ in a relatively familiar way. However, the new system, $(\mathbb{H}, +, \cdot)$, where $\mathbb{H} = \mathbb{R}^4$ is the set of *quaternions*—sometimes called *Hamilton numbers* after their inventor, William Rowan Hamilton (1805–1865)—is not going to be a field; all properties will hold with the exception that multiplication is not commutative any more. We can go one step further and define the *octonions*—or *Cayley numbers* after Arthur Cayley (1821–1895)—as the elements of $\mathbb{R}^8$ with specific addition and multiplication—the octonion system $(\mathbb{O}, +, \cdot)$, however, fails both commutativity and associativity of multiplication. Nevertheless, both the quaternions and the octonions are *normed division algebras*—systems with some attractive and applicable properties (but which we will not define here).

One of the breakthroughs of twentieth century mathematics was the following theorem.

Theorem 21.31 (Hurwitz's Theorem) *The only normed division algebras over* $\mathbb{R}$ *are* $\mathbb{R}$, $\mathbb{C}$, $\mathbb{H}$, *and* $\mathbb{O}$.

Well, we have already been "out of order" after $\mathbb{R}$, and now we cannot even continue conforming to the "norm," so it is high time to end this chapter.

Problems

1. Recall that we have set $2 = S(1)$, $3 = S(2)$, etc.

   (a) Use the relevant definitions to verify that $2 \cdot 3 = 6$.
   (b) Given natural numbers n_1 and n_2, provide a definition for $n_1^{n_2}$. (Cf. Problem 4 (d) of Chapter 2.)
   (c) Use your definition from part (b) to verify that $2^3 = 8$.

2. In this problem we prove Proposition 21.9. The proof is rather delicate; in particular, the order in which the various properties are established has to be chosen carefully.

   (a) Prove that addition of natural numbers is associative.
       (Hint: Show that, for each $n_1, n_2 \in \mathbb{N}$, the set

$$M(n_1, n_2) = \{n_3 \in \mathbb{N} \mid (n_1 + n_2) + n_3 = n_1 + (n_2 + n_3)\}$$

   is 1-inductive.)
   (b) Prove that addition of natural numbers is commutative.
       (Hints: Prove that, for each $n_1 \in \mathbb{N}$, the set

$$M(n_1) = \{n_2 \in \mathbb{N} \mid n_1 + n_2 = n_2 + n_1\}$$

   is 1-inductive. To show that $1 \in M(n_1)$, use induction to prove that $1 + n = n + 1$ holds for all $n \in \mathbb{N}$. In both these claims, use the fact that addition of natural numbers is associative.)
   (c) Prove that 1 is a multiplicative identity in $\mathbb{N}$.
   (d) Prove that multiplication of natural numbers is distributive with respect to addition.
       (Hint: Since we have not yet established commutativity of multiplication, two identities need to be proved.)
   (e) Prove that multiplication of natural numbers is commutative.
   (f) Prove that multiplication of natural numbers is associative.

3. Suppose that $(N, 1, S)$ is a Peano system. Prove each of the following propositions.

   (a) **Proposition 21.32** *For any elements m_1, m_2, and n of N, the equation $m_1 + n = m_2 + n$ implies $m_1 = m_2$.*
       (Hints: Let $m_1 \in N$, $m_2 \in N$, and suppose that $m_1 \neq m_2$. Consider the set

$$A(m_1, m_2) = \{n \in N \mid m_1 + n \neq m_2 + n\}.$$

   Our claim follows from showing that $A(m_1, m_2) = N$.)
   (b) **Proposition 21.33** *For any elements m and n of N, exactly one of the following three statements holds:*

- $m = n;$
- $\exists k \in N, m = n + k;$
- $\exists k \in N, n = m + k.$

(Hints: Let $n \in N$ be arbitrary, and define

$$A(n) = \{m \in N \mid \exists k \in N, m = n + k\},$$

$$B(n) = \{m \in N \mid \exists k \in N, n = m + k\},$$

and

$$C(n) = \{n\} \cup A(n) \cup B(n).$$

Prove that $C(n)$ is 1-inductive, thus $C(n) = N$. This implies that for all $m, n \in N$, at least one of the three statements holds.
To prove that no more than one of the three statements holds, first prove that, for an arbitrary $n \in N$,

$$\{m \in N \mid m \neq n + m\} = N$$

holds.)

(c) **Proposition 21.34** *For any elements m_1, m_2, and n of N, the equation $m_1 \cdot n = m_2 \cdot n$ implies $m_1 = m_2$.*
(Hints: Proceed indirectly, and use Proposition 21.33.)

4. Verify that Definitions 21.12 and 21.13 are valid (they do not depend on which elements we choose).

5. Let $(Z, +, \cdot, 1, P)$ be an integral system, and let $S : Z \to Z$ denote the function given by $S(z) = z + 1$. Prove that $(P, 1, S)$ is a Peano system.

6. Make the development of the rational number system precise (cf. page 359) following the outline below.

   (a) Give a careful definition of a *rational system* $(Q, +, \cdot, P)$ listing a categorical set of axioms.

   (b) Construct the rational number system $(\mathbb{Q}, +, \cdot, P_\mathbb{Q})$ using the relation given in Problem 7 (b) of Chapter 16.
   (Hints: Prove (again) that the relation is an equivalence relation; in order to prove transitivity (without using division!), state and prove a cancellation property for integer multiplication, similar to Proposition 21.34. Define $\mathbb{Q}, +, \cdot$, and $P_\mathbb{Q}$; show that your definitions are independent of the equivalence class representatives chosen. Prove that $(\mathbb{Q}, +, \cdot, P_\mathbb{Q})$ satisfies the characteristic properties of part (a).)

7. (a) Prove that the sum of two real numbers is a real number.
   (b) Prove that the sum of two positive real numbers is a positive real number.

(c) Prove that the additive inverse of a real number, defined in Definition 21.20, is a real number.

(d) Prove that the sum of a real number and its additive inverse is $0_\mathbb{R}$.

(e) Prove that for any real number $x_\mathbb{R}$, exactly one of the following holds: $x_\mathbb{R}$ is positive, $x_\mathbb{R}$ is negative, or $x_\mathbb{R} = 0_\mathbb{R}$.

(f) Prove that the product of two real numbers is a real number.

(g) Prove that the product of two positive real numbers is a positive real number.

(h) Prove that every nonzero real number has a multiplicative inverse.

(i) Prove that a real number $x_\mathbb{R}$ is less than a real number $y_\mathbb{R}$ if, and only if, $x_\mathbb{R} \subset y_\mathbb{R}$.

(j) Prove that every nonempty bounded set of real numbers has a supremum. (Hint: Consider their union.)

8. Define the set

$$\sqrt{3}_\mathbb{R} = \{q \in \mathbb{Q} \mid q \leq 0 \text{ or } q^2 < 3\}.$$

(a) Use Definition 21.15 to prove that $\sqrt{3}_\mathbb{R}$ is a real number.

(b) Use Definition 21.16 (and Proposition 21.17) to prove that $\sqrt{3}_\mathbb{R}$ is an irrational real number.

(c) Use Definition 21.22 (and Proposition 21.23) to prove that $\sqrt{3}_\mathbb{R} \cdot \sqrt{3}_\mathbb{R} = 3_\mathbb{R}$.

9. We say that an ordered field F has the *Cantor Property* if every infinite chain of closed, bounded, and nonempty intervals

$$I_1 \supseteq I_2 \supseteq I_3 \supseteq \cdots$$

in F has an element in common. In other words, suppose that $(a_n)_{n=1}^\infty$ is an increasing sequence in F and that $(b_n)_{n=1}^\infty$ is a decreasing sequence in F for which $a_n < b_n$ for every $n \in \mathbb{N}$; the Cantor Property of F guarantees an element $f \in F$ for which $a_n \leq f \leq b_n$ holds for all n.

(a) Prove that $\mathbb{R}$ has the Cantor Property. Prove also that if we further assume that $\lim(b_n - a_n) = 0$, then the element $f \in \mathbb{R}$ is unique.

(b) Does $\mathbb{Q}$ have the Cantor Property?

(c) Does $\mathbb{R}(x)$ have the Cantor Property? (Recall that $\mathbb{R}(x)$ denotes the ordered field of rational functions; cf. Problem 9 (d) of Chapter 13.)

(d) Is a version of the Cantor Property true in $\mathbb{R}$ where instead of a chain of closed intervals we are given a chain of open intervals? In other words, can we claim that there is a real number f for which $a_n < f < b_n$ holds for all n?

(e) Prove that the Archimedean Property of $\mathbb{R}$ (cf. Theorem 18.3) and the Cantor Property of $\mathbb{R}$ together (as well as the ordered field axioms) imply the Completeness Axiom (cf. Axiom 17.7).

Remark According to this result, we could have said that the categorical axioms of $\mathbb{R}$ are the ordered field axioms together with the Archimedean Property and the Cantor Property.

(Hints: Let S be a nonempty and bounded subset of $\mathbb{R}$. We will prove that S has a supremum; the proof that S also has an infimum can be done similarly (cf. also Theorem 17.13). We will construct a chain of intervals

$$I_1 \supseteq I_2 \supseteq I_3 \supseteq \cdots$$

recursively, as follows. Since S is bounded, we can choose a lower bound $a_1 \in \mathbb{R}$ for S and an upper bound $b_1 \in \mathbb{R}$ for S, and we set $I_1 = [a_1, b_1]$. Given $I_n = [a_n, b_n]$ with a midpoint of $c_n = (a_n + b_n)/2$ for some $n \geq 1$, we set $I_{n+1} = [a_{n+1}, b_{n+1}]$ equal to $[a_n, c_n]$ if c_n is an upper bound of S and to $[c_n, b_n]$ otherwise. Use the Archimedean and the Cantor Properties of $\mathbb{R}$ to prove that there is a unique real number f that is in $\cap_{n=1}^{\infty} I_n$, then prove that $f = \sup S$.)

10. (a) Define the set

$$\mathbb{Q}(\sqrt{2}) = \{a + b\sqrt{2} \mid a \in \mathbb{Q}, b \in \mathbb{Q}\}.$$

   i. Prove that $\mathbb{Q}(\sqrt{2})$ is a field.
   ii. Prove that $\mathbb{Q}(\sqrt{2})$ is an algebraic extension of $\mathbb{Q}$.
   iii. Is $\mathbb{Q}(\sqrt{2})$ algebraically closed?

(b) Define the set

$$\mathbb{Q}(\sqrt[3]{5}) = \{a + b\sqrt[3]{5} + c\sqrt[3]{25} \mid a \in \mathbb{Q}, b \in \mathbb{Q}, c \in \mathbb{Q}\}.$$

   i. Prove that $\mathbb{Q}(\sqrt[3]{5})$ is a field.
   ii. Prove that $\mathbb{Q}(\sqrt[3]{5})$ is an algebraic extension of $\mathbb{Q}$.
   iii. Is $\mathbb{Q}(\sqrt[3]{5})$ algebraically closed?

11. As we have mentioned before, Euler's number e is transcendental; that is, there is no not-identically-zero polynomial f with integer coefficients for which $f(e) = 0$. The fact that there is no such linear polynomial is equivalent to saying that e is irrational, and we proved this in Theorem 18.15. Prove that e is not the root of any quadratic polynomial with integer coefficients either.

(Hints: Suppose, indirectly, that there are integers a, b, and c for which $a \neq 0$ and

$$ae^2 + be + c = 0;$$

we then have

$$ae + b + \frac{c}{e} = 0.$$

Use the infinite series for e and for $1/e$; cf. page 282.)

12. In this problem we briefly discuss the elements of *algebraic number theory*, including some famous results, open questions, and even an April Fool's hoax.

(a) A complex number α is called a *quadratic integer* if it is the root of the polynomial $x^2 + ux + v$ for some integers u and v.

In each part below, a pair of numbers is given so that one number is a quadratic integer and the other is not. Decide which is which.

   i. 11 and $\frac{3}{2}$

   ii. $7 + \sqrt{2}$ and $7 + \sqrt[3]{4}$

   iii. $\frac{3}{2} + \frac{11}{2}\sqrt{3}$ and $\frac{3}{2} + \frac{11}{2}\sqrt{5}$

   iv. $7 + 3i$ and $7 + 3\sqrt{i}$

   v. $\frac{3}{2} + \frac{11}{2}\sqrt{-3}$ and $\frac{3}{2} + \frac{11}{2}\sqrt{-5}$

(b) Prove that the roots of the polynomial $x^2 + ux + v$ are integers if, and only if, $u^2 - 4v$ is a square number (including 0).

(c) Let u and v be integers for which $u^2 - 4v$ is not a square number; we define a field associated with the polynomial $x^2 + ux + v$, as follows. Let

$$m = \frac{u^2 - 4v}{k^2},$$

where k^2 is the largest square number divisor of $u^2 - 4v$. (Note that 1 is a square, and thus every positive or negative integer has a unique largest square divisor.) We thus see that m is a *square-free integer*: its largest square divisor is 1.

Define the set

$$Q(\sqrt{m}) = \{u + b\sqrt{m} \mid u \in \mathbb{Q}, b \in \mathbb{Q}\};$$

as Problem 10 (a) above suggests, $Q(\sqrt{m})$ is a field for the usual operations. For example, we see that the field associated with the polynomial $x^2 + 3x + 1$ is $Q(\sqrt{5})$. (The same field is associated with $x^2 + 3kx + k^2$ for any nonzero integer k.)

Furthermore, we define $I(m)$ as the set of quadratic integers in $Q(\sqrt{m})$; $I(m)$ is then an integral domain.

Prove that

$$
I(m) = \begin{cases} \{a + b\sqrt{m} \mid a \in \mathbb{Z}, b \in \mathbb{Z}\} & \text{if } m \not\equiv 1 \bmod 4, \\[2mm] \{\frac{a}{2} + \frac{b}{2}\sqrt{m} \mid a \in \mathbb{Z}, b \in \mathbb{Z}, a \equiv b \bmod 2\} & \text{if } m \equiv 1 \bmod 4. \end{cases}
$$

(d) Analogously to the integral domain of the integers, we can develop the number theory of $I(m)$. We make the following definitions.

Definition 21.35 *For quadratic integers* $\alpha, \beta \in I(m)$, *we say that* α *divides* β, *and write* $\alpha|\beta$, *if there is a* $\gamma \in I(m)$ *for which* $\alpha \cdot \gamma = \beta$.

Note that Definition 21.35 is analogous to Definition 2.2.

Definition 21.36 *We say that an element* μ *of* $I(m)$ *is a* unit element *if* μ *divides every element of* $I(m)$.

It is easy to see that μ is a unit element if, and only if, it has a multiplicative inverse in $I(m)$. While we only had two unit elements among the integers, this is not necessarily the case among quadratic integers.

   Prove each of the following statements.

   i. $I(-1)$ has four unit elements: ± 1 and $\pm i$.
   ii. $I(-3)$ has six unit elements: ± 1 and $\pm\frac{1}{2} \pm \frac{1}{2}\sqrt{-3}$.
   iii. When $m < 0$ and $m \neq -1, -3$, then $I(m)$ has two unit elements: ± 1.
   iv. $I(2)$ has infinitely many unit elements.
      (Hint: Verify that $(3 + 2\sqrt{2})^n$ is a unit element for all $n \in \mathbb{N}$.)

(e) Next, we define primes in $I(m)$, but we need to be careful: in contrast to the integers, Definitions 2.1a and 2.1b (see Chapter 2) are not equivalent in general. Therefore, we distinguish between irreducible elements and primes, as follows.

Definition 21.37 *An element of* $I(m)$ *is called* irreducible *if it is not a unit element, and it cannot be factored into a product of two other elements of* $I(m)$ *without one of them being a unit element.*

Definition 21.38 *An element of* $I(m)$ *is called* prime *if it is not 0 and not a unit element, and it cannot divide a product of two other elements of* $I(m)$ *without dividing at least one of them.*

Prove that each of the numbers

$$
2, \ 3, \ 1 + \sqrt{-5}, \ 1 - \sqrt{-5}
$$

is irreducible in $I(-5)$ but that none of them are prime.

(Hint: Note that the product of the first two equals the product of the last two.)

(f) The most important number-theoretic question in any integral domain is whether the Fundamental Theorem of Arithmetic (FTA) holds; that is, if every nonzero and non-unit element factors into a product of irreducible elements in an essentially unique way. (For irreducible elements $\alpha_1, \ldots, \alpha_r; \beta_1, \ldots, \beta_s \in I(m)$, we say that the irreducible factorization $\alpha_1 \cdots \alpha_r$ is essentially the same as the irreducible factorization $\beta_1 \cdots \beta_s$, if each factor in one factorization is a unit times a factor in the other factorization; more precisely, if $r = s$ and there are unit elements $\mu_1, \ldots, \mu_r$ for which the multisets $[\mu_1\alpha_1, \ldots, \mu_r\alpha_r]$ and $[\beta_1, \ldots, \beta_r]$ are the same.)

Prove that the FTA fails in $I(-5)$.

(Hint: Use part (e).)

Remarks The problem of finding all square-free integers m for which the FTA holds in $I(m)$ has a long and fascinating history. In the case in which m is negative, we have known the answer since 1952: the FTA holds if, and only if,

$$m = -1, -2, -3, -7, -11, -19, -43, -67, \text{ or } -163.$$

These nine integers are known as Heegner numbers, named after the German mathematician Kurt Heegner (1893–1965) who first established the result.

Regarding the case in which $m > 0$, at the present time, we know that the FTA holds in $I(m)$ for

$$m = 2, 3, 5, 6, 7, 11, 13, 14, 17, 19, 21, 22, 23, 29, 31, 33, 37, 38, 41, 43, 46, 47, \ldots;$$

but the complete characterization seems quite elusive.

Astute observers may notice that the last six Heegner numbers are in one-to-one correspondence with the six lucky numbers of Euler (see Problem 3 in Chapter 3). In particular, one can prove that the existence of another such value k would be equivalent to the FTA holding in $I(m)$ with $m = 1 - 4k$; this fact rests on the factorization

$$n^2 - n + k = \left(\frac{2n-1}{2} + \frac{1}{2}\sqrt{1-4k}\right) \cdot \left(\frac{2n-1}{2} - \frac{1}{2}\sqrt{1-4k}\right).$$

The Heegner numbers give us an opportunity to mention two other famous individuals: Srinivasa Ramanujan (1887–1920) and Martin Gardner (1912–2010). Ramanujan was a self-trained genius who, at the beginning of the twentieth century in poor and rural India, made a range of amazing mathematical discoveries, some of which are still not fully understood today. Gardner was a brilliant American mathematics and science writer, most famous for his long-running column in *Scientific American* magazine.

As an April Fool's joke, Gardner wrote in 1975 that the conjecture of Ramanujan regarding the value of

$$e^{\pi\sqrt{163}}$$

has been proven: the value is exactly an integer! Gardner soon afterwards admitted that his article was a hoax: Ramanujan made no such conjecture and, in fact, as we mentioned on page 337, the value is transcendental! However, it is indeed true that Ramanujan's constant, as the number is now being referred to, is very close to being an integer:

$$e^{\pi\sqrt{163}} \approx 262537412640768743.99999999999925\ldots.$$

In fact, the other Heegner numbers yield near-integers as well; for example,

$$e^{\pi\sqrt{19}} \approx 12^3 \cdot (3^2 - 1)^3 + 744 - 0.222\ldots,$$

$$e^{\pi\sqrt{43}} \approx 12^3 \cdot (9^2 - 1)^3 + 744 - 0.000223\ldots,$$

$$e^{\pi\sqrt{67}} \approx 12^3 \cdot (21^2 - 1)^3 + 744 - 0.00000134\ldots,$$

$$e^{\pi\sqrt{163}} \approx 12^3 \cdot (231^2 - 1)^3 + 744 - 0.00000000000075\ldots.$$

For an explanation, one needs to turn to another fast-developing branch of mathematics: *analytic number theory*.

Chapter 22
Games Are Valuable!

In this chapter we return to our very first adventure in this book: the analysis of games. As an illustrative example, we evaluate our good old game *Arion* of Chapter 1; namely, we show that it has exactly a "one half move advantage" for player A (we will, of course, make this notion precise). But more generally—and a lot more abstractly—we will discuss a fascinating novel branch of mathematics that has both far-reaching theoretical significance and very practical applicability. In the process, we revisit a remarkable variety of material from previous chapters and thus take advantage of the opportunity to deepen and extend our understanding.

Our main focus in this chapter will be on a large collection Γ of two-person games that includes a variety of games we have already seen. We will introduce a very natural equivalence relation on Γ and thus create equivalence classes; we then consider these equivalence classes of games to be our collection of "surreal numbers" $\mathbb{S}$. For example, we will see that all fair games (cf. Problem 1 of Chapter 7) form a single equivalence class, and we identify this class with the surreal number $0_{\mathbb{S}}$. (In a sense, the "further away" a game is from $0_{\mathbb{S}}$, the more "unfair" it is.) We will then learn that $\mathbb{S}$ is an ordered field for certain addition, multiplication, and order. The field $\mathbb{S}$ of surreal numbers is, in fact, the largest ordered field in the sense that it contains an isomorphic copy of every ordered field. In particular, $\mathbb{S}$ will have a clearly identifiable subfield that is isomorphic to $\mathbb{R}$; furthermore, it will also contain (copies of) Cantor's infinite ordinal numbers (ever-increasing notions of infinite "lengths" of sets) as well as infinitesimals (positive values that are below all the positive reals).

This branch of mathematics, sometimes referred to as *combinatorial game theory* (to be distinguished from *economic game theory*), was invented and first developed by John Horton Conway (1937–2020) in the 1970s; the beautiful and deep theory is undoubtedly one of the top inventions in the history of mathematics. (The term of "surreal number" was coined by Donald Knuth in a charming book—*Surreal Numbers*, Addison–Wesley, 1974—that explains "how two ex-students" on a remote island discovered the concept "and found total happiness.")

© Springer Nature Switzerland AG 2020
B. Bajnok, *An Invitation to Abstract Mathematics*, Undergraduate Texts in Mathematics, https://doi.org/10.1007/978-3-030-56174-1_22

The collection of games that we are studying here are played by two players—whom we call Left and Right—who take turns selecting one of the options available to them at the time. At each stage of the game, all options as well as all subsequent options—both those available for Left and those available for Right—are known to both players in advance. (This condition rules out card games that build on hidden information and dice games where the options are determined by chance while the game is being played.) Note that the number of options may be finite or infinite. The game ends when one player is unable to move; this player is then declared the loser and the player able to move last the winner. We require that every game ends by one of the players winning after a finite number of steps. (So here we exclude games that can be drawn out forever or may end in a tie.) We let Ω denote the class of games satisfying these very general conditions. The collection of games Γ mentioned above is a certain subclass of Ω; while all games in Ω are in a certain definite sense "value-able," it is the games in Γ to which we will assign numeric (though "surreal") values. We will, of course, explain all this precisely below.

Note that, since we have full information, every game G is completely determined by the set of options $\mathcal{L}$ and $\mathcal{R}$ available for players Left and Right, respectively, and, therefore, we may identify G with the ordered pair $(\mathcal{L}, \mathcal{R})$. For example, consider the game *Arion* of Chapter 1 given by Figure 22.1.

Fig. 22.1 The game *Arion*

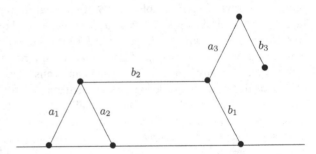

Recall that players A and B take turns to remove edges available to them; if the removal of an edge disconnects a part of the diagram from the "ground," then that component becomes unavailable. Let L_1, L_2, and L_3 denote the results of player Left exercising options a_1, a_2 and a_3, respectively, and let R_1, R_2, and R_3 denote the results of player Right exercising options b_1, b_2 and b_3, respectively—see Figures 22.2 and 22.3.

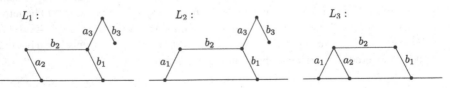

Fig. 22.2 The games L_1, L_2, and L_3, resulting after Left exercises options a_1, a_2 and a_3, respectively

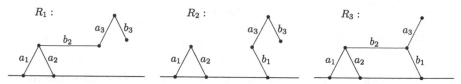

Fig. 22.3 The games R_1, R_2, and R_3, resulting after Right exercises options b_1, b_2 and b_3, respectively

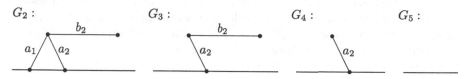

Fig. 22.4 A play of *Arion* after L_3

We then have

$$Arion = (\{L_1, L_2, L_3\}, \{R_1, R_2, R_3\}).$$

Of course, the elements of $\mathcal{L} = \{L_1, L_2, L_3\}$ and $\mathcal{R} = \{R_1, R_2, R_3\}$ are games themselves, and thus could each be presented as ordered pairs of their respective left- and right-options.

When playing a game, each player, when it is his or her turn, chooses an option and thereby transforms the game to another—"simpler"—game. The process continues, and each play of a particular game G is essentially a finite sequence of—ever "simpler"—games $G_0 = G, G_1, G_2, G_3$, etc., ending in a game G_n that has no available options for the player whose turn it would be to move. For instance, a particular play of *Arion* is the sequence $(G_0, G_1, G_2, G_3, G_4, G_5)$ where $G_0 = Arion$, $G_1 = L_3$ (see above), and the games G_2, G_3, G_4, and G_5 are as in Figure 22.4. In particular, in this play of the game, Left makes the final move and thus wins the game. This list of games reveals pertinent information: for example, if Left makes the initial move in G as well as the final (winning) move, as in the example above, then

- n is odd;
- G_i is a left-option of G_{i-1} for $i = 1, 3, 5, \ldots, n$ and a right-option of G_{i-1} for $i = 2, 4, 6, \ldots, n - 1$; and
- G_n has no right-options.

(Analogous claims could be made for the cases in which Right starts the game or when Right wins.)

After this heuristic introduction, let's turn to a precise development. In the treatment that we choose here, the collection Ω of games will not be given a definition; instead, we consider it a primitive, satisfying—and, in the sense of Chapter 21, categorically determined by—the following two axioms. We think of

Fig. 22.5 The game $G(0)$

Ω as a collection of ordered pairs of sets; that is, it consists of elements of the form $(\mathcal{L}, \mathcal{R})$ where $\mathcal{L}$ and $\mathcal{R}$ are sets. (We need to choose our terminology carefully: while we insist on $\mathcal{L}$ and $\mathcal{R}$ being sets, Ω will only be a "collection" as it will be too large to be a set—cf. Appendix C.) So that we can state our axioms more concisely, we will use the following term: We say that a collection of ordered pairs of sets is *inductive* if it has the property that whenever $\mathcal{L}$ and $\mathcal{R}$ are *subsets* of the collection, the ordered pair $(\mathcal{L}, \mathcal{R})$ is an *element* of the collection.

Our two axioms for Ω are as follows:

Axiom 22.1 *The collection Ω is inductive.*

Axiom 22.2 *The only inductive subcollection of Ω is Ω itself.*

Axioms 22.1 and 22.2 resemble the Peano Axioms (cf. Chapter 21); while the objective of Axiom 22.1 is to ensure that Ω is large enough to contain all games we would want to play, Axiom 22.2 assures that it doesn't contain other— unnecessary—elements.

It may seem strange at first sight that—in contrast to the Peano Axioms—we do not have an explicit axiom stating that Ω is nonempty. Note, however, that the empty-set clearly satisfies $\emptyset \subseteq \Omega$ (even if Ω were to be empty itself), thus $(\emptyset, \emptyset) \in \Omega$ by Axiom 22.1; adding this as an axiom is therefore not necessary. We let

$$G(0) = (\emptyset, \emptyset);$$

this (very uninteresting) game can be thought of as the *Hackenbush* game of Figure 22.5, where neither player has any options. We say that $G(0)$ appears in "generation 0" ($G(0)$ is the only game in generation 0).

Once we have $G(0)$, we can use it to define three new games:

$$G(1) = (\{G(0)\}, \emptyset),$$

$$G(-1) = (\emptyset, \{G(0)\}),$$

and

$$N(1) = (\{G(0)\}, \{G(0)\});$$

these are the games of "generation 1"; these three games are indeed in Ω by Axiom 22.1. The *Hackenbush* representations of these games are in Figure 22.6. (We mark edges available for Left by solid lines, those available by Right by dotted lines, and those available to both by dash lines.) For example, to see that the *Hackenbush* game marked "$G(-1)$" above is indeed what we defined $G(-1)$ to be, note that

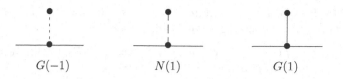

Fig. 22.6 The three games of Ω_1

Left has no options, and Right has one option that reduces the game to $G(0)$, thus we have $(\emptyset, \{G(0)\}) = G(-1)$.

Our four games so far provide examples for the four different possible outcomes that a game may have. Namely, by definition, the second player wins $G(0)$ (as the first player immediately loses), Left wins $G(1)$ (as Right will run out of moves first, regardless of who starts), Right wins $G(-1)$ (again, regardless of who starts to play it), and the first player to move wins $N(1)$.

It is helpful to think of the recursive construction procedure of games in terms of these "generations"; the collection of games appearing in generations 0, 1, 2, etc., will be denoted by Ω_0, Ω_1, Ω_2, and so on, and are said to have *birthdays* 0, 1, 2, etc., respectively. (We will see that not all games appear in a finite generation, thus we have games with infinite birthdays indexed not just by natural numbers, but by infinite ordinals.) So far, we have seen that

$$\Omega_0 = \{G(0)\}$$

and

$$\Omega_1 = \{G(-1), N(1), G(1)\}.$$

In order to create a game of birthday 2, observe that we may choose any subset of $\Omega_0 \cup \Omega_1$ for both the set of left-options and the set of right-options, as long as not both are subsets of Ω_0 (that is, equal to $\emptyset$ or $\{G(0)\}$), therefore

$$|\Omega_2| = 2^4 \cdot 2^4 - 4 = 252.$$

Clearly, for each $n \in \mathbb{N} \cup \{0\}$, $|\Omega_n|$ is finite, but grows very rapidly as n increases. In particular, we see that

$$|\Omega_n| = 2^{\sum_{i=0}^{n-1} |\Omega_i|} \cdot 2^{\sum_{i=0}^{n-1} |\Omega_i|} - \sum_{i=0}^{n-1} |\Omega_i|;$$

for example, $|\Omega_3| = 4^{256} - 256$—a number that has 154 decimal digits!

Our next goal is to find a way to compare two games from the point of view of the two players. Namely, we will introduce a "less than" relation on the collection of all games: we will say that the game G_1 is less than the game G_2, if G_1 is "less desirable" for Left than G_2 is (and thus G_2 is "more desirable" for Right than G_1 is). More precisely, we make the following definition.

Definition 22.3 *Let $G = (\mathcal{L}_G, \mathcal{R}_G) \in \Omega$ and $H = (\mathcal{L}_H, \mathcal{R}_H) \in \Omega$. We say that G is less than or similar to H and write $G \lesssim H$ if, and only if,*

* *there is no $L \in \mathcal{L}_G$ for which $H \lesssim L$, and*
* *there is no $R \in \mathcal{R}_H$ for which $R \lesssim G$.*

If G is not less than or similar to H we write $G \not\lesssim H$.

Note that we have a recursive definition without an initial condition given. However, as before, we see that the definition applies to the game $G(0) = (\emptyset, \emptyset)$: we trivially have $G(0) \lesssim G(0)$. More generally, we have the following obvious fact.

Proposition 22.4 *For all sets of games $\mathcal{X}$ and $\mathcal{Y}$, we have $(\emptyset, \mathcal{X}) \lesssim (\mathcal{Y}, \emptyset)$.*

By Proposition 22.4, we have $G(0) \lesssim G(0)$, $G(0) \lesssim G(1)$, $G(-1) \lesssim G(0)$, and $G(-1) \lesssim G(1)$. By Definition 22.3, $G(0) \lesssim G(0)$ implies that $G(1) \not\lesssim G(0)$ and $G(0) \not\lesssim G(-1)$; furthermore, $G(-1) \lesssim G(0)$ implies that $G(1) \not\lesssim G(-1)$. With a bit more work, we can verify each entry in Table 22.1—see Problem 1.

In addition to the $\lesssim$ relation on games just defined, we introduce the following relations.

Definition 22.5 *Suppose that $G \in \Omega$ and $H \in \Omega$. We say that*

* *G is similar to H, and write $G \approx H$, if $G \lesssim H$ and $H \lesssim G$;*
* *G is greater than or similar to H, and write $G \gtrsim H$, if $H \lesssim G$;*
* *G is less than H, and write $G < H$, if $G \lesssim H$ and $G \not\approx H$; and*
* *G is greater than H, and write $G > H$, if $H < G$.*

For example, according to the table above, each of the four games of generations 0 and 1 is similar to itself; furthermore, we have

$$G(-1) < G(0) < G(1)$$

and

$$G(-1) < N(1) < G(1).$$

Table 22.1 The "less than or similar to" relation on the games of Ω_0 and Ω_1

Is $G \lesssim H$?	$H = G(-1)$	$H = G(0)$	$H = N(1)$	$H = G(1)$
$G = G(-1)$	Yes	Yes	Yes	Yes
$G = G(0)$	No	Yes	No	Yes
$G = N(1)$	No	No	Yes	Yes
$G = G(1)$	No	No	No	Yes

(Without assuming that the "less than" relation is transitive—luckily, it is!—it would have been more clear to state these claims via separate inequalities; instead of the somewhat ambiguous $G(-1) < G(0) < G(1)$, write $G(-1) < G(0)$, $G(0) < G(1)$, and $G(-1) < G(1)$.)

However, we see that $\lesssim$ is not a total order relation on Ω: we have $G(0) \not\lesssim N(1)$ and $N(1) \not\lesssim G(0)$; that is, $G(0)$ and $N(1)$ are incomparable! We should also note that, while reflexivity and transitivity of $\lesssim$ hold, antisymmetry fails: we will soon see examples in which each of two distinct games is less than or similar to the other. (This explains why we prefer to use the $\lesssim$ sign and not the $\leq$ sign that suggests antisymmetry.)

For all practical purposes, two similar games are indeed similar (pun intended): exchanging one by the other will not make a difference. We will make this notion more precise later; for now, we just state the following:

Proposition 22.6 *Two games that have similar left-options and similar right-options are themselves similar. More precisely, let $G = (\mathcal{L}_G, \mathcal{R}_G) \in \Omega$ and $H = (\mathcal{L}_H, \mathcal{R}_H) \in \Omega$, and suppose that*

- $\forall L_G \in \mathcal{L}_G, \exists L_H \in \mathcal{L}_H, L_G \approx L_H$;
- $\forall L_H \in \mathcal{L}_H, \exists L_G \in \mathcal{L}_G, L_H \approx L_G$;
- $\forall R_G \in \mathcal{R}_G, \exists R_H \in \mathcal{R}_H, R_G \approx R_H$; *and*
- $\forall R_H \in \mathcal{R}_H, \exists R_G \in \mathcal{R}_G, R_H \approx R_G$;

Then $G \approx H$.

We leave the easy proof to Problem 2.

We now define a subclass of Ω that we are most interested in: the class of numeric games.

Definition 22.7 *The collection of numeric games Γ consists of games $(\mathcal{L}, \mathcal{R})$ where $\mathcal{L} \subseteq \Gamma$, $\mathcal{R} \subseteq \Gamma$, and there is no $L \in \mathcal{L}$ and $R \in \mathcal{R}$ for which $R \lesssim L$.*

Definition 22.7 calls for a few comments. First of all, note that, as before, while we seem to have a recursive definition without a starting point, it is not necessary: $G(0) = (\emptyset, \emptyset)$ is trivially a numeric game.

Also, observe that in Definition 22.7, we used the predicate

$$\neg(\exists L \in \mathcal{L}, \exists R \in \mathcal{R}, R \lesssim L)$$

instead of the more direct form

$$\forall L \in \mathcal{L}, \forall R \subset \mathcal{R}, L < R.$$

The reason is that while, as we prove later, the two forms are indeed equivalent for numeric games, we did not want to make this assumption in our definition. (In the larger class Ω of not-necessarily-numeric games the two forms are not equivalent: as we have seen, we have neither $G(0) \lesssim N(1)$ nor $N(1) < G(0)$.)

Before continuing with our development, it may be worth briefly recalling our discussion of real numbers from Chapter 21. We constructed real numbers using Dedekind cuts; that is, "initial segments" of rational numbers (cf. Definition 21.15). For example, we let

$$0_\mathbb{R} = \{q \in \mathbb{Q} \mid q < 0_\mathbb{Q}\},$$

$$2_\mathbb{R} = \{q \in \mathbb{Q} \mid q < 2_\mathbb{Q}\},$$

and

$$\sqrt{2}_\mathbb{R} = \{q \in \mathbb{Q} \mid q \leq 0_\mathbb{Q} \text{ or } q^2 < 2_\mathbb{Q}\}.$$

Perhaps a more pleasing variation of this approach would have been to let real numbers denote ordered pairs of sets of rational numbers of the form $(\mathcal{L}, \mathcal{R})$, where the "left segment" $\mathcal{L}$ denotes a Dedekind cut and the "right segment" denotes its complement $\mathbb{Q} \setminus \mathcal{L}$. For example, we could write

$$0_\mathbb{R} = (\{q \in \mathbb{Q} \mid q < 0_\mathbb{Q}\}, \{q \in \mathbb{Q} \mid q \geq 0_\mathbb{Q}\}),$$

$$2_\mathbb{R} = (\{q \in \mathbb{Q} \mid q < 2_\mathbb{Q}\}, \{q \in \mathbb{Q} \mid q \geq 2_\mathbb{Q}\}),$$

and

$$\sqrt{2}_\mathbb{R} = (\{q \in \mathbb{Q} \mid q \leq 0_\mathbb{Q} \text{ or } q^2 < 2_\mathbb{Q}\}, \{q \in \mathbb{Q} \mid q > 0_\mathbb{Q} \text{ and } q^2 > 2_\mathbb{Q}\}).$$

(As we have shown, there is no rational number q for which $q^2 = 2_\mathbb{Q}$.)

This formulation would align well with our intuition of a "cut": the ordered pair $(\mathcal{L}, \mathcal{R})$ is simply a partition of the (previously defined) set of rational numbers with certain additional properties, most importantly, that $\mathcal{L}$ is "to the left" of $\mathcal{R}$, that is,

$$\forall l \in \mathcal{L}, \forall r \in \mathcal{R}, l < r.$$

(We should note that $<$ here refers to the previously defined order relation among the rational numbers.)

Conway's brilliant idea was that the ordered pair $(\mathcal{L}, \mathcal{R})$ can be considered in general as a game. Thus, games are generalizations of Dedekind cuts; in fact, as we already mentioned and will see below, the definition of games is so general that it will lead us to rediscover not just real numbers, but many other types of "numbers," such as Cantor's infinite ordinal numbers, as well as infinitesimally small positive numbers.

So let us return to Definition 22.7 and see some specific numeric games. We have already seen that

$$G(0) = (\emptyset, \emptyset)$$

is trivially numeric; it is also easy to see that

$$G(1) = (\{G(0)\}, \emptyset)$$

and

$$G(-1) = (\emptyset, \{G(0)\})$$

are as well. In fact, we have the following obvious generalization.

Proposition 22.8 *For all sets of numeric games $\mathcal{X}$ and $\mathcal{Y}$, we have $(\emptyset, \mathcal{X}) \in \Gamma$ and $(\mathcal{Y}, \emptyset) \in \Gamma$.*

However, $N(1) = (\{G(0)\}, \{G(0)\}) \notin \Gamma$, since, by Proposition 22.4, $G(0) \lesssim G(0)$, contrary to our requirement in Definition 22.7.

Letting $\Gamma_n = \Gamma \cap \Omega_n$ denote the collection of numeric games with birthday n, we thus see that

$$\Gamma_0 = \{G(0)\}$$

and

$$\Gamma_1 = \{G(-1), G(1)\}.$$

Let us now attempt to exhibit Γ_2. As it turns out, it is a bit easier to consider all games that have birthday at most 2 (that is, we include the games of Γ_0 and Γ_1). We have three numeric games of birthdays at most 1 available, $G(-1)$, $G(0)$, and $G(1)$. As we have already seen, we have

$$G(-1) < G(0) < G(1).$$

Each of the games in $\Gamma_0 \cup \Gamma_1 \cup \Gamma_2$ is of the form $(\mathcal{L}, \mathcal{R})$ for some sets

$$\mathcal{L} \subseteq \{G(-1), G(0), G(1)\}$$

and

$$\mathcal{R} \subseteq \{G(-1), G(0), G(1)\};$$

we also must uphold the condition that no element of $\mathcal{R}$ is less than or similar to any element of $\mathcal{L}$. When creating such a game $(\mathcal{L}, \mathcal{R})$, we may first decide how many of the games $G(-1)$, $G(0)$, and/or $G(1)$ should appear in either $\mathcal{L}$ or $\mathcal{R}$ (no game may appear in both). If we want k of them to appear ($k \in \{0, 1, 2, 3\}$), then we have $\binom{3}{k}$ choices for selecting them. Next, we need to decide which of them will be in $\mathcal{L}$ and which in $\mathcal{R}$; we can put the l smallest ones (with $0 \leq l \leq k$) into $\mathcal{L}$ and the rest into $\mathcal{R}$.

Therefore, we have a total of

$$\sum_{k=0}^{3}\sum_{l=0}^{k}\binom{3}{k} = 20$$

numeric games that get created by generation 2; they are as follows:

$k = 0 : l = 0 : (\emptyset, \emptyset)$

$k = 1 : l = 0 : (\emptyset, \{G(-1)\})$ $(\emptyset, \{G(0)\})$ $(\emptyset, \{G(1)\})$

$\quad\quad l = 1 : (\{G(-1)\}, \emptyset)$ $(\{G(0)\}, \emptyset)$ $(\{G(1)\}, \emptyset)$

$k = 2 : l = 0 : (\emptyset, \{G(-1), G(0)\})$ $(\emptyset, \{G(-1), G(1)\})$ $(\emptyset, \{G(0), G(1)\})$

$\quad\quad l = 1 : (\{G(-1)\}, \{G(0)\})$ $(\{G(-1)\}, \{G(1)\})$ $(\{G(0)\}, \{G(1)\})$

$\quad\quad l = 2 : (\{G(-1), G(0)\}, \emptyset)$ $(\{G(-1), G(1)\}, \emptyset)$ $(\{G(0), G(1)\}, \emptyset)$

$k = 3 : l = 0 : (\emptyset, \{G(-1), G(0), G(1)\})$

$\quad\quad l = 1 : (\{G(-1)\}, \{G(0), G(1)\})$

$\quad\quad l = 2 : (\{G(-1), G(0)\}, \{G(1)\})$

$\quad\quad l = 3 : (\{G(-1), G(0), G(1)\}, \emptyset)$

Since one of these, $G(0) = (\emptyset, \emptyset)$, was already created in generation 0, and two others, $G(-1) = (\emptyset, \{G(0)\})$ and $G(1) = (\{G(0)\}, \emptyset)$, were created in generation 1, we get a total of seventeen new games in Γ_2. Here we introduce notations for four of them:

$$G(-2) = (\emptyset, \{G(-1), G(0)\}),$$

$$G\left(-\tfrac{1}{2}\right) = (\{G(-1)\}, \{G(0))\},$$

$$G\left(\tfrac{1}{2}\right) = (\{G(0)\}, \{G(1)\}),$$

$$G(2) = (\{G(0), G(1)\}, \emptyset);$$

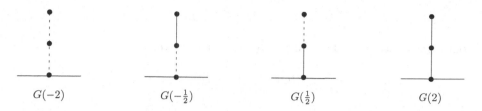

$G(-2)$ $G(-\frac{1}{2})$ $G(\frac{1}{2})$ $G(2)$

Fig. 22.7 Four games of Γ_2

the *Hackenbush* representation of these games can be seen in Figure 22.7. For example, to verify that the *Hackenbush* game marked "$G(\frac{1}{2})$" is indeed the game

$$G\left(\tfrac{1}{2}\right) = (\{G(0)\}, \{G(1)\}),$$

note that both Left and Right have a single option; exercising their options, Left would reduce the game to $G(0)$, and Right would reduce it to $G(1)$. Similarly, in the game marked "$G(2)$," Left has two options, to reduce the game to $G(0)$ or to $G(1)$, and Right has no options, so this is indeed the game $(\{G(0), G(1)\}, \emptyset)$.

As we are about to show, these four games are pivotal among the seventeen games of Γ_2: every game in Γ_2 is similar to one of these four or to one of the games of an "earlier" generation.

Proposition 22.9 *Among the seventeen games of Γ_2,*

- *$(\{G(-1)\}, \emptyset)$, $(\emptyset, \{G(1)\})$, and $(\{G(-1)\}, \{G(1)\})$ are similar to $G(0)$;*
- *$(\{G(-1), G(0)\}, \emptyset)$ is similar to $G(1)$;*
- *$(\emptyset, \{G(0), G(1)\})$ is similar to $G(-1)$;*
- *$(\{G(1)\}, \emptyset)$, $(\{G(-1), G(1)\}, \emptyset)$, and $(\{G(-1), G(0), G(1)\}, \emptyset)$ are similar to $G(2)$;*
- *$(\emptyset, \{G(-1)\})$, $(\emptyset, \{G(-1), G(1)\})$, and $(\emptyset, \{G(-1), G(0), G(1)\})$ are similar to $G(-2)$;*
- *$(\{G(-1), G(0)\}, \{G(1)\})$ is similar to $G(\frac{1}{2})$; and*
- *$(\{G(-1)\}, \{G(0), G(1)\})$ is similar to $G(-\frac{1}{2})$.*

We should point out that, as Proposition 22.9 illustrates, similar games may have different birthdays!

Proof We only prove our first claim; the others can be proved similarly. (Also, all claims other than our first claim follow directly from the Domination Theorem; see Theorem 22.31.)

We start by noting that, by Proposition 22.4, we have

$$(\{G(-1)\}, \emptyset) \gtrsim G(0)$$

and

$$(\emptyset, \{G(1)\}) \lesssim G(0).$$

Furthermore, we know that $G(0) \not\lesssim G(-1)$, and, therefore,

$$(\{G(-1)\}, \emptyset) \lesssim G(0)$$

and

$$(\{G(-1)\}, \{G(1)\}) \lesssim G(0).$$

Similarly, $G(1) \not\lesssim G(0)$, and thus

$$G(0) \lesssim (\emptyset, \{G(1)\})$$

and

$$G(0) \lesssim (\{G(-1)\}, \{G(1)\}).$$

Therefore, we see that

$$(\{G(-1)\}, \emptyset) \approx G(0),$$

$$(\emptyset, \{G(1)\}) \approx G(0),$$

and

$$(\{G(-1)\}, \{G(1)\}) \approx G(0)$$

by Definition 22.5.   □

While the size of Γ_n is not nearly as large as that of Ω_n, it still increases very rapidly with n; for example, Γ_3 has 7,143,404 elements (cf. Problem 3). Thus exhibiting each game of Γ—even of Γ_3—is hopeless; however, as we will see, there is no need to do that as all of them will be similar to one of only fifteen games: seven of them,

$$G(-2), \ G(-1), \ G(-\tfrac{1}{2}), \ G(0), \ G(\tfrac{1}{2}), \ G(1), \ G(2),$$

we met before, and we will define eight others,

$$G(-3), \ G(-\tfrac{3}{2}), \ G(-\tfrac{3}{4}), \ G(-\tfrac{1}{4}), \ G(\tfrac{1}{4}), \ G(\tfrac{3}{4}), \ G(\tfrac{3}{2}), \ G(3).$$

In other words, Γ_3 can be partitioned into fifteen equivalence classes based on similarity. To see this, we first turn to a more general discussion of the beautiful theory of combinatorial games.

Since games are defined recursively, it is natural to expect that many of our proofs will be inductive; but, since we have more than just countably many games, it is also clear that we need a version of induction that is stronger than regular induction (which is suitable only for statements about countable sets). We have already mentioned transfinite induction (cf. Theorem 17.21) that can handle such proofs. The version of transfinite induction we employ here can be stated as follows:

Theorem 22.10 (Conway's Induction Principle) *Suppose that $P(G)$ is a predicate that becomes a statement for all $G \in \Omega$. If the implication*

$$\left(\bigwedge_{K \in \mathcal{L} \cup \mathcal{R}} P(K) \right) \Rightarrow P(H)$$

holds for all $H = (\mathcal{L}, \mathcal{R}) \in \Omega$, then $P(G)$ is true for every $G \in \Omega$.

It is easy to see that Conway's Induction Principle follows from Axiom 22.2.

So, to use Conway induction to prove that a predicate $P(G)$ is true for every game G, one just needs to verify that, for every game $H \in \Omega$, $P(H)$ holds whenever $P(L)$ and $P(R)$ hold for all left-options L and right-options R of H. It might be strange at first sight that in Conway induction there is no need for a "base case." Note, however, that the inductive hypothesis guarantees that $P(G(0))$ is true: indeed, since $G(0) = (\emptyset, \emptyset)$, the assumption in the inductive hypothesis is trivially true for $H = G(0)$.

As an easy application, we use Conway induction to prove that every game (even those with infinitely many options) comes to an end after a finite number of moves. More generally, we prove the following:

Theorem 22.11 *There is no infinite sequence of games $(G_n)_{n=1}^{\infty}$ so that for each $n \in \mathbb{N}$, G_{n+1} is a left- or right-option of G_n.*

Proof Let $P(G)$ be the predicate that there is no infinite sequence of games $(G_n)_{n=1}^{\infty}$ so that for each $n \in \mathbb{N}$, G_{n+1} is a left- or right-option of G_n, and G_1 is a left- or right-option of G. We need to prove that $P(G)$ is true for every $G \in \Omega$.

Let $H = (\mathcal{L}, \mathcal{R}) \in \Omega$, and suppose that $P(K)$ is true for all $K \in \mathcal{L} \cup \mathcal{R}$. If $P(H)$ were to be false, then there would be an infinite sequence of games $(H_n)_{n=1}^{\infty}$ so that for each $n \in \mathbb{N}$, H_{n+1} is a left- or right-option of H_n, and H_1 is a left- or right-option of H. But this means that we also have an infinite sequence starting with H_1, making $P(H_1)$ false, but $H_1 \in \mathcal{L} \cup \mathcal{R}$, so $P(H_1)$ is true by assumption. Therefore, $P(H)$ must be true, and our claim follows by Conway induction.   □

Theorem 22.11 is not only a consequence of Axiom 22.2, but it is equivalent to it—see Problem 4.

Our next goal is to use Conway induction to prove that the relation $\lesssim$ (and, therefore, $\approx$) is reflexive and transitive. We start by showing that $\lesssim$ is reflexive.

Proposition 22.12 *For all games G, we have $G \lesssim G$.*

Proof We use Conway induction. Let $H = (\mathcal{L}, \mathcal{R}) \in \Omega$, and assume that $L \lesssim L$ and $R \lesssim R$ hold for all $L \in \mathcal{L}$ and $R \in \mathcal{R}$. We need to prove that $H \lesssim H$.

We suppose, indirectly, that $H \not\lesssim H$. By definition, this implies that either there exists an $L \in \mathcal{L}$ for which $H \lesssim L$, or there exists an $R \in \mathcal{R}$ for which $R \lesssim H$. Let's assume that the first situation holds and we have $H \lesssim L$ for some $L \in \mathcal{L}$. (The second situation can be handled similarly.) But this, by definition, implies that there is no $L' \in \mathcal{L}$ for which $L \lesssim L'$; in particular, we have $L \not\lesssim L$, which is a contradiction with our inductive hypothesis above.   □

In order to prove that the $\lesssim$ relation is transitive, we need a generalized version of Conway induction. First we introduce a notation.

For a natural number n, let $\mathbf{H} = (H_1, \ldots, H_n)$ be an ordered n-tuple of games; suppose that $H_i = (\mathcal{L}_i, \mathcal{R}_i)$. We then define $\Theta(\mathbf{H})$ to be the set of n-tuples $\mathbf{M} = (M_1, \ldots, M_n)$ where there is a unique index $i \in \{1, \ldots, n\}$ for which $M_i \in \mathcal{L}_i \cup \mathcal{R}_i$; for all other indices j, $M_j = H_j$. Finally, for each n-tuple $\mathbf{M} = (M_1, \ldots, M_n)$ of $\Theta(\mathbf{H})$, we generate the $n!$ possible permutations of the n components; the collection of all these n-tuples is then denoted by $\Psi(\mathbf{H})$.

For example, for $n = 1$, we have simply $\Psi(\mathbf{H}) = \mathcal{L}_1 \cup \mathcal{R}_1$. For $n = 2$, $\Psi(\mathbf{H})$ consists of ordered pairs of the form

$$(L_1, H_2) \ (R_1, H_2) \ (H_1, L_2) \ (H_1, R_2)$$
$$(H_2, L_1) \ (H_2, R_1) \ (L_2, H_1) \ (R_2, H_1)$$

with $L_i \in \mathcal{L}_i$, $R_i \in \mathcal{L}_i$, $i \in \{1, 2\}$. For $n = 3$ (which is the instance we will need for our proof of transitivity below), $\Psi(\mathbf{H})$ consists of $2^3 \cdot 3! = 48$ kinds of ordered triples.

Theorem 22.13 (Conway's Multiple Induction Principle) *Let n be a natural number, and suppose that $P(\mathbf{G})$ is a predicate that becomes a statement for all n-tuples $\mathbf{G} = (G_1, \ldots, G_n)$ with $G_1, \ldots, G_n \in \Omega$. If the implication*

$$\left(\bigwedge_{\mathbf{K} \in \Psi(\mathbf{H})} P(\mathbf{K}) \right) \Rightarrow P(\mathbf{H})$$

holds for all n-tuples $\mathbf{H} = (H_1, \ldots, H_n)$, then $P(\mathbf{G})$ is true for every n-tuple of games $\mathbf{G} = (G_1, \ldots, G_n)$.

The template of Conway's Multiple Induction Principle looks complicated but is, in fact, rather simple: to prove that $P(G_1, \ldots, G_n)$ holds for all n-tuples of games $(G_1, \ldots, G_n)$, one needs to verify that $P(H_1, \ldots, H_n)$ holds whenever the predicate is assumed to be true in every case in which the n games are permuted arbitrarily and any one term is replaced by any of its left- or right-options. Note that, for $n = 1$, Conway's Multiple Induction Principle reduces exactly to Conway's Induction Principle; for $n = 2$, this is the "game" equivalent of the double induction template of Problem 3 (e) of Chapter 11. Conway triple induction (the case $n = 3$ of Theorem 22.13) is used to prove that $\lesssim$ is a transitive relation on games.

Proposition 22.14 *Let G, H, and K be games, and suppose that* $G \lesssim H$ *and* $H \lesssim K$. *Then we also have* $G \lesssim K$.

Proof We use Conway's Multiple Induction Principle for the predicate $P(G_1, G_2, G_3)$ denoting the implication

$$(G_1 \lesssim G_2) \wedge (G_2 \lesssim G_3) \Rightarrow (G_1 \lesssim G_3).$$

Let $H_1 = (\mathcal{L}_1, \mathcal{R}_1)$, $H_2 = (\mathcal{L}_2, \mathcal{R}_2)$, and $H_3 = (\mathcal{L}_3, \mathcal{R}_3)$, and assume that $H_1 \lesssim H_2$ and $H_2 \lesssim H_3$. We need to prove that $H_1 \lesssim H_3$.

Indirectly, suppose that $H_1 \not\lesssim H_3$. We assume that there is an $L \in \mathcal{L}_1$ for which $H_3 \lesssim L$ as the case in which there is an $R \in \mathcal{R}_3$ for which $R \lesssim H_1$ can be treated similarly. By our inductive hypothesis, we know that $P(H_2, H_3, L)$ holds; that is, $H_2 \lesssim H_3$ and $H_3 \lesssim L$ imply $H_2 \lesssim L$. But, since $H_1 \lesssim H_2$, we also have $H_2 \not\lesssim L$, which is a contradiction. □

The fact that $\lesssim$ is transitive immediately implies the following useful variations (cf. Problem 5).

Proposition 22.15 *Let* $G_1, G_2, G_3 \in \Omega$ *with* $G_1 \lesssim G_2 \lesssim G_3$, *and suppose that* $G_1 < G_2$ *or* $G_2 < G_3$. *In this case we have* $G_1 < G_3$.

Proposition 22.16 *Let* $G_1, G_2, H_1, H_2 \in \Omega$, *and suppose that* $G_1 \approx G_2$, $H_1 \approx H_2$. *In this case, if* $G_1 \lesssim H_1$, *then* $G_2 \lesssim H_2$, *and if* $G_1 < H_1$, *then* $G_2 < H_2$.

Proposition 22.17 *Let* $G_1, G_2 \subset \Omega$, *and suppose that* $G_1 \approx G_2$. *Then* $G_1 > G(0)$ *if, and only if,* $G_2 > G(0)$.

We now turn to an important property of the $\lesssim$ relation that says that any two numeric games are comparable; that is, for any pair of numeric games G and H, we have at least one of $G \lesssim H$ or $H \lesssim G$. Indeed, as we prove below, for any pair of numeric games G and H, we have exactly one of $G \approx H$, $G < H$, or $G > H$. It is important to note, however, that these properties do not hold in the larger class of (not necessarily numeric) games Ω!

First we prove the following useful proposition. (As we remarked earlier, the recursive nature of Γ means that we will frequently need to rely on Conway's Induction Principle and Conway's Multiple Induction Principle. From now on, in these situations we will simply write "by induction.")

Proposition 22.18 *Suppose that* $G = (\mathcal{L}, \mathcal{R}) \in \Gamma$, *and let* $L \in \mathcal{L}$, $R \in \mathcal{R}$. *Then we have*

$$L < G < R.$$

According to Proposition 22.18, given an arbitrary numeric game G, Left's move in G will always result in a game L that is less than G, and Right's move will always be to a game R that is greater than G, reflecting the sentiment that having to make a move is considered undesirable.

Proof We will only prove $L < G$ here as $G < R$ can be shown similarly. Let $L = (\mathcal{L}_L, \mathcal{R}_L)$.

We first show that $L \lesssim G$. By definition, this means that (i) we must have no $H \in \mathcal{R}$ for which $H \lesssim L$, and (ii) we must have no $K \in \mathcal{L}_L$ for which $G \lesssim K$. Since $G \in \Gamma$, (i) holds. To show (ii), let $K \in \mathcal{L}_L$ be arbitrary. By our inductive hypothesis, we have $K < L$ and therefore $K \lesssim L$. But this implies that we cannot have $G \lesssim K$, proving (ii).

To complete our proof, we need to verify that $G \not\lesssim L$, but this follows immediately from the definition of $\lesssim$, since by reflexivity we have $L \lesssim L$. $\qquad\square$

Theorem 22.19 *For any pair of numeric games G and H, we have exactly one of $G \approx H$, $G < H$, or $G > H$.*

Proof It is easy to see that Definition 22.5 guarantees that no $G, H \in \Gamma$ can satisfy more than one of the above relations. For example, if we were to have $G < H$ and $H < G$, then we would have $G \lesssim H$ and $H \lesssim G$, thus $G \approx H$, as well, contradicting both $G < H$ and $H < G$.

To prove that at least one of the relations holds, we proceed indirectly, and assume that $G \not\approx H$, $G \not< H$, and $H \not> G$. Therefore, $G \not\lesssim H$ and $H \not\lesssim G$ (since, if we had, say, $G \lesssim H$, then $G \not\approx H$ would mean that $G < H$).

Let $G = (\mathcal{L}_G, \mathcal{R}_G)$ and $H = (\mathcal{L}_H, \mathcal{R}_H)$. Our inequality $G \not\lesssim H$ means that

   (i) there exists an $L_G \in \mathcal{L}_G$ for which $H \lesssim L_G$, or
  (ii) there exists an $R_H \in \mathcal{R}_H$ for which $R_H \lesssim G$.
       Similarly, $H \not\lesssim G$ means that
  (i*) there exists an $L_H \in \mathcal{L}_H$ for which $G \lesssim L_H$, or
 (ii*) there exists an $R_G \in \mathcal{R}_G$ for which $R_G \lesssim H$.

We show that none of the four pairs of conditions (i) and (i*), (i) and (ii*), (ii) and (i*), or (ii) and (ii*) can hold.

The case in which (i) and (ii*) are both assumed to be true leads to a contradiction immediately: by transitivity, we have $R_G \lesssim L_G$, contradicting Proposition 22.18. Similarly, (ii) and (i*) cannot both hold.

To show that (i) and (i*) cannot both hold is only one step more complicated: using Proposition 22.18, we get

$$L_H < H \lesssim L_G < G \lesssim L_H,$$

from which $L_H < L_H$, a contradiction. The proof that (i) and (ii*) cannot both hold is similar. $\qquad\square$

Turning now to the similarity relation of games, we immediately see from Propositions 22.12 and 22.14 that $\approx$ satisfies the reflexive and transitive properties; since, by definition, it is also symmetric, we arrive at the following:

Corollary 22.20 *The relation $\approx$ of similarity is an equivalence relation on the collection of games Ω.*

By Corollary 22.20, the similarity relation partitions the collection Ω of all games into equivalence classes. (A disclaimer is in order: since Ω is not a actually a set, the terms "partition" and "equivalence class" would need to be qualified; as this slight imprecision will not cause problems for us here, we will not make the distinction.) Below, we will only be interested in the equivalence classes within the collection Γ of numeric games. We denote the equivalence class of a game G by $[G]$; that is,

$$[G] = \{H \in \Gamma \mid H \approx G\}.$$

Definition 22.21 *The equivalence classes of the similarity relation in Γ are called* surreal numbers. *The collection of surreal numbers is denoted by* $\mathbb{S}$.

As we mentioned earlier, $\mathbb{S}$ is an ordered field for certain addition, multiplication, and order; we are about to make this more precise. We start with the binary operation of addition of games.

Definition 22.22 *Let $G = (\mathcal{L}_G, \mathcal{R}_G) \in \Omega$ and $H = (\mathcal{L}_H, \mathcal{R}_H) \in \Omega$. We define the* sum of G and H recursively, as

$$G + H = ((G + \mathcal{L}_H) \cup (\mathcal{L}_G + H), (G + \mathcal{R}_H) \cup (\mathcal{R}_G + H)).$$

Here, as customary, the sum of a game and a collection of games is simply the collection of the appropriate sums; for example,

$$G + \mathcal{L}_H = \{G + L_H \mid L_H \in \mathcal{L}_H\}.$$

The way we defined the addition of games captures our intuitive notion of combining two games. Consider two players (Left and Right) playing the two games G and H simultaneously, so that, when it is his or her turn, each player is to make a move in exactly one of the games (the players can freely decide whether to make a move in G or H). Clearly, Left's choices are to move from G "plus" H to either G plus L_{II} for some $L_{II} \in \mathcal{L}_{II}$ (when choosing to move in H) or to L_G plus H for some $L_G \in \mathcal{L}_G$ (when deciding to move in G); Right's choices are similar. Thus, our definition of the addition of two games aligns with the situation in which the players play two games simultaneously.

As a matter of fact, it is quite often the case that the players only play a single game, but that, at some point during play, the game naturally decomposes into two (or more) separate games. For example, when playing *Arion*, one of Right's initial options is to remove b_2, thereby decomposing the game into two components (cf. R_2 on page 383). Thus, it is helpful to introduce and study the binary operation of addition of games.

Examining the properties of game addition, we arrive at the following results.

Theorem 22.23 *The addition of games has the following properties.*

1. Addition is a closed operation: For all $G \in \Omega$ and $H \in \Omega$, we have $G + H \in \Omega$.
2. Addition is commutative: For all $G \in \Omega$ and $H \in \Omega$, we have

$$G + H = H + G.$$

3. *Addition is associative: For all $G \in \Omega$, $H \in \Omega$, and $K \in \Omega$, we have*

$$(G + H) + K = G + (H + K).$$

4. *Addition has the identity property: For all $G \in \Omega$, we have*

$$G + G(0) = G.$$

5. *The additive inverse property holds up to similarity: For all $G = (\mathcal{L}, \mathcal{R}) \in \Omega$, we have*

$$G + (-G) \approx G(0),$$

where $-G$ is defined recursively as $-G = (-\mathcal{R}, -\mathcal{L})$.

Thus, Theorem 22.23 says that $(\Omega, +)$ is close to being an Abelian group for addition ($\approx$ becomes equality in the last property only in the case in which $G = G(0)$). We leave the rather easy proof to Problem 9.

Additionally, we can prove that addition interacts with order, as expected.

Theorem 22.24 *For all games $G_1 \in \Omega$, $G_2 \in \Omega$, and $H \in \Omega$, we have*

$$G_1 \lesssim G_2 \iff G_1 + H \lesssim G_2 + H.$$

The proof of Theorem 22.24 is carried out in Problem 10. As a useful corollary, we can prove that game addition preserves similarity.

Corollary 22.25 *If games G_1, G_2, H_1, and H_2 satisfy $G_1 \approx G_2$ and $H_1 \approx H_2$, then*

$$G_1 + H_1 \approx G_2 + H_2.$$

Proof Since $G_1 \approx G_2$ and $H_1 \approx H_2$ imply $G_1 \lesssim G_2$ and $H_1 \lesssim H_2$, by transitivity, we have

$$G_1 + H_1 \lesssim G_1 + H_2 \lesssim G_2 + H_2;$$

the other direction is similar. $\square$

Next, we define the multiplication of games.

Definition 22.26 *Let $G = (\mathcal{L}_G, \mathcal{R}_G) \in \Omega$ and $H = (\mathcal{L}_H, \mathcal{R}_H) \in \Omega$. We define the product of G and H recursively, as*

$$G \cdot H = ((G \cdot \mathcal{L}_H + \mathcal{L}_G \cdot H - \mathcal{L}_G \cdot \mathcal{L}_H) \cup (G \cdot \mathcal{R}_H + \mathcal{R}_G \cdot H - \mathcal{R}_G \cdot \mathcal{R}_H),$$

$$(G \cdot \mathcal{R}_H + \mathcal{L}_G \cdot H - \mathcal{L}_G \cdot \mathcal{R}_H) \cup (G \cdot \mathcal{L}_H + \mathcal{R}_G \cdot H - \mathcal{R}_G \cdot \mathcal{L}_H)).$$

As usual, subtraction simply denotes the addition of the negative; furthermore, unless parentheses dictate otherwise, multiplication is to be performed before addition and subtraction.

Multiplication of games is not nearly as applicable as addition is; furthermore, the rule we just defined leads to some unsettling situations, such as the fact that we may have a pair of similar games that, when multiplied by the same game, result in non-similar products (cf. Problem 11). However, one can prove that, when restricted to numeric games, multiplication satisfies all the usual properties. For example, assuming that

$$L_G < G < R_G$$

and

$$L_H < H < R_H$$

imply

$$(G - L_G) \cdot (H - L_H) > G(0),$$

$$(R_G - G) \cdot (H - L_H) > G(0),$$

$$(G - L_G) \cdot (R_H - H) > G(0),$$

and

$$(R_G - G) \cdot (R_H - H) > G(0),$$

the usual arithmetic operations yield

$$\left. \begin{array}{c} G \cdot L_H + L_G \cdot H - L_G \cdot L_H \\ \\ G \cdot R_H + R_G \cdot H - R_G \cdot R_H \end{array} \right\} < G \cdot H < \left\{ \begin{array}{c} G \cdot R_H + L_G \cdot H - L_G \cdot R_H \\ \\ G \cdot L_H + R_G \cdot H - R_G \cdot L_H \end{array} \right.$$

—providing some much-needed illumination for Definition 22.26.

We now return to the collection $\mathbb{S}$ of surreal numbers. We make the following definitions.

Definition 22.27 *Let $x_\mathbb{S}$ and $y_\mathbb{S}$ be surreal numbers, and suppose that $G \in x_\mathbb{S}$ and $H \in y_\mathbb{S}$. We define surreal addition and multiplication as*

- $x_\mathbb{S} + y_\mathbb{S} = [G + H]$ *and*
- $x_\mathbb{S} \cdot y_\mathbb{S} = [G \cdot H]$.

Furthermore, we say that $x_\mathbb{S}$ is positive *if $G > G(0)$ and that it is* negative *if $G < G(0)$; we let $P_\mathbb{S}$ denote the collection of positive surreal numbers.*

Note that, by Corollary 22.25 and Proposition 22.17, addition and order are well defined; the fact that multiplication of surreal numbers is also independent of the chosen representative can be established as well (we omit the lengthy proof).

We are now ready for one of our main results:

Theorem 22.28 *The surreal number system $(\mathbb{S}, +, \cdot, P_\mathbb{S})$ is an ordered field.*

Proof First, we need to prove that the sum of any two numeric games is a numeric game; that is, if $G = (\mathcal{L}_G, \mathcal{R}_G) \in \Gamma$ and $H = (\mathcal{L}_H, \mathcal{R}_H) \in \Gamma$, then $G + H \in \Gamma$. This is indeed the case; in fact, we can prove the stronger claim that

$$\left. \begin{array}{c} G + L_H \\ L_G + H \end{array} \right\} < G + H < \left\{ \begin{array}{c} G + R_H \\ R_G + H \end{array} \right.$$

holds for all $L_G \in \mathcal{L}_G, R_G \in \mathcal{R}_G, L_H \in \mathcal{L}_H$, and $R_H \in \mathcal{R}_H$. For example, to show that

$$G + L_H < G + H,$$

we can use Proposition 22.18 and Theorem 22.24 to write

$$G + L_H \lesssim G + H.$$

Conversely,

$$G + H \lesssim G + L_H$$

would imply $H \lesssim L_H$, which cannot be. This establishes that surreal number addition is closed.

Commutativity and associativity of addition follow immediately from Theorem 22.23, as does the fact that $0_\mathbb{S}$ is the additive identity.

To verify the additive inverse property, we first prove that the negative of a numeric game is numeric as well; that is, if $G = (\mathcal{L}, \mathcal{R}) \in \Gamma$, then $-G = (-\mathcal{R}, -\mathcal{L}) \in \Gamma$ as well. Indeed, for any $L \in \mathcal{L}$ and $R \in \mathcal{R}$, we have $L < R$ and thus

$$-R = -R + G(0) \approx -R + L + (-L) \lesssim -R + R + (-L) \approx G(0) + (-L) = -L,$$

and thus $-R \lesssim -L$; to see that we cannot have $-L \lesssim -R$, observe that this would imply

$$R \approx R + (-L) + L \lesssim R + (-R) + L \approx L,$$

which contradicts $G \in \Gamma$. Therefore, for every surreal number $x_\mathbb{S}$, $-x_\mathbb{S}$ is surreal as well, and we have

$$x_\mathbb{S} + (-x_\mathbb{S}) = 0_\mathbb{S}.$$

The order axiom that for every surreal number $x_\mathbb{S}$, exactly one of $x_\mathbb{S} = 0_\mathbb{S}$, $x_\mathbb{S} \in P_\mathbb{S}$, or $-x_\mathbb{S} \in P_\mathbb{S}$ holds, follows immediately from Theorem 22.19.

To prove that $P_\mathbb{S}$ is closed for addition, suppose that $G = (\mathcal{L}_G, \mathcal{R}_G) \in \Gamma$ and $H = (\mathcal{L}_H, \mathcal{R}_H) \in \Gamma$, $G > G(0)$, and $H > G(0)$. Therefore, by Theorem 22.24, we have

$$G + H \gtrsim G + G(0) \gtrsim G(0) + G(0) = G(0);$$

if we also had $G + H \approx G(0)$, then this would imply

$$H = H + G(0) \lesssim H + G \approx G(0),$$

which is a contradiction.

Verifying that each of the axioms involving multiplication holds as well is remarkably complicated and will be omitted here. (However, we prove the multiplicative identity property in Problem 11.)   □

The surreal number field is, in fact, a very special ordered field. Recall that the rational number field is the smallest ordered field; that is, for every ordered field system $(F, +_F, \cdot_F, P_F)$, there exists a subfield $Q \subseteq F$ for which the systems $(\mathbb{Q}, +, \cdot, P_\mathbb{Q})$ and $(Q, +_Q, \cdot_Q, P_Q)$ are isomorphic. (Here $+_Q$, $\cdot_Q$, and P_Q are the restrictions of $+_F$, $\cdot_F$, and P_F to Q.) The surreal number system is at the other end of the spectrum:

Theorem 22.29 *The surreal number system* $(\mathbb{S}, +, \cdot, P_\mathbb{S})$ *is the largest ordered field; that is, for every ordered field system* $(F, +_F, \cdot_F, P_F)$*, there exists a subfield* $S \subseteq \mathbb{S}$ *for which the systems* $(F, +_F, \cdot_F, P_F)$ *and* $(S, +_S, \cdot_S, P_S)$ *are isomorphic.*

We will soon see that, in accordance with Theorem 22.29, the field of surreal numbers contains (an isomorphic copy of) the rational numbers as well as the real numbers, and one can show that it contains (an isomorphic copy of) the ordered field $\mathbb{R}(x)$ as well.

Let us now return to games. When analyzing a particular game, it is helpful to "simplify" the game as much as possible. As we are about to explain, this simplification procedure is quite efficient: with the application of only two types of techniques, one can reduce every numeric game to a unique member of a well-understood family of games. In order to make this precise, we introduce some terminology.

Definition 22.30 *Let $G \in \Gamma$ and $H \in \Gamma$. We say that*

- *H is a* restriction *of G if*

    – *every left-option of H is also a left-option of G, and*
    – *every right-option of H is also a right-option of G;*

- *H is a* dominating restriction *of G if it is a restriction so that*

    – *every left-option of G is less than or similar to some left-option of H, and*
    – *every right-option of G is greater than or similar to some right-option of H;*

- *H is a* compromise *of G if*

    – *every left-option of G is less than H, and*
    – *every right-option of G is greater than H;*

- *H is a* simple compromise *of G if it is a compromise of G, but*

    – *none of the left-options of H are compromises of G, and*
    – *none of the right-options of H are compromises of G.*

Let's consider some examples. The game

$$G = (\{G(0), (\emptyset, \{G(1)\})\}, \{G(1), G(2)\})$$

has 16 restrictions, since with any

$$\mathcal{L} \subseteq \{G(0), (\emptyset, \{G(1)\})\}$$

and

$$\mathcal{R} \subseteq \{G(1), G(2)\},$$

$H = (\mathcal{L}, \mathcal{R})$ is a restriction of G. For H to be a dominating restriction, we must take at least one of $G(0)$ or $(\emptyset, \{G(1)\})$ to be in $\mathcal{L}$ (note that $G(0) \approx (\emptyset, \{G(1)\})$), and we must have $G(1) \in \mathcal{R}$ (but may or may not want to take $G(2)$). Therefore, G has $3 \cdot 2 = 6$ dominating restrictions.

By verifying that

$$(\{G(-2)\}, \{G(2)\}) \approx G(0),$$

we can also see that

$$G(0) = (\emptyset, \emptyset),$$

$$(\emptyset, \{G(1)\}),$$

$$(\{G(-1)\}, \emptyset),$$

and

$$(\{G(-1)\}, \{G(1)\})$$

are all compromises of the game

$$G = (\{G(-2)\}, \{G(2)\});$$

but only $G(0)$ is a simple compromise, since neither

$$G(1) = (\{G(0)\}, \emptyset)$$

nor

$$G(-1) = (\emptyset, \{G(0)\})$$

is a compromise of G.

We should note that every game G has a dominating restriction and a simple compromise; in fact, by Proposition 22.18, G itself serves as an example for both.

We will now prove that all dominating restrictions of a game G are similar to G (and thus to each other), and all simple compromises are similar to G (and thus to each other).

Theorem 22.31 (The Domination Theorem) *Let $G \subset \Gamma$. If H is a dominating restriction of G, then $G \approx H$.*

Proof $G = (\mathcal{L}_G, \mathcal{R}_G)$ and $H = (\mathcal{L}_H, \mathcal{R}_H)$. To show that $G \approx H$, we will need to prove the following four statements:

 (i) there is no $L_G \in \mathcal{L}_G$ for which $H \lesssim L_G$;
 (ii) there is no $R_H \in \mathcal{R}_H$ for which $R_H \lesssim G$;
 (iii) there is no $L_H \in \mathcal{L}_H$ for which $G \lesssim L_H$; and
 (iv) there is no $R_G \in \mathcal{R}_G$ for which $R_G \lesssim H$.

Now Proposition 22.18 implies that for every $L_G \in \mathcal{L}_G$ and $R_G \in \mathcal{R}_G$ we have

$$L_G < G < R_G;$$

since $\mathcal{L}_H \subseteq \mathcal{L}_G$ and $\mathcal{R}_H \subseteq \mathcal{R}_G$, we have (ii) and (iii).

Furthermore, since H is a dominating restriction of G, for every $L_G \in \mathcal{L}_G$ and for every $R_G \in \mathcal{R}_G$, we can find options $L_H \in \mathcal{L}_H$ and $R_H \in \mathcal{R}_H$ for which $L_G \lesssim L_H$ and $R_H \lesssim R_G$. Therefore, by Proposition 22.18, we have

$$L_G \lesssim L_H < H < R_H \lesssim R_G,$$

and our claims (i) and (iv) follow by Proposition 22.15. $\qquad\square$

The Domination Theorem is quite useful in simplifying games. In particular, it immediately implies the last six parts of Proposition 22.9 (which classifies the seventeen games of Γ_2): For example, to see that

$$(\{G(-1), G(0), G(1)\}, \emptyset)$$

is similar to

$$G(2) = (\{G(0), G(1)\}, \emptyset),$$

note that, by the Domination Theorem, each game is similar to

$$(\{G(1)\}, \emptyset).$$

In order to classify numeric games with birthdays 3 or more, we need the following sibling of the Domination Theorem.

Theorem 22.32 (The Simplicity Theorem) *Let $G \in \Gamma$. If H is a simple compromise of G, then $G \approx H$.*

Proof Let $G = (\mathcal{L}_G, \mathcal{R}_G)$ and $H = (\mathcal{L}_H, \mathcal{R}_H)$. Since H is a compromise of G, we have

$$\forall L_G \in \mathcal{L}_G, \forall R_G \in \mathcal{R}_G, L_G < H < R_G;$$

furthermore, since H is a simple compromise of G, the same condition does not hold for any left- or right-option of H; that is,

- there is no $L_H \in \mathcal{L}_H$ for which $\forall L_G \in \mathcal{L}_G, \forall R_G \in \mathcal{R}_G, L_G < L_H < R_G$, and
- there is no $R_H \in \mathcal{R}_H$ for which $\forall L_G \in \mathcal{L}_G, \forall R_G \in \mathcal{R}_G, L_G < R_H < R_G$.

We need to prove that $G \lesssim H$ and $H \lesssim G$. Here we only show $G \lesssim H$ as the other claim is similar.

To show that $G \lesssim H$, we need to verify that

(i) there is no $L_G \in \mathcal{L}_G$ for which $H \lesssim L_G$, and
(ii) there is no $R_H \in \mathcal{R}_H$ for which $R_H \lesssim G$.

Clearly, (i) holds by our hypothesis that $\forall L_G \in \mathcal{L}_G, L_G < H$.

To verify (ii), we proceed indirectly and assume that $R_H \lesssim G$ for some $R_H \in \mathcal{R}_H$. Therefore, we know that there is no $R_G \in \mathcal{R}_G$ for which $R_G \lesssim R_H$; that is, $R_H < R_G$ holds for all $R_G \in \mathcal{R}_G$. Furthermore, by Proposition 22.18, we have $H < R_H$; therefore, by assumption and by transitivity, $L_G < R_H$ holds for all $L_G \in \mathcal{L}_G$. We thus have a right-option R_H of H for which

$$\forall L_G \in \mathcal{L}_G, \forall R_G \in \mathcal{R}_G, L_G < R_H < R_G,$$

and this is a contradiction with our hypothesis. $\square$

The Domination Theorem and the Simplicity Theorem allow us to simplify games and make their evaluation easier. As an example, here we classify all 7,143,404 numeric games with birthday 3.

Given $G \in \Gamma_3$, we first find a dominating restriction H of G. In order to do so, we will employ the set of games

$$S = \{G(-2),\ G(-1),\ G(-\tfrac{1}{2}),\ G(0),\ G(\tfrac{1}{2}),\ G(1),\ G(2)\};$$

we can verify that the seven games above are listed in increasing order. We have the following three possibilities for H:

(1) $H \approx (\{L\}, \emptyset)$ for some $L \in S$;
(2) $H \approx (\emptyset, \{R\})$ for some $R \in S$; or
(3) $H \approx (\{L\}, \{R\})$ for some $L, R \in S$ with $L < R$.

We may examine case (1) further to find that

(a) if $L < G(0)$, then $H \approx G(0)$;
(b) if $L = G(0)$, then $H \approx G(1)$;
(c) if $L = G(\tfrac{1}{2})$, then $H \approx G(1)$;
(d) if $L = G(1)$, then $H \approx G(2)$;
(e) if $L = G(2)$, then $H \approx G(3)$, where $G(3)$ is defined as

$$G(3) = (\{G(0),\ G(1),\ G(2)\},\ \emptyset).$$

Similarly, in case (2), we find that H is similar to one of the games in S or to

$$G(-3) = (\emptyset,\ \{G(-2),\ G(-1),\ G(0)\}).$$

Finally, in case (3), we may use the Domination Theorem and the Simplicity Theorem to verify that

(a) if $L < G(0)$ and $R > G(0)$, then $H \approx G(0)$;
(b) if $L, R \in \{G(0),\ G(\tfrac{1}{2}),\ G(1),\ G(2)\}$, then

   (i) if $L = G(0)$ and $R = G(\tfrac{1}{2})$, then $H \approx G(\tfrac{1}{4})$, where

$$G(\tfrac{1}{4}) = (\{G(0)\},\ \{G(\tfrac{1}{2}),\ G(1)\});$$

   (ii) if $L = G(0)$ and $R = G(1)$, then $H \approx G(\tfrac{1}{2})$;
   (iii) if $L = G(0)$ and $R = G(2)$, then $H \approx G(1)$;
   (iv) if $L = G(\tfrac{1}{2})$ and $R = G(1)$, then $H \approx G(\tfrac{3}{4})$, where

$$G(\tfrac{3}{4}) = (\{G(0),\ G(\tfrac{1}{2})\},\ \{G(1)\});$$

   (v) if $L = G(\tfrac{1}{2})$ and $R = G(2)$, then $H \approx G(1)$; and

(vi) if $L = G(1)$ and $R = G(2)$, then $H \approx G(\frac{3}{2})$, where

$$G(\tfrac{3}{2}) = (\{G(0), G(1)\}, \{G(2)\}).$$

(c) if $L, R \in \{G(-2),\ G(-1),\ G(-\frac{1}{2}),\ G(0)\}$, then H is similar to $G(-1)$, $G(-\frac{1}{2})$ or to

$$G(-\tfrac{3}{2}) = (\{G(-2)\}, \{G(-1), G(0)\}),$$

$$G(-\tfrac{3}{4}) = (\{G(-1)\}, \{G(-\tfrac{1}{2}), G(0)\}),$$

or

$$G(-\tfrac{1}{4}) = (\{G(-1), G(-\tfrac{1}{2})\}, \{G(0)\}).$$

In summary, we can state our results as follows. Let

$$\Delta_0 = \{G(0)\},$$

$$\Delta_1 = \{G(-1),\ G(1)\},$$

$$\Delta_2 = \{G(-2),\ G\left(-\tfrac{1}{2}\right),\ G\left(\tfrac{1}{2}\right),\ G(2)\},$$

and

$$\Delta_3 = \{G(-3),\ G\left(-1\tfrac{1}{2}\right),\ G\left(-\tfrac{3}{4}\right),\ G\left(-\tfrac{1}{4}\right),\ G\left(\tfrac{1}{4}\right),\ G\left(\tfrac{3}{4}\right),\ G\left(1\tfrac{1}{2}\right),\ sG(3)\};$$

with these notations we have the following:

Proposition 22.33 *Every numeric game of birthday 3 (or less) is similar to one of the fifteen games in*

$$\Delta_0 \cup \Delta_1 \cup \Delta_2 \cup \Delta_3.$$

(Similarly, the abbreviated summary of Proposition 22.9 is that each game in Γ_2 is similar to a game in $\Delta_0 \cup \Delta_1 \cup \Delta_2$.)

The fifteen games in Proposition 22.33 can be visualized by *Hackenbush* games; more precisely, for each game, there is a *Hackenbush* string H so that $G \approx H$. (A *Hackenbush string* is a *Hackenbush* game whose graph is a path.) The *Hackenbush* string representation of the fifteen games of Proposition 22.33 are given in Figure 22.8 (the label q corresponds to the game $G(q)$).

For example, we can see that the *Hackenbush* string marked $-\frac{3}{4}$ provides only one option for Left (to move the game to $G(-1)$) and two options for Right (moves to $G(-\frac{1}{2})$ or $G(0)$). Thus, the game is indeed

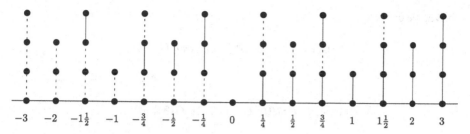

Fig. 22.8 The fifteen games of Proposition 22.33

$$G(-\tfrac{3}{4}) = (\{G(-1)\}, \{G(-\tfrac{1}{2}), G(0)\}),$$

as defined above. As the chart indicates, the birthday of each *Hackenbush* string corresponds to its "height" (the number of edges in it). Furthermore, the eight elements of Δ_3 alternate with the seven elements of $\Delta_0 \cup \Delta_1 \cup \Delta_2$; see Problem 6 for generalizations and proofs.

It is time for us now to explain how we chose the labels for our games thus far. Notice that every game in

$$\Delta_0 \cup \Delta_1 \cup \Delta_2 \cup \Delta_3$$

is of the form $G(q)$ for some real number q whose binary form is finite. For example, we have

$$\Delta_3 = \{G(-3), G\left(-1\tfrac{1}{2}\right), G\left(-\tfrac{3}{4}\right), G\left(-\tfrac{1}{4}\right), G\left(\tfrac{1}{4}\right), G\left(\tfrac{3}{4}\right), G\left(1\tfrac{1}{2}\right), G(3)\},$$

corresponding, in order, to the binary forms

$$-11, \ -1.1, \ -.11, \ -.01, \ .01, \ .11, \ 1.1, \ 11.$$

Recall from Problem 3 (e) of Chapter 12 that $\mathbb{Z}[\tfrac{1}{2}]$ is the set of all real numbers with finite binary forms; as we have seen there, $\mathbb{Z}[\tfrac{1}{2}]$ is an integral domain, and it consists of the set of integers together with rational numbers of the form

$$\pm \left(n + \frac{a}{2^k}\right)$$

where n is a nonnegative integer, k is a positive integer, and a is an odd integer with $1 \leq a < 2^k$ (this form is unique).

We can recursively define the game $G(q)$ for each $q \in \mathbb{Z}[\tfrac{1}{2}]$; for example, having defined the elements of

$$\Delta_0 \cup \Delta_1 \cup \Delta_2 \cup \Delta_3$$

above, we can set

$$G(\tfrac{5}{8}) = (\{G(0), G(\tfrac{1}{2})\}, \{G(\tfrac{3}{4}), G(1)\})$$

and

$$G(4) = (\{G(0), G(1), G(2), G(3)\}, \emptyset).$$

We will carry out this development in Problem 6; here we only point out that our notations are in line with the way we denoted natural numbers: for example, on page 351 we had the analogous

$$4_{\mathbb{N}} = \{\emptyset, 1_{\mathbb{N}}, 2_{\mathbb{N}}, 3_{\mathbb{N}}\}.$$

The set of games

$$\Delta = \{G(q) \mid q \in \mathbb{Z}[\tfrac{1}{2}]\}$$

play an important role in the evaluation of games. The games of Δ, often called *dyadic games*, behave as expected when it comes to addition, multiplication, and order: they are consistent with our "labels." Namely, we have the following:

Proposition 22.34 *For any $q_1, q_2 \in \mathbb{Z}[\tfrac{1}{2}]$, the following hold:*

- $G(q_1) + G(q_2) \approx G(q_1 + q_2)$;
- $-G(q_1) \approx G(-q_1)$;
- $G(q_1) \cdot G(q_2) \approx G(q_1 \cdot q_2)$; *and*
- $G(q_1) < G(q_2)$ *if, and only if, $q_1 < q_2$.*

In Problem 12, we prove that Proposition 22.34 holds when $q_1, q_2 \in \mathbb{N}$.

Observe that there are only countably many games of finite birthday but, as we mentioned earlier, the collection of all numeric games Γ is so large that it's not even a proper set; therefore, "most" games have infinite birthdays. So let us now meet some games with infinite birthdays.

Two such games are the *infinite game* $G(\omega)$ and the *infinitesimal game* $G(\epsilon)$, defined as

$$G(\omega) = (\{G(0), G(1), G(2), G(3), \dots\}, \emptyset)$$

and

$$G(\epsilon) = \left(\{G(0)\}, \left\{G(1), G\left(\tfrac{1}{2}\right), G\left(\tfrac{1}{4}\right), \dots\right\}\right).$$

These two games are examples of games of Γ_ω, the games created in "generation ω" and thus have birthday ω. (Here ω stands for the first infinite ordinal number;

see page 344.) Γ_ω contains a large variety of games beyond $G(\omega)$ and $G(\epsilon)$; most importantly, it contains games corresponding to non-dyadic rational numbers and, in fact, all real numbers. For example, given the binary representations

$$\tfrac{1}{3} = 0.010101010101010101\ldots,$$

$$\sqrt{2} = 1.011010100000100111\ldots,$$

and

$$\pi = 11.001001000011111101\ldots$$

(cf. Table 18.1), we can set

$$G\left(\tfrac{1}{3}\right) = \left(\left\{G\left(\tfrac{1}{4}\right), G\left(\tfrac{5}{16}\right), G\left(\tfrac{21}{64}\right), \ldots\right\}, \left\{G\left(\tfrac{1}{2}\right), G\left(\tfrac{3}{8}\right), G\left(\tfrac{11}{32}\right), \ldots\right\}\right),$$

$$G(\sqrt{2}) = \left(\left\{G\left(1\right), G\left(1\tfrac{1}{4}\right), G\left(1\tfrac{3}{8}\right), \ldots\right\}, \left\{G\left(1\tfrac{1}{2}\right), G\left(1\tfrac{7}{16}\right), G\left(1\tfrac{29}{64}\right), \ldots\right\}\right),$$

and

$$G\left(\pi\right) = \left(\left\{G(3), G\left(3\tfrac{1}{8}\right), G\left(3\tfrac{9}{64}\right), \ldots\right\}, \left\{G\left(3\tfrac{1}{2}\right), G\left(3\tfrac{1}{4}\right), G\left(3\tfrac{3}{16}\right), \ldots\right\}\right).$$

(Note the resemblance of these ordered pairs to the corresponding Dedekind cuts.) Of course, to establish that these games are well defined, one would need to verify that their left options are all less than each of their right options, cf. Problem 8.

In fact, one can go further and create games in $\Gamma_{\omega+1}$, etc., such as

$$G(\omega + 1) = (\{G(\omega)\}, \emptyset),$$

$$G(\omega + 2) = (\{G(\omega + 1)\}, \emptyset),$$

$$G(\omega \cdot 2) = (\{G(\omega), G(\omega + 1), G(\omega + 2), \ldots\}, \emptyset),$$

$$G(\omega^2) = (\{G(\omega), G(\omega \cdot 2), G(\omega \cdot 3), \ldots\}, \emptyset),$$

$$G(\omega^\omega) = (\{G(\omega), G(\omega^2), G(\omega^3), \ldots\}, \emptyset),$$

and on and on (cf. page 343 for a better understanding of these notations). Similarly, one could define ever-decreasing infinitesimal games, such as

$$G\left(\tfrac{\epsilon}{2}\right) = (\{G(0)\}, \{G(\epsilon)\}),$$

$$G\left(\tfrac{\epsilon}{4}\right) = (\{G(0)\}, \{G\left(\tfrac{\epsilon}{2}\right)\}),$$

$$G(\epsilon^2) = (\{G(0)\}, \{G\left(\tfrac{\epsilon}{2}\right), G\left(\tfrac{\epsilon}{4}\right), G\left(\tfrac{\epsilon}{8}\right), \dots\}),$$

etc. Furthermore, the infinitesimal games can be "shifted" next to other games; for example,

$$G(3+\epsilon) = \left(\{G(0), G(1), G(2), G(3)\}, \left\{G(4), G\left(3\tfrac{1}{2}\right), G\left(3\tfrac{1}{4}\right), \dots\right\}\right)$$

is a game "infinitely close to," but larger than, $G(3)$. And, of course, this is just the beginning; there are too many games to exhibit explicitly!

It is important to note that our requirement that each game ends after a finite number of moves applies to games with infinite birthdays as well. For example, we see that $G(\omega)$ ends immediately if Right starts and after a single move if Left starts. Indeed, if Left moves

$$G(\omega) = (\{G(0), G(1), G(2), G(3), \dots\}, \emptyset)$$

to, say,

$$G(3) = (\{G(0), G(1), G(2)\}, \emptyset),$$

then Right will be unable to respond. Thus, $G(\omega)$ is a win for Left. Similarly, we see that $G(\epsilon)$ ends by Left winning after a single move if Left starts, and after two moves (one round) if Right starts. For example, if Right chooses to move from

$$G(\epsilon) = \left(\{G(0)\}, \left\{G(1), G\left(\tfrac{1}{2}\right), G\left(\tfrac{1}{4}\right), \dots\right\}\right)$$

to

$$G\left(\tfrac{1}{4}\right) = \left(\{G(0)\}, \{G\left(\tfrac{1}{2}\right), G(1)\}\right),$$

then Left will move to $G(0)$; that leaves no options for Right. So $G(\epsilon)$ is also a win for Left regardless of which player starts. Among the many intriguing properties, we will see that $G(\epsilon)$ is similar to the reciprocal of $G(\omega)$; indeed, denoting the surreal numbers corresponding to $G(\omega)$, $G(\epsilon)$, and $G(1)$ by ω_S, ϵ_S, and 1_S, respectively, we can show that

$$\omega_S \cdot \epsilon_S = 1_S$$

(see Problem 13).

We close this chapter—and, indeed, our book—with a brief discussion of the analysis and evaluation of games. When playing a game, we are mostly interested, of course, in whether we are able to win it or are destined to lose it. In fact, there are two questions about every game: who wins the game if Left makes the initial move, and who wins the game if Right starts? When we say, for example, that a certain game G is a win for Left if Left starts, what we mean is that Left can conduct a winning play of G no matter what Right's moves are. (A thorough analysis of these kinds of statements was presented in Chapter 7.) So we are uninterested in what happens if the players don't play optimally. Note also that the existence of a winning strategy does not necessarily mean that such a strategy is known. (For example, in Problem 1 of Chapter 12 we proved that the *Divisor* game is always won by the first player, but a winning strategy is not yet known.)

We introduce the following notations:

- $\Omega_{L \to L}$: the collection of games that Left can win if Left starts;
- $\Omega_{L \to R}$: the collection of games that Right can win if Left starts;
- $\Omega_{R \to L}$: the collection of games that Left can win if Right starts; and
- $\Omega_{R \to R}$: the collection of games that Right can win if Right starts.

For example, as we have seen earlier, $G(0)$ is in $\Omega_{L \to R}$ and also in $\Omega_{R \to L}$; the games $G(1)$, $G(\omega)$, and $G(\epsilon)$ are all in $\Omega_{L \to L}$ and $\Omega_{R \to L}$; and $G(-1)$ is in $\Omega_{L \to R}$ and $\Omega_{R \to R}$. It may seem quite obvious that every game is either in $\Omega_{L \to L}$ or $\Omega_{L \to R}$, but not both; similarly, each game is in $\Omega_{R \to L}$ or $\Omega_{R \to R}$, but not both. This is indeed the case:

Theorem 22.35 (The Fundamental Theorem of Combinatorial Game Theory)
The sets $\{\Omega_{L \to L}, \Omega_{L \to R}\}$ and $\{\Omega_{R \to L}, \Omega_{R \to R}\}$ are both partitions of the collection of all games Ω; that is,

- *every $G \in \Omega$ is a win for either Left or Right (but not both) if Left starts the game; and*
- *every $G \in \Omega$ is a win for either Left or Right (but not both) if Right starts the game.*

Proof For a game $G = (\mathcal{L}, \mathcal{R}) \in \Omega$, let $P(G)$ be the predicate that

$$G \in [(\Omega_{L \to L} \setminus \Omega_{L \to R}) \cup (\Omega_{L \to R} \setminus \Omega_{L \to L})] \cap [(\Omega_{R \to L} \setminus \Omega_{R \to R}) \cup (\Omega_{R \to R} \setminus \Omega_{R \to L})];$$

our two claims are equivalent to proving that $P(G)$ holds for all $G \in \Omega$.

We will use induction, so let us assume that $P(G)$ holds for all $L \in \mathcal{L}$ and all $R \in \mathcal{R}$. Below we prove only that

$$G \in (\Omega_{L \to L} \setminus \Omega_{L \to R}) \cup (\Omega_{L \to R} \setminus \Omega_{L \to L});$$

the proof that we also have

$$G \in (\Omega_{R \to L} \setminus \Omega_{R \to R}) \cup (\Omega_{R \to R} \setminus \Omega_{R \to L})$$

is similar.

We will consider two cases depending on whether $\mathcal{L} \cap \Omega_{R \to L}$ is the empty-set.

Case 1: If $\mathcal{L} \cap \Omega_{R \to L} \neq \emptyset$, then let $L \in \mathcal{L} \cap \Omega_{R \to L}$. Then, by our inductive hypothesis, $L \notin \Omega_{R \to R}$. Therefore, if Left moves first in G, choosing option L will assure that Left wins and Right loses; we thus have $G \in \Omega_{L \to L} \setminus \Omega_{L \to R}$.

Case 2: If $\mathcal{L} \cap \Omega_{R \to L} = \emptyset$, then, according to our inductive hypothesis,

$$\mathcal{L} \subseteq (\Omega_{R \to R} \setminus \Omega_{R \to L}).$$

Therefore, no matter what Left's first move is in G, the game will result in a win for Right and a loss for Left, so $G \in \Omega_{L \to R} \setminus \Omega_{L \to L}$.   □

Let us introduce some additional notations:

$$\Omega_L = \Omega_{L \to L} \cap \Omega_{R \to L},$$

$$\Omega_R = \Omega_{L \to R} \cap \Omega_{R \to R},$$

$$\Omega_I = \Omega_{L \to L} \cap \Omega_{R \to R},$$

and

$$\Omega_{II} = \Omega_{L \to R} \cap \Omega_{R \to L}.$$

We call these four collections *outcome classes* as they correspond to the four possible outcomes that a game may have. For example, Ω_L denotes the collection of games that Left can win (regardless of who starts), and Ω_{II} stands for the collection of those games that are won by the second player (whether that is Left or Right).

The Fundamental Theorem of Combinatorial Game Theory immediately implies the following:

Corollary 22.36 *The four outcome classes Ω_L, Ω_R, Ω_I, and Ω_{II} are pairwise disjoint and their union is Ω.*

It turns out that we can say quite a bit more about the outcome classes within the collection of numeric games Γ. Defining $\Gamma_L, \Gamma_R, \Gamma_I$, and Γ_{II} by $\Gamma_L = \Gamma \cap \Omega_L$, and so on, we have the following important theorem.

Theorem 22.37 *We have the following characterization of numeric games.*

1. *$G \in \Gamma_L$ if, and only if, $G > G(0)$;*
2. *$G \in \Gamma_R$ if, and only if, $G < G(0)$;*
3. *$G \in \Gamma_{II}$ if, and only if, $G \approx G(0)$; and*
4. *$\Gamma_I = \emptyset$.*

In particular, we see that only three of the four outcome classes within Γ are nonempty: no numeric game is always a win for the first player (if both players play optimally).

Proof By Theorem 22.19, it suffices to prove that the predicate $P(G)$, defined as

$$[(G > G(0)) \wedge (G \in \Gamma_L)] \vee [(G < G(0)) \wedge (G \in \Gamma_R)] \vee [(G \approx G(0)) \wedge (G \in \Gamma_{II})],$$

holds for every game $G \in \Gamma$. We will use induction.

Let $G = (\mathcal{L}, \mathcal{R}) \in \Gamma$, and assume first that $G > G(0)$. Since then $G \not\lesssim G(0) = (\emptyset, \emptyset)$, we must have a left-option $L \in \mathcal{L}$ for which $G(0) \lesssim L$. By our inductive hypothesis $P(L)$ holds, thus $L \in \Gamma_L \cup \Gamma_{II}$, and, therefore, $L \in \Gamma_{R \to L}$. This yields $G \in \Gamma_{L \to L}$. Similarly, we could show that $G < G(0)$ implies $G \in \Gamma_{R \to R}$.

Assume next that $G \gtrsim G(0)$. Then, by definition, there are no right-options $R \in \mathcal{R}$ for which $R \lesssim G(0)$; that is, $G(0) < R$ holds for all $R \in \mathcal{R}$. By induction, we get that $\mathcal{R} \subseteq \Gamma_L$, which yields $G \in \Gamma_{R \to L}$. A similar argument would show that $G \lesssim G(0)$ implies $G \in \Gamma_{L \to R}$.

In summary, we have shown that, if $G > G(0)$ (in which case we also have $G \gtrsim G(0)$), then $G \in \Gamma_{L \to L} \cap \Gamma_{R \to L}$ and so $G \in \Gamma_L$; if $G < G(0)$, then $G \in \Gamma_R$; and, if $G \approx G(0)$ (in which case $G \lesssim G(0)$ and $G \gtrsim G(0)$), then $G \in (\Gamma_{R \to L}) \cap (\Gamma_{L \to R})$ and so $G \in \Gamma_{II}$. Therefore, $P(G)$ must hold for all $G \in \Gamma$. $\square$

In closing, let us return one more time to our game *Arion*. In our next proposition, we evaluate the game in two different ways. Our first method, while shorter here, relies on extensive work we have done in Problem 1 of Chapter 7; it also has the disadvantage that we had to first guess the value of the game before we could verify it. Our second approach is more general and, in fact, can be adapted to evaluate any numeric game.

Proposition 22.38 *The surreal number value of* Arion *equals* $\left(\frac{1}{2}\right)_{\mathbb{S}}$.

Proof I We can use Theorem 22.37, combined with our other previous results, to show that *Arion* $\approx G\left(\frac{1}{2}\right)$. Indeed, we have already seen that the game *Arion* $+ G\left(-\frac{1}{2}\right)$ is in Γ_{II} (cf. Problem 1 (b) of Chapter 7), and thus

$$Arion + G\left(-\tfrac{1}{2}\right) \approx G(0);$$

this then yields

$$Arion = Arion + G(0) \approx Arion + G\left(-\tfrac{1}{2}\right) + G\left(\tfrac{1}{2}\right) \approx G(0) + G\left(\tfrac{1}{2}\right) \approx G\left(\tfrac{1}{2}\right),$$

from which our claim follows. $\square$

Proof II Of the three left-options L_1, L_2, and L_3, and the three right-options R_1, R_2, and R_3 (cf. page 383), here we evaluate two: L_3 and R_2 (admittedly, these are the two easiest ones). Using Proposition 22.34, we immediately get

$$R_2 = G(2) + G\left(-\tfrac{3}{4}\right) \approx G\left(1\tfrac{1}{4}\right).$$

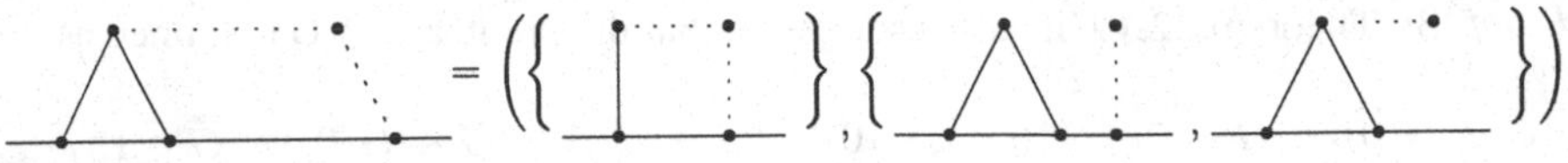

Fig. 22.9 The options for L_3

The options of L_3 are given in Figure 22.9. Therefore, using some of the *Hackenbush* values exhibited earlier, together with Proposition 22.34, the Domination Theorem, the Simplicity Theorem, cases (3) (b) and (3) (c) on page 403, and Proposition 22.6, we have

$$L_3 = \left(\left\{ \left(\{G(-2)\}, \{G(-1) + G(1), G(\tfrac{1}{2})\} \right) \right\}, \left\{ \left(\{G(-1) + G(1)\}, \{G(2)\} \right), \left(\{G(\tfrac{1}{2})\}, \{G(2)\} \right) \right\} \right)$$

$$= \left(\left\{ \left(\{G(-2)\}, \{G(0), G(\tfrac{1}{2})\} \right) \right\}, \left\{ \left(\{G(0)\}, \{G(2)\} \right), \left(\{G(\tfrac{1}{2})\}, \{G(2)\} \right) \right\} \right)$$

$$\approx \left(\left\{ \left(\{G(-2)\}, \{G(0)\} \right) \right\}, \left\{ \left(\{G(0)\}, \{G(2)\} \right), \left(\{G(\tfrac{1}{2})\}, \{G(2)\} \right) \right\} \right)$$

$$\approx \left(\{G(-1)\}, \{G(1)\} \right)$$

$$\approx G(0).$$

Similar calculations show that $L_1 = L_2 \approx G\left(-\tfrac{1}{2}\right)$, $R_1 \approx G\left(\tfrac{3}{2}\right)$, and $R_3 \approx G(1)$; see Problem 15. Thus, we arrive at

$$Arion = (\{L_1, L_2, L_3\}, \{R_1, R_2, R_3\})$$

$$\approx \left(\{G\left(-\tfrac{1}{2}\right), G(0)\}, \{G\left(\tfrac{3}{2}\right), G\left(1\tfrac{1}{4}\right), G(1)\} \right)$$

$$\approx \left(\{G(0)\}, \{G(1)\} \right)$$

$$= G\left(\tfrac{1}{2}\right),$$

as claimed.   □

We opened our book with the introduction of the game *Arion*, and it is a pleasing way to close it with its complete analysis. Our journey introduced us to many different branches of (abstract) mathematics and had us meet many of its classical results, recent developments, and future goals. We trust that the journey

was informative, instructive, and enjoyable. And, most importantly, we hope that
our book served as an invitation to engage in further studies of mathematics.

Problems

1. Verify the assertions in Table 22.1.
2. Prove Proposition 22.6.
3. Verify that $|\Gamma_3| = 7, 143, 404$.
4. Prove that Theorem 22.11 implies Axiom 22.2.
5. Prove Propositions 22.15, 22.16, and 22.17.
6. In this problem we analyze dyadic games (games corresponding to real numbers
   with finite binary forms).

   (a) Provide a definition for the dyadic game $G(q)$ for every $q \in \mathbb{Z}\left[\frac{1}{2}\right]$. Verify
       also that your definition yields numeric games.
       (Hints: Separate the cases in which $q \in \mathbb{Z}$ and in which $q \notin \mathbb{Z}$. Proceed
       recursively in both cases.)
   (b) Find a formula for the birthday of the dyadic game $G(q)$ for every $q \in \mathbb{Z}\left[\frac{1}{2}\right]$.
   (c) Let Δ_n be the set of dyadic games with birthday $n \in \mathbb{N}$. Find the elements
       of Δ_4 explicitly.
   (d) Extend the chart on page 407 to include all *Hackenbush* strings of height 4.
       (Hint: Start with the *Hackenbush* strings of height 3.)
   (e) Prove that, for every nonnegative integer n, $|\Delta_n| = 2^n$.
   (f) Prove that, for every positive integer n, $|\cup_{i=0}^{n-1} \Delta_i| = 2^n - 1$.
   (g) Let $n \in \mathbb{N}$. Suppose that $G_1, G_2, \ldots, G_{2^n}$ is the full list of games in Δ_n in
       increasing order, that is,

       $$G_1 < G_2 < \cdots < G_{2^n},$$

       and that $G'_1, G'_2, \ldots, G'_{2^n-1}$ is the full list of games in $\cup_{i=0}^{n-1} \Delta_i$, also in
       increasing order. Prove that

       $$G_1 < G'_1 < G_2 < G'_2 < \cdots < G'_{2^n-1} < G_{2^n}.$$

   (h) Let $q_1, q_2 \in \mathbb{Z}\left[\frac{1}{2}\right]$ with $q_1 < q_2$. Find (in terms of q_1 and q_2) the unique
       element $q \in \mathbb{Z}\left[\frac{1}{2}\right]$ for which

       $$(\{G(q_1)\}, \{G(q_2)\}) \approx G(q).$$

   (i) Prove Theorem 22.39.

Fig. 22.10 The *Hackenbush*
string corresponding to the
sequence (L, R, L)

Theorem 22.39 *Any numeric game with a finite birthday is similar to a dyadic game. More precisely, if $G \in \Gamma_n$ for some nonnegative integer n, then there exists a unique integer $m \in \{0, 1, \ldots, n\}$ and a unique dyadic game $G(q) \in \Delta_m$ for which $G \approx G(q)$.*

(Hint: Use the Domination Theorem and the Simplicity Theorem.)

7. In this problem we learn a rule for the evaluation of *Hackenbush* strings. Note that each *Hackenbush* string can be identified with a sequence of letters, each being L (for edges available for Left) or R (for edges available for Right); so that the first term in the sequence corresponds to the edge at the bottom, the next term to the edge directly on top of it, and so on. For example, the *Hackenbush* string corresponding to the sequence (L, R, L) is the game in Figure 22.10.

    Suppose that a *Hackenbush* string G is given by the sequence

$$(X_1, X_2, \ldots, X_n)$$

(here $n \in \mathbb{N}$ and $X_i \in \{L, R\}$ for $i = 1, 2, \ldots, n$); assume that $X_1 = L$. (A similar calculation can be given if $X_1 = R$.)

Define $q \in \mathbb{Z}\left[\frac{1}{2}\right]$, as follows. If $X_i = L$ for all i, then let $q = n$. Otherwise, let

$$q = m + \sum_{i=k}^{n} \frac{s_i}{2^{i-m}},$$

where k is the smallest index i for which $X_i = R$, $m = k - 1$, and s_i equals 1 when $X_i = L$ and -1 when $X_i = R$ (here $i = k, k+1, \ldots, n$). For example, if G is given by the sequence (L, R, L), then $n = 3, k = 2, m = 1, s_2 = -1$, and $s_3 = 1$, so

$$q = 1 - \frac{1}{2^1} + \frac{1}{2^2} = \frac{3}{4}.$$

Prove that $G = G(q)$.

Remark This rule was discovered by Thea van Roode who was a Canadian undergraduate student at the time.

8. In this problem we investigate infinite *Hackenbush* strings.

   (a) Prove each of the following identities.

$$\frac{1}{4} + \frac{1}{16} + \frac{1}{64} + \frac{1}{256} + \cdots = \frac{1}{3},$$

$$\frac{1}{2} - \frac{1}{8} - \frac{1}{32} - \frac{1}{128} - \cdots = \frac{1}{3},$$

and

$$1 - \frac{1}{2} - \frac{1}{4} + \frac{1}{8} - \frac{1}{16} + \frac{1}{32} - \frac{1}{64} + - \cdots = \frac{1}{3}.$$

   (Hint: Use the formula for geometric series from Problem 11 (b) of Chapter 18.)

   (b) Recall that we defined $G\left(\frac{1}{3}\right)$ as

$$G\left(\tfrac{1}{3}\right) = \left(\left\{G\left(\tfrac{1}{4}\right), G\left(\tfrac{5}{16}\right), G\left(\tfrac{21}{64}\right), \ldots\right\}, \left\{G\left(\tfrac{1}{2}\right), G\left(\tfrac{3}{8}\right), G\left(\tfrac{11}{32}\right), \ldots\right\}\right).$$

   Prove that $G\left(\frac{1}{3}\right) \in \Gamma$.

   (Hints: We need to prove that each of the left-options of $G\left(\frac{1}{3}\right)$ are less than all of the right-options. Use the first two identities of part (a).)

   (c) Find the (infinite) *Hackenbush* string representation of $G\left(\frac{1}{3}\right)$.
   (Hint: Use the third identity of part (a).)

   (d) Find the *Hackenbush* string representation of $G\left(\frac{2}{7}\right)$.
   (Hint: Use the identity

$$\frac{2}{7} = \frac{1 - \frac{1}{2} - \frac{1}{4}}{1 - \frac{1}{8}}$$

   to write $\frac{2}{7}$ as the limit of an infinite series.)

   (e) Find the *Hackenbush* string representation of $G\left(\frac{1}{7}\right)$.

   (f) Find the *Hackenbush* string representation of $G\left(\frac{1}{5}\right)$.

   (g) Let p be any prime number that is greater than 2. Prove that the game $G(\frac{1}{p})$ has a *periodic Hackenbush* string representation; that is, there is a nonnegative integer k and a positive integer r for which the *Hackenbush* string representing $G(\frac{1}{p})$ has the form

$$(X_1, \ldots, X_k, X_{k+1}, \ldots, X_{k+r}, X_{k+1}, \ldots, X_{k+r}, X_{k+1}, \ldots, X_{k+r}, \ldots).$$

(Hints: Note that p and 2 are relatively prime, so by Fermat's Little Theorem, we can then find a positive integer a for which

$$2^{p-1} - 1 = a \cdot p$$

or, equivalently,

$$\frac{1}{p} = \frac{a}{2^{p-1}} \cdot \frac{1}{1 - 1/2^{p-1}}.$$

We also see that a is odd and

$$1 \leq a \leq 2^{p-1} - 1.$$

Use the formula for infinite geometric series.)

(h) Find the *Hackenbush* string representation of $G(\epsilon)$.

(i) Find the infinite *Hackenbush* string representations of $G(-\omega)$, $G(\omega + 3)$, $G(\omega \cdot 2)$, and $G(\omega^2)$.

(Hints: The well-orderings of $\mathbb{N}$ on page 343 may be helpful.)

(j) Prove that

$$G(\omega + 3) + G(-\omega) + G(-3)$$

is similar to $G(0)$ by verifying that it is a win for the second player.

9. Prove Theorem 22.23.

10. Prove Theorem 22.24.

11. (a) Prove that for every game $G \in \Omega$, we have

$$G \cdot G(0) = G(0).$$

(b) Prove that for every game $G \in \Omega$, we have

$$G \cdot G(1) = G.$$

(c) Find an example for games $G_1, G_2, H \in \Omega$ for which $G_1 \approx G_2$ but

$$G_1 \cdot H \not\approx G_2 \cdot H.$$

12. Suppose that m and n are positive integers. Prove each of the following statements.

(a) $G(m) + G(n) \approx G(m + n)$

(b) $-G(m) = G(-m)$

(c) $G(m) \cdot G(n) \approx G(m \cdot n)$

(Hint: Prove that for every $i \in \{0, 1, \ldots, n-1\}$ and $j \in \{0, 1, \ldots, m-1\}$ we have

$$0 \leq m \cdot i + n \cdot j - i \cdot j \leq m \cdot n - 1,$$

with equality possible at both ends.)

13. Let $\omega_S = [G(\omega)]$, etc. Prove the following identities.

   (a) $\omega_S + 1_S = (\omega + 1)_S$;
   (b) $\omega_S + \omega_S = (\omega \cdot 2)_S$;
   (c) $2_S \cdot \omega_S = (\omega \cdot 2)_S$; and
   (d) $\omega_S \cdot \epsilon_S = 1_S$.

14. (a) Can two distinct games be mutual restrictions of each other?
   (b) Can two distinct games be mutual compromises of each other?
   (c) Can two distinct games be simple compromises of each other?

15. (a) Referring to the games listed on page 383, prove that $L_1 \approx G\left(-\frac{1}{2}\right)$, $R_1 \approx G\left(\frac{3}{2}\right)$, and $R_3 \approx G(1)$.
   (b) Evaluate the *Hackenbush* game of Problem 1 (a) of Chapter 1.
   (c) Evaluate the *Hackenbush* game of Problem 1 (b) of Chapter 1.

16. In this problem we evaluate the game *Cutcake* (cf. Problem 5 of Chapter 1 and Problem 9 of Chapter 7). Suppose that Vertical plays as Left and Horizontal plays as Right.

   (a) Verify that the *Cutcake* game $C(m, n)$ for $1 \leq m \leq 8$ and $1 \leq n \leq 8$ is similar to the dyadic game given by the entry in row m and column n in Table 22.2 (for $G(q)$ we only mark "q").
   (Hints: Use the ideas of Problem 9 of Chapter 7. Start with $C(1, n)$ and $C(m, 1)$, and proceed recursively. For example, for $C(2, 3)$, note that

   $$C(2, 3) = (\{C(2, 1) + C(2, 2)\}, \{C(1, 3) + C(1, 3)\})$$
   $$\approx (\{G(-1) + G(0)\}, \{G(2) + G(2)\})$$

Table 22.2 Values of small *Cutcake* games

	1	2	3	4	5	6	7	8
1	0	1	2	3	4	5	6	7
2	−1	0		1		2		3
3	−2							
4	−3	−1		0				1
5	−4							
6	−5	−2						
7	−6							
8	−7	−3		−1				0

Fig. 22.11 The *Cutcake*
game $C(8, 8)$ after one move
by each player

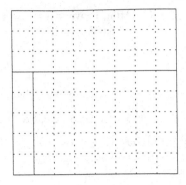

$$\approx (\{G(-1)\}, \{G(4)\})$$

$$\approx G(0).$$

For higher values of m and n you will also need to use the Domination
Theorem.)

(b) As Table 22.2 indicates (and as we already found in Problem 9 of
Chapter 7), the *Cutcake* game $C(8, 8)$ is a win for the second player.
Suppose that Horizontal starts this game by cutting the cake into a 3-by-
8 piece and a 5-by-8 piece, to which Vertical (mistakenly!) responds by
cutting the 5-by-8 piece into a 5-by-1 piece and a 5-by-7 piece. The status
of the game at this point is as given in Figure 22.11.

Prove that Horizontal can now win the game, but only by cutting the 5-by-1
piece next.

(Hint: Use the table of values above.)

(c) Find a formula, in terms of m and n, for the value of $C(m, n)$ for all positive
integers m and n.

Appendix E
Graphic Content

In Chapter 16 we introduced graphs by way of relations: we defined a *graph relation* as an irreflexive and symmetric relation on a given set (see Definition 16.4). Graphic relations occur naturally in many situations. A typical example that we have seen is "knowing one another" in a group of people (cf. Ramsey numbers in Definition 12.7), but one can easily think of many others. A graph relation is indeed one of the most ubiquitous concepts, not just in mathematics, but in a wide variety of applications in the physical and social sciences, computer science, and engineering. Here we attempt to provide a brief introduction to the vibrant and extensive field of *graph theory*.

As the name implies, a graph relation R on a set S can be represented by a drawing in the plane in which points in the plane represent the elements of S and two points u and v are connected by a line (or curve) whenever $(u, v) \in R$. In the resulting drawing, the points in the plane representing S are called *vertices*, and the connections are called *edges*. It usually does not matter how the points in the plane are arranged and whether connections are represented by straight line segments or other curves. For example, the relations

$$\{(a, b) \in \{1, 2, 3, 4, 5, 6\}^2 \mid |a - b| \in \{1, 5\}\}$$

and

$$\{(a, b) \in \{1, 2, 3, 4, 5, 6\}^2 \mid |a - b| \in \{2, 4\}\}$$

are both easily seen to be graph relations with six vertices and six edges; they can be drawn as shown in Figure E.1.

Of course, a graph has many different drawings; for example, the two triangles in the second graph of Figure E.1 could be drawn next to one another so that they do not intersect (neither do they need to have straight line edges, etc.). Given two graph drawings, it is an interesting, but often very difficult, question to decide whether

© Springer Nature Switzerland AG 2020

B. Bajnok, *An Invitation to Abstract Mathematics*, Undergraduate Texts in Mathematics, https://doi.org/10.1007/978-3-030-56174-1

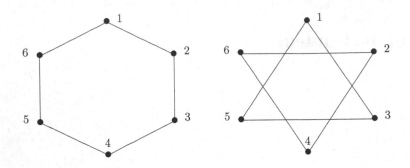

Fig. E.1 Two graphs on six vertices and six edges

the two drawings represent the same relation. To be precise, we would need to differentiate between a graph relation R on a set S and its drawing G in the plane; for simplicity, however, we will just use the term *graph* for both—hoping that it is clear from the context what we mean. (A further source of confusion may come from the fact that the graphs in this context are different from the graphs introduced on page 246; for this reason, we sometimes refer to graphs described here as combinatorial graphs and the graphs given in a coordinate system as *Cartesian graphs*, named after the French philosopher and mathematician René Descartes (1596–1650).)

Let us introduce some terminology. We say that two vertices in a graph are *adjacent* if they are connected by an edge and that an edge is *incident* to a vertex if the vertex is one of its endpoints. The number of edges that are incident to a vertex is called the *degree* of the vertex. For example, each vertex in the two graphs above has degree 2.

Suppose that u and v are vertices in a graph G. By a (u, v)-*trail* in G we mean an alternating sequence of vertices and edges, starting with u and ending with v, so that any two consecutive vertices in the sequence are distinct, and each edge in the sequence is incident to the two vertices next to it in the sequence. We say that G is *connected* if there is a (u, v)-trail in G between any two distinct vertices u and v. For example, the first graph above is connected, but the second is not: there is no trail in the graph from (for example) vertex 1 to vertex 2.

We say that the (u, v)-*trail* is a *path* if all vertices on the trail are distinct; if $u = v$ but all other vertices are distinct on the trail, then we say that the trail is a *cycle*.

We say that a graph is a *forest* if it contains no cycles, and we say that it is a *tree* if it is a connected forest. Decision trees, introduced in Chapter 1, are examples of trees. Trees have the characteristic property that there is a unique path between any two of their vertices. This and other properties make trees very useful in applications, particularly in computer science.

Another important type of graph is the planar graph, defined as follows. We say that a drawing of a graph is a *plane drawing* if no two edges intersect each other (other than, perhaps, at one of their endpoints), and we say that a graph is *planar* if it has at least one plane drawing. For example, the first drawing on page 424 is a plane drawing but the second is not; however, both graphs are planar, since the second graph can be re-drawn without any edge crossings. The regions in the plane that are created by a plane drawing are called *faces*. (Somewhat more precisely, the faces of a plane drawing G are the connected components of $\mathbb{R}^2 \setminus G$.) If the graph has finitely many vertices, then all but one of the faces in the plane drawing are bounded.

Planar graphs are reminiscent of planar maps, introduced in Appendix B. Indeed, each planar map corresponds to a planar graph G where the "countries" of the map form the vertex set, and where two vertices are adjacent whenever the corresponding countries share a boundary (of nonzero length). Conversely, it is also easy to see that every planar graph arises this way. Recall that the Four-Color Theorem states that the chromatic number of planar maps is at most 4. Equivalently, we can define the *chromatic number of a graph* G to be the minimum number of colors needed to color the vertices of G so that adjacent vertices receive different colors; according to the Four-Color Theorem, the chromatic number of every planar graph is at most 4. (In Problem 5, we prove a weaker statement that the chromatic number of a planar graph is at most 6.)

Graph theory is a highly applicable and fascinating branch of mathematics; we explore some of its main results and applications in the problems and assignments below.

Problems and Assignments

1. Suppose that G is a graph on V vertices and E edges; for $i = 1, 2, \ldots, V$, let $\deg(v_i)$ denote the degree of the vertex v_i. Prove that we have

$$E = \frac{1}{2} \sum_{i=1}^{V} \deg(v_i).$$

2. Let T be a tree with V vertices and E edges.

   (a) Prove that T has a vertex of degree 1.
   (b) Prove that $E = V - 1$.
       (Hints: Use induction on V and part (a).)

3. Prove the following famous result.

   Theorem E.1 (Euler's Formula) *Let G be a connected planar graph with V vertices, E edges, and F faces. Then $V - E + F = 2$.*

   (Hints: Use induction on the number of faces. If $F = 1$, then G is a tree; use Problem 2 part (b). For the inductive step, choose an edge e of G that is on a cycle; consider the graph G' that results from G by removing e from it.)

4. Let G be a connected planar graph with V vertices, E edges, and F faces, and
   suppose that $V \geq 3$.

   (a) Prove that $E \geq 3F/2$.
   (b) Prove that if no face of G is a triangle (i.e., a face bounded by three edges),
       then $E \geq 2F$.
   (c) Prove that $E \leq 3V - 6$.
       (Hint: Use Euler's Formula and part (a).)
   (d) Prove that if no face of G is a triangle, then $E \leq 2V - 4$.

5. (a) Prove that every planar graph contains a vertex that has degree at most 5.
       (Hint: Use Problems 1 and 4 (c).)
   (b) Prove that every planar graph has chromatic number at most 6.
       (Hint: Use part (a).)

   Remark Of course, the Four-Color Theorem states a much stronger result. To
   prove that six colors are not needed is trickier, and to prove that five colors are
   not needed requires substantial computer usage.

6. Regular polyhedra generalize the notion of regular polygons to three dimensions. A *regular polyhedron* is a polyhedron whose faces are bounded by
   congruent regular polygons and whose polyhedral angles are congruent (and
   therefore, the same number of edges are incident at each vertex). Interestingly,
   while regular polygons exist for any number of (at least three) vertices, there
   are only five regular polyhedra. Prove the following classification theorem.

   Theorem E.2 *There are exactly five different regular polyhedra: the regular
   tetrahedron, the regular octahedron, the cube, the icosahedron, and the dodec-
   ahedron.*

   (Hints: For any polyhedron P one can associate a planar graph $G(P)$ so that
   the vertices, edges, and faces of P correspond to the vertices, edges, and faces
   of $G(P)$. Note that all but one face of $G(P)$ is bounded. For example, if P is
   the cube, then $G(P)$ is the graph of Figure E.2.

Fig. E.2 The planar graph of
the cube, often denoted by
Q_3

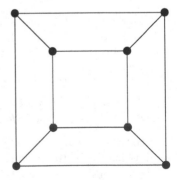

Now let k be the degree of each vertex of the regular polyhedron P, and let l be the number of edges surrounding each surface. Prove that $V = 2E/k$, $F = 2E/l$, then use Euler's Formula.)

7. (a) For a positive integer n, the *complete graph* K_n is defined to be the graph on n vertices where every pair of (distinct) vertices is adjacent. For example, K_5 is the graph shown in Figure E.3.

Fig. E.3 The complete graph K_5

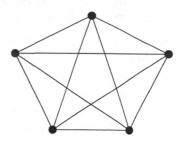

Prove the following:

Proposition E.3 *The complete graph K_n is planar if, and only if, $n \le 4$.*

In particular, K_5 is not a planar graph.
(Hints: To prove that K_n is not planar for $n \ge 5$, use Problem 4 (c).)

(b) For positive integers m and n, the *complete bipartite graph* $K_{m,n}$ is defined to be the graph on two disjoint sets on vertices, one of size m and the other of size n, where every vertex from the first set of vertices is adjacent to every vertex in the second set of vertices, but there are no other edges. For example, $K_{3,3}$ is the graph shown in Figure E.4. (The top row of vertices is one set, and the bottom row is the other set.)

Fig. E.4 The complete bipartite graph $K_{3,3}$

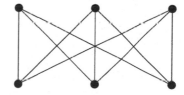

Prove the following:

Proposition E.4 *The complete bipartite graph $K_{m,n}$ is planar if, and only if, $m \le 2$ or $n \le 2$.*

In particular, $K_{3,3}$ is not a planar graph.
(Hint: Use Problem 4 (d).)

Remarks By Propositions E.3 and E.4, if K_5 or $K_{3,3}$ is a *subgraph* of a graph G, then G cannot be planar. Obviously, the same holds if edges of these two graphs are replaced by paths having the same endpoints but passing through some other vertices; that is, if G contains *subdivisions* of K_5 or $K_{3,3}$. The famous theorem of the Polish mathematician Kazimierz Kuratowski of 1930 says that this condition of planarity is sufficient as well.

Theorem E.5 (Kuratowski's Theorem) *A graph is planar if, and only if, it contains no subgraphs that are subdivisions of K_5 or $K_{3,3}$.*

8. In this problem we return to the scheduling problems of Problem 8 of Appendix B.

   (a) Construct the graph G corresponding to Problem 8 (a) of Appendix B; that is, the vertices of G will correspond to the courses, with two of them adjacent whenever the corresponding courses have students in common. Prove that G is planar.
   (Hint: Find an explicit plane drawing for G.)
   (b) Construct the graph G corresponding to Problem 8 (b) of Appendix B. Prove that G is not planar.
   (Hint: Use Proposition E.4.)

9. Graph theory has applications in many different areas of science and industry. We have mentioned *scheduling problems* in Appendix B and Problem 8 above; another area with both theoretical and practical importance concerns *transportation problems*. Review different kinds of transportation problems, such as the *traveling salesman problem* (that deals with *Hamiltonian graphs*) and the *Königsberg Bridge Problem* (that exemplifies *Eulerian graphs*).

10. As we mentioned above, graph theory has close connections with several different branches of mathematics. Review some of these connections with combinatorics, matrix theory, group theory, and topology.

Appendix F
All Games Considered

Let us return to the beautiful and intriguing topic of combinatorial games. In Chapter 22 we learned how to evaluate numeric games; we now attempt to do the same for all games. As it turns out, several of our results for numeric games carry through just fine, but some others need to be adjusted. In particular, while this much larger collection of games is still "value-able," the values do not form an ordered field any more.

As an example, consider the *Hackenbush* flower F of Figure F.1, played by A and B: options a_1 and a_2 are available to A, b_1 is available to B, and c_1 is available to both A and B.

It is quite clear that it is the first player who has a winning strategy (remove c_1), so we have $F \in \Omega_I$, and the value of F is not zero, positive, or negative—it is said to be *fuzzy*. However, one can show that F is comparable to all positive and to all negative games: it is "between" them: it is greater than any negative game, and less than any positive game. To see, for example, that $F < G(\frac{1}{2})$, consider the sum of F and $G(-\frac{1}{2})$, as shown in Figure F.2.

This game is clearly won by B (B's initial move should be c_1), so has a negative value; therefore, $F < G(\frac{1}{2})$. The same argument would show that $F < G(c)$;

© Springer Nature Switzerland AG 2020
B. Bajnok, *An Invitation to Abstract Mathematics*, Undergraduate Texts in Mathematics, https://doi.org/10.1007/978-3-030-56174-1

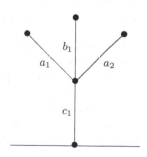

Fig. F.1 A *Hackenbush* flower

b_1

a_1 a_2

c_1

Fig. F.2 The *Hackenbush* flower F with $G(\frac{1}{2})$

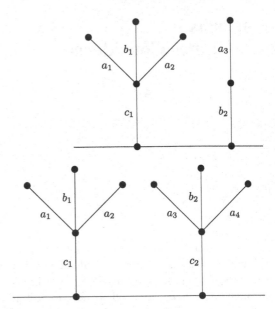

Fig. F.3 The bouquet of flowers F and F

indeed, F is less than any positive game. We could similarly show that F is greater than any negative game. Yet, $F \not\approx G(0)$!

We can see, however, that the game has some advantage for A, as there are more options that are available for A. Let us look at the game added to itself (Figure F.3).

It is not hard to see that the sum has a positive value, as A can play in such a way that B will be forced to remove one of the c edges, after which A can remove the other c edge and win the game. So $F + F > G(0)$, but $F \not> G(0)$—this is, literally, beyond surreal!

Restricting our attention, however, to a certain family of games, the so-called *impartial games*, will result in a subclass that rivals in elegance the class of numeric games; in particular, the values attached to impartial games form a field just like the surreal numbers do. In fact, the theory of impartial games is more developed—and, in many ways, simpler—than the theory of numeric games.

Definition F.1 *A game* $G = (\mathcal{L}, \mathcal{R})$ *with* $\mathcal{L} = \mathcal{R}$ *is called* impartial.

Impartial games have a beautiful theory that dates back to the works of Charles Bouton, Patrick Grundy, and Roland Sprague during the first half of the twentieth century, and has been further developed by John H. Conway, Richard Guy, and many others since then.

We have seen a variety of impartial games in this book: *Capture*, *Acrostic Twins*, *Turning Corners*, and *Nim* among them. The simplest family of impartial games is probably the *star games* $N(n)$, defined recursively, as

$$N(n) = (\{N(m) \mid m < n\}, \{N(m) \mid m < n\}).$$

(The reason for the name "star" is that $N(n)$ is sometimes denoted by $*n$ in the literature.) For example, we have

$$N(0) = (\emptyset, \emptyset),$$

$$N(1) = (\{N(0)\}, \{N(0)\}),$$

$$N(2) = (\{N(0), N(1)\}, \{N(0), N(1)\}),$$

and so on. Of course, we have $N(0) = G(0)$, and $N(1) = (\{G(0)\}, \{G(0)\})$, as we have defined earlier.

It is helpful to illustrate star games using *Hackenbush* diagrams. As we have already seen, the *Hackenbush* diagram of $N(1) = (\{G(0)\}, \{G(0)\})$ consists of a single edge that is available to both players: when removing this edge, each player reduces the game to $G(0)$. This game is clearly won by the player who starts the game. Similarly, we see that $N(n)$ corresponds to a *Hackenbush* string of "height" n where each edge is available to both players.

One fundamental property of star games is the following:

Proposition F.2 *Let*

$$G = (\{N(n_1), \ldots, N(n_m)\}, \{N(n_1), \ldots, N(n_m)\}),$$

and define n as the smallest nonnegative integer that is not an element of $\{n_1, \ldots, n_m\}$. Then $G \approx N(n)$.

Proposition F.2 is analogous to the Simplicity Theorem for numeric games. For example, we have

$$(\{N(0), N(3)\}, \{N(0), N(3)\}) \approx N(1),$$

$$(\{N(1), N(2), N(3)\}, \{N(1), N(2), N(3)\}) \approx N(0) = G(0),$$

and, of course,

$$(\{N(0), N(1), \ldots, N(n-1)\}, \{N(0), N(1), \ldots, N(n-1)\}) = N(n).$$

The star family plays a crucial role in the theory of games; for example, we can easily see that the one-heap *Nim* game $N(n)$ (cf. Problem 6 of Chapter 1) is exactly the just-introduced star $N(n)$, hence the same notation. Indeed, when playing *Nim* with a single heap of size n, each player can reduce the heap to any size less than n, thus

$$N(n) = (\{N(m) \mid m < n\}, \{N(m) \mid m < n\}).$$

In particular, we see that $N(n) \in \Omega_I$ for any $n > 0$ and, of course, $N(0) \in \Omega_{II}$.

It is quite easy to see that *Capture* is also governed by star games. Let $C(n, k)$ stand for the general version of the game *Capture* (cf. Problem 4 (d) of Chapter 1); for example, $C(n, 3)$ is the case of (c). We can analyze $C(n, 3)$, as follows.

Clearly, neither player has a move in $C(0, 3)$, so $C(0, 3) = N(0)$. In $C(1, 3)$, each player has only one option: to reduce the game to $C(0, 3)$. Thus

$$C(1, 3) = (\{N(0)\}, \{N(0)\}) = N(1).$$

Similarly, $C(2, 3) = N(2)$, and $C(3, 3) = N(3)$. Next, we have

$$C(4, 3) = (\{C(1, 3), C(2, 3), C(3, 3)\}, \{C(1, 3), C(2, 3), C(3, 3)\})$$

$$= (\{N(1), N(2), N(3)\}, \{N(1), N(2), N(3)\});$$

by Proposition F.2 above, we thus have $C(4, 3) \approx N(0)$. Moving on to $C(5, 3)$, we have

$$C(5, 3) = (\{C(2, 3), C(3, 3), C(4, 3)\}, \{C(2, 3), C(3, 3), C(4, 3)\})$$

$$= (\{N(2), N(3), N(0)\}, \{N(2), N(3), N(0)\})$$

$$= N(1),$$

and so on. An easy inductive argument thus leads the following:

Proposition F.3 *For fixed positive integers n and k, let r be the remainder of n when divided by $k + 1$. Then the game* Capture *$C(n, k)$ is similar to $N(r)$. Consequently, the second player has a winning strategy for $C(n, k)$ when n is divisible by $k + 1$, and the first player wins the game if n is not divisible by $k + 1$.*

The following classical and striking result states that Proposition F.3 can be generalized to any impartial game.

Theorem F.4 (The Sprague–Grundy Theorem) *Every impartial game is similar to $N(n)$ for some ordinal number n.*

Thus, analyzing impartial games is easy in theory: the second player to move has a winning strategy when the number n in Theorem F.4 is 0, and the first player can win if it is not 0. In reality, however, it is often very difficult (or even unknown how) to find this value. We will analyze *Acrostic Twins*, *Turning Corners*, and *Nim* below.

Combinatorial game theory is a vibrant and relatively new field of study, and it has an interdisciplinary flavor that attracts mathematicians, computer scientists, professional players, and amateurs alike. The assignments below offer a wide variety of questions for everyone with an interest in pursuing research projects on this topic.

Problems and Assignments

1. Prove the claim made on page 211 that for every nonnegative integer m, the set $\{0, 1, 2, \ldots, 2^{2^m} - 1\}$, equipped with *Nim* addition and *Nim* multiplication, is a field.

2. We have already seen in Problem 12 of Chapter 22 that for all $m, n \in \mathbb{N}$, we have

$$G(m) + G(n) \approx G(m + n).$$

   (a) Prove that for all $m, n \in \mathbb{N}$, we have

$$N(m) + N(n) \approx N(m \oplus n)$$

   (here $\oplus$ denotes *Nim* addition).

   (b) Is there a nice formula for $G(n) + N(m)$?

3. Prove Proposition F.2.

4. Prove Theorem F.4.

5. Prove each of the following claims made in Problem 12 of Chapter 2.

   (a) The *Nim* game $N(n_1, n_2, \ldots, n_m)$ is a win for Second if, and only if,

$$n_1 \oplus n_2 \oplus \cdots \oplus n_m = 0$$

   (and thus for First if, and only if, the *Nim* sum is not 0).

   (b) The game *Acrostic Twins* on an m-by-n board is a win for Second if, and only if,

$$\bigoplus_{i=0}^{m-1} \bigoplus_{j=0}^{m-1} (i \oplus j) = 0$$

   (and First can win if the *Nim* sum is not 0).

   (c) The game *Turning Corners* on an m-by-n board is a win for Second if, and only if,

$$\bigoplus_{i=0}^{m-1} \bigoplus_{j=0}^{m-1} (i \otimes j) = 0$$

   (and First can win if the *Nim* sum is not 0).

6. Design an algorithm that evaluates any finite impartial *Hackenbush* game.

7. In the game *Subtraction* $S(n; A)$, we are given a heap of size n and a finite set of positive integers A. The game is played by two players taking turns choosing

an element $a \in A$ and removing a chips from the heap. As usual, the first player who is unable to move loses. Note that $S(n; A)$ is a generalization of the *Capture* game $C(n, k)$ (with $A = \{1, 2, \ldots, k\}$).

(a) Prove that

$$S(n; \{2, 5, 6\}) \approx \begin{cases} N(0) \text{ when } n \equiv 0, 1, 4, \text{ or } 8 \bmod 11; \\ N(1) \text{ when } n \equiv 2, 3, 6, \text{ or } 10 \bmod 11; \\ N(2) \text{ when } n \equiv 5 \text{ or } 9 \bmod 11; \\ N(3) \text{ when } n \equiv 7 \bmod 11. \end{cases}$$

In particular,

$$S(n; \{2, 5, 6\}) \in \begin{cases} \Omega_{II} \text{ when } n \equiv 0, 1, 4, \text{ or } 8 \bmod 11; \\ \Omega_I \text{ otherwise.} \end{cases}$$

(b) Prove that $S(n; A)$ is always periodic; that is, there exists a positive integer p (assumed to be the smallest) for which $S(n; A)$ has the same outcome as $S(n + p; A)$ does for all "large enough" n. What can one say about p? When is $S(n; A)$ purely periodic (that is, when does periodicity hold for all $n \in \mathbb{N}$)?

(c) What if A is allowed to be infinite in $S(n; A)$?

(d) In the (non-impartial) generalization $S(n; A, B)$ of $S(n; A)$, Left and Right must choose the number of chips from sets $A \subset \mathbb{N}$ and $B \subset \mathbb{N}$, respectively. Prove that

$$S(n; \{1, 2\}, \{2, 3\}) \in \begin{cases} \Omega_L \text{ when } n \equiv 1 \bmod 4; \\ \Omega_{II} \text{ when } n \equiv 0 \bmod 4; \\ \Omega_I \text{ when } n \equiv 2 \text{ or } 3 \bmod 4. \end{cases}$$

(e) Analyze $S(n; A, B)$ in general.

8. An entertaining yet deep and influential book on combinatorial games containing, indeed, the complete foundation to the field, is *Winning Ways*, by Elwyn Berlekamp, John Conway, and Richard Guy (A K Peters, Second edition, 2001–2004). The four-volume set describes hundreds of games, discusses what is known about them, and lists specific open questions about them, thereby serving as an excellent source for research projects. By browsing through the books, choose your favorite game(s) and analyze them as thoroughly as possible (outcome class, similarity class, winning strategy, etc.).

9. The theoretical foundation of combinatorial game theory is laid out in the book *On Numbers and Games*, written by John Conway (A K Peters, Second edition, 2001). The book builds up the entire field starting with the axioms and including the deep theory of algebra and analysis on surreal numbers and other, more

general, numbers. While the book is not an easy read, those interested in a fuller understanding of the field are encouraged to study it.

10. The very last theorem in Conway's book *On Numbers and Games* says the following:

Theorem F.5 *This is the last theorem in this book.*

Prove Theorem F.5.

Appendix G
A Top Forty List of Math Theorems

This is the author's (very subjective) list of the top forty greatest theorems of mathematics—or, at least, the top forty greatest theorems in this book—arranged in approximate chronological order. For each theorem on our list, we provide the reference number for where it can be found in this book.

© Springer Nature Switzerland AG 2020
B. Bajnok, *An Invitation to Abstract Mathematics*, Undergraduate Texts in Mathematics, https://doi.org/10.1007/978-3-030-56174-1

Index

© Springer Nature Switzerland AG 2020
B. Bajnok, *An Invitation to Abstract Mathematics*, Undergraduate Texts in Mathematics, https://doi.org/10.1007/978-3-030-56174-1

Printed in the United States
by Baker & Taylor Publisher Services